Fuels and Petrochemicals Division 2017

Core Programming Area at the AIChE Annual Meeting

Minneapolis, Minnesota, USA
29 October - 3 November 2017

ISBN: 978-1-5108-5803-9

Printed from e-media with permission by:

Curran Associates, Inc.
57 Morehouse Lane
Red Hook, NY 12571

Some format issues inherent in the e-media version may also appear in this print version.

Printed by Curran Associates, Inc. (2018)

For permission requests, please contact AIChE
at the address below.

AIChE
120 Wall Street, FL 23
New York, NY 10005-4020

Phone: (800) 242-4363
Fax: (203) 775-5177

www.aiche.org

Additional copies of this publication are available from:

Curran Associates, Inc.
57 Morehouse Lane
Red Hook, NY 12571 USA
Phone: 845-758-0400
Fax: 845-758-2633
Email: curran@proceedings.com
Web: www.proceedings.com

Fuels and Petrochemicals Division 2017

Core Programming Area at the AIChE Annual Meeting

Minneapolis, Minnesota, USA
29 October - 3 November 2017

AIChE
The Global Home of Chemical Engineers

TABLE OF CONTENTS

(64b) Imaging Single Molecules by Scanning Probe Microscopy 1
Bruno Schuler

(64c) Non-Contact Atomic Force Microscopy (NC-AFM) Imaging of Model Compound Bispyrene Molecules 2
Yunlong Zhang, Michael R. Harper, Doug Kushnerick

(64d) Molecular Structure of Heavy Oil Revealed - 1 3
Michael Moir

(64e) Molecular Structure of Heavy Oil Revealed - 2: The Effect of Severe Hydroprocessing 14
Frans Van den Berg

(64f) Molecular Structure of Heavy Oil Revealed - 3: High Resolution Atomic Force Microscopy - Perspectives and Examples in Heavy Oil Characterization 24
Bruno Schuler

(64g) Molecular Structure of Heavy Oil Revealed - 4: Molecular Structure of Heavy Oil Revealed with Non-Contact Atomic Force Microscopy 25
Yunlong Zhang, Michael R. Harper, Doug Kushnerick

(64h) Atom Explicit Compositional Models of Heavy Oils: Making and Using State-of-the Art Measurements 26
Michael T. Klein

(64i) Perspectives on Catalyst Design for Complex Feedstocks 27
Matthew Neurock

(64j) New Approaches to Big Chemistry: Rapid Screening and Quantification of Complex Hydrocarbon Mixtures 28
Paul J. Dauenhauer, Omar A. Abdelrahman

(152a) Digital Alchemy for Assembly Engineeringc 29
Sharon C. Glotzer

(152b) Improved Algebraic, Numerical, and Graphical Representations in Fluid Mechanics 30
Stuart W. Churchill, James C. Hill

(152c) The Scaling of Turbulence Near the Wall and the Churchill Turbulent Flux Correlation: Insights with Lagrangian Simulations 31
Dimitrios V. Papavassiliou, Quoc T. Nguyen, Chiranth Srinivasan

(152d) Flow Boiling Using a Piranha Pin Fin Heat Sink 32
Cory Woodcock, Xiangfei Yu, Yoav Peles, Joel L. Plawsky

(152e) Transport Problems in the Spirit of Stuart Churchill for Teaching and Research at the University of Michigan 33
Ronald G. Larson, Claudio Vilas Boas Favero

(175a) A Framework for Optimization-Based Design of Heat-Integrated Crude Oil Distillation Units Using a Surrogate Model and Support Vector Machine 34
Dauda Ibrahim, Megan Jobson, Jie Li, Gonzalo Guillén-Gosálbez

(175b) Optimization-Based Design of Energy-Efficient Crude Oil Distillation Systems with Pre-Separation Units 35
Minerva Ledezma-Martinez, Megan Jobson, Robin Smith

(175c) Vacuum Distillation: Slop Wax Disposition 50
Andrew W. Sloley

(175d) Acid Regenerator Optimization for Increased Unit Throughput 67
Kayla Erickson

(175e) Vacuum Distillation: Fractionating Wash Zones 68
Andrew W. Sloley

(175f) Advances in Salt Point Management Practices: Mitigating Overhead Corrosion with Big Data 87
Collin Cross

(175g) Fouling Modelling for Crude Oil Heat Exchanger Networks 98
J. Loyola-Fuentes, Robin Smith, Megan Jobson

(176a) Solve this! Fundamental Approach to Problem Solving in Industrial Processes I (Invited Talks) 111
Zdravko Stefanov, Paul Chauvel, Eldad Herceg, Dana A. Livingston

(242a) Mitigating Risks for Refiners Processing Opportunity Crudes 112
Tim Olsen

(242e) Impact of Opportunity Crudes on Desalter Operation and Wastewater Treatment Performance in a Refinery 113
Somnath Basu
(242c) CFD Applied to SHALE OIL Pyrolisys 125
Carina Stahnke, Marcela Kotsuka Silva, Leonardo M. Rosa, Dirceu Noriler, Henry F. Meier, Jaci C. S. C. Bastos
(242d) Future Oil Sands Production Costs and GHG Emissions Based on Emerging Technologies 126
Experience I. Nduagu, Alpha Sow, Evar Umeozor, Dinara Millington
(242b) Asphaltene Precipitation in Bitumen-Diluent Blends 135
Wattana Chaisoontornyotin, Yuan Yang, Jingzhou Zhang, Samson Ng, Michael P. Hoepfner
(242f) A Tool for Understanding Foaming Tendencies, Coke Morphology, and Quench Effectiveness - Real Time Visualization into Pilot Delayed Coker Drum 136
Amaka Waturuocha, Dwijen Banerjee, Michael Volk, Glixon Mavarez Nava, Keith Wisecarver
(242g) Shale Oil As Steam Cracking Feedstock: GC x GC Characterization and COILSIM1D Modeling 137
Nenad D. Ristic, Marko R. Djokic, I. Amghizar, Kevin M. Van Geem, Guy B. Marin
(242h) Characteristics of Unconventional and Heavy Oils: Their Economic and Environmental Impacts in Production and Processing 138
M. R. Riazi
(279b) Numerical Evaluation of a Fluid Catalytic Cracking Unit with Internal Baffles 139
Tania Sanchez Martinez, Leonardo M. Rosa, Jonathan Utzig, Waldir Pedro Martignoni, Henry F. Meier
(279d) A Group Contribution Method for the Prediction of the Mid Infrared (MIR) Absorption Spectra of Species Involved in Fluid Catalytic Cracking (FCC) 140
Sandra Milena Lopez-Zamora, Hugo de Lasa, Alejandro Molina
(279f) Fundamental research on phase equilibrium of extraction desulfurization 158
Yuhao Zhang, Liang Zhao, Jinsen Gao, Chunming Xu
(347a) The Effect of New Nanocomposites on Fluidity of Waxy Crude Oil in Low Temperature 161
Huirong Huang, Wei Wang, Yanfen Ding, Zeheng Peng, Jing Gong
(347b) Co-Processing of Liquid Phase Pyrolysis Oil and Refinery Intermediates in a Continuous Hydrodeoxygenation Reactor 171
Klara Treusch, Nikolaus Schwaiger, Roland Nagl, Berndt Hammerschlag, Julia Ausserleitner, Anna Huber, Peter Pucher, Matthaeus Siebenhofer
(347d) Static and Dynamic Catalytic Adsorptive Desulfurization (CADS) of Real Diesel Using Low-Cost TiO_2/Fumed Silica 172
Xiaoling Ren, Zewei Liu, Lei Dong, Jing Xiao
(347e) [O]-Induced Reactive Adsorptive Desulfurization of Liquid Fuel over $Ag_{X}o$@SBA-15 Under Ambient Conditions 173
Guang Miao, Liqiong Wu, Ying Wu, Feiyan Ye, Jing Xiao
(417a) Energy Saving from Process Design - a Service Oriented Architecture (SOA) Methodology 174
Jia Li
(417b) Mathematical Modeling and Optimization of Commercial-Scale Catalytic Two-Stage, Ebullated Bed, Direct Coal Liquefaction Reactors 175
Yuan Jiang, Debangsu Bhattacharyya
(417c) Optimal Design of Pipeline System for Flare Minimization in Multiple Chemical Plants 176
Yiling Xu, Tianxing Cai, Wang Zhenlei, Qiang Xu
(417d) Development of a Natural Gas to Industrial Chemicals Bioprocess Platform 177
Bryan Yeh
(417e) Comparison of Conventional and Newly Proposed Solvent Deasphalting Processes: Numerical Simulation Based on Energy and Cost Analysis 178
Junwoo Park, Soo Ik Im, Ki Bong Lee, Kang Seok Go, Nam Sun Nho
(463b) Evaluation of Aromatics Formation in Thermal Cracking of Triglycerides 179
Ramon F. Beims, L. Ender, Vanderleia Botton, Dilamara R. Sharf, E. L. Simionatto, Henry F. Meier, Vinicyus R. Wiggers
(463c) Tuning the Structure and Electronic Properties of Manganese Nitrides for Ammonia Production from First-Principles Modeling 180
Bin Liu, Nannan Shan, Peter H. Pfromm
(463d) Rapid and Extended Analysis of Syngas Using Micro GC Fusion 181
Shawn Wilson
(463f) Dual-Alcohol Blending Effects on Gasoline Properties 182
Saeid Aghahossein Shirazi, Thomas D. Foust, Kenneth F. Reardon
(484a) NO Uptake and Desorption on Fe-Modified Pd/ZSM-5: Impact of Pd and Fe Loading 183
Kyle Karinshak, Michael Harold, Kiran Premchand

(484b) Insights into Influencing Factors on Oxygen Storage Characteristics for Ceria-Zirconia Catalysts 185
Jason Wu, Giovanni Cavataio

(484c) Syntheses of Pd Doped $CaCo_xZr_{1-X}O_{3-\delta}$ novel Perovskite-Type Redox Materials for Automotive Emissions Control 186
Qinghe Zheng, Marty Lail

(484d) Probing the Oxidation and Reduction Halves of the Cu Redox Cycle during Standard NH_3-Selective Catalytic Reduction of NO_x on Cu-SSZ-13 199
Ishant Khurana, Christopher Paolucci, Atish A. Parekh, Arthur J. Shih, Jonatan D. Albarracin Caballero, Aleksey Yezerets, W. Nicholas Delgass, Jeffrey T. Miller, William F. Schneider, Rajamani Gounder, Fabio H. Ribeiro

(484e) Sulfur Oxidation Studies with Bimetallic Pd/Pt Catalysts 200
Monique S. Wilburn, William Epling

(484f) Nature of SO_2 Poisoned Cu-SSZ-13 Catalysts Under Ammonia Selective Catalytic Reduction (NH_3-SCR) Conditions 201
Arthur J. Shih, Hui Li, Ashok Kumar, Juan M. Gonzalez, Ishant Khurana, Christopher Paolucci, Jonatan D. Albarracin Caballero, Atish A. Parekh, A. L. Villa, W. Nicholas Delgass, Rajamani Gounder, Aleksey Yezerets, William F. Schneider, Jeffrey T. Miller, Fabio H. Ribeiro

(484g) Measurement of the Oxygen Storage Capacity and Its Role in the Catalytic Oxidation 202
Ping Li, Chang Yup Seo, Xiaoyin Chen, Yongdan Li, Johannes W. Schwank

(484h) Activation of Pd/SSZ-13 in Low Temperature NO Adsorption for Cold Start Application 203
Youngseok Ryou, J. Lee, Hyokyoung Lee, Changhwan Kim, Do Heui Kim

(578a) Natural Gas Liquids (NGL)- Fractionate or Not to Fractionate at the Gas Processing Plant 204
Bijal Gangar, Ali A. Pilehvari

(578b) LNG Market and Technology 212
Jeffrey Zhang

(578c) Visual Measurements of Hydrocarbon Freeze-out Temperature in LNG Mixtures: Experimental Design 222
Arman Siahvashi, Saif Al Ghafri, Eric May

(578d) Plant-Wide Modeling, Techno-Economic Analysis and Optimization of the Shale-Gas to Dimethyl Ether (DME) Process Via Direct and Indirect Synthesis Route 234
Chirag Mevawala, Yuan Jiang, Debangsu Bhattacharyya

(578e) New Thermophysical Properties Measurements of Complex Mixtures Relevant to Liquefied Natural Gas (LNG) Processing 235
Saif Al Ghafri, Masoumeh Akhfash, Arash Arami-Niya, Sofia Mylona, Kumarini Seneviratne, Fuyu Jiao, Jordan Oakley, Thomas Hughes, Michael L. Johns, Eric May

(578f) Numerical Investigation for Heat Transfer of Supercritical Methane Heated in a Horizontal Circular Tube 236
Binhui Ruan, Wensheng Lin

(578g) Study on Ice Storage System Based on Cold Energy Utilization in Lcng Fueling Stations 237
Siyue Zhao, Wensheng Lin, Wei Qiu

(584a) The Structural Characteristics of Oil Shale Kerogens Changed with Their Humic Degree 238
Qian Wang, Qing Liu, Fan Yang, Yucui Hou, Weize Wu

(584c) Development and Application of a Fuel Property Database for Mono-Alcohols As Fuel Blend Components for Spark Ignition Engines 239
Saeid Aghahossein Shirazi, Kenneth F. Reardon

(584e) Study on Crude Oil Extraction Method from Sludge Including Oil 240
Tadashi Sano, Masatomo Watanabe, Hideaki Kurokawa, Kei Hayashida, Hisashi Isogami, Yojiro Hayashi

(584f) Insight into Coal Structure Based on Benzene Carboxylic Acids from Coal Via Oxidation 241
Fan Yang, Yucui Hou, Muge Niu, Shuhang Ren, Weize Wu

(584h) Production of Carboxylic Acids from Lignite with Two-Stage Alkali-O_2 Oxidation 242
Yucui Hou

(584i) Methane Production from Crude Solid Residue: A Minimal Organic Waste Strategy 243
Aditi David, Saurabh Dhiman, Glenn Johnson, Rajesh Sani

(584j) Production of Carboxylic Acids from Lignite with Two-Stage Alkali-Oxygen Oxidation 244
Yucui Hou, Wenbin Li, Fan Yang, Shuhang Ren, Weize Wu

(584k) CFD Modeling of Immiscible Liquid-Liquid Flow in a Large Scale Crude Oil Storage Tank Equipped with Side-Entering Mixers 245
Reynaldo Fonseca, Diener Volpin Ribeiro Fontoura, Nicolas Spogis, J. R. Nunhez

(584l) Reaction Kinetics of 1-Methylnapthalene Hydrocracking over Metal/Beta Catalyst to BTX 246
Tao Wu, Sheng-Li Chen, Guimei Yuan, Jie Xu, Ling-xiang Huang, Ying-qian Cao

(584m) CO_2 Micro Foam and Its Application in Enhanced Oil Recovery 247
Shuangxing Liu

(584o) Kinetic Modelling of the Combustion of Aliphatic Hydrocarbons 248
Okoh Elechi
(584p) Predicting the Properties of Petroleum Blends 249
Hessa AlMulla, Tareq Albahri
(584q) On-Sun Demonstration of Hydrogen Production Via Solar Thermochemical Water Splitting 250
Samantha L. Millican, Amanda Hoskins, Caitlin Czernik, Mark Wallace, Ibraheam Al-Shankiti, Judy Netter, Charles B. Musgrave, Alan W. Weimer
(584r) Optimal Aspect Ratio and Recirculation Rate for LNG Storage Tanks in a Regasification Terminal 251
Mohd S. Khan, Surya Effendy, S. Farooq, Iftekhar A. Karimi
(584t) Update: Gas Production from the SCOOP 252
Richard L. Long
(584u) The Relationship between Olefin Hydrogenation and Octane Number Loss in FCC Gasoline Hydrodesulfurization 262
Lixia Dong, Liang Zhao, Kaiwei Luo, Yuhao Zhang, Jinsen Gao, Chunming Xu
(584w) Inherent Safety Assessment in Natural Gas Liquefaction Process 263
Dae-Hyun Kim, Hye-Ri Gye, Chul-Jin Lee
(690a) A Comparative Assessment of Sol-Gel and Solid State Syntheses Derived Redox Materials for H_2 Production *Via* Thermochemical Water-Splitting Process 264
Vinod S. Amar, Rajesh V. Shende, Jan A. Puszynski
(690b) Overview of Impact of Electrode Catalyst Loadings on Thin Flexible Fuel Cell (TFFC) Performance 265
Matthew Mayer, Seyed Reza Mahmoodi, Ronald S. Besser
(690c) NiMo-Ceria-Zirconia Catalytic Reforming Layer for Solid Oxide Fuel Cells Running on Isooctane 266
Kai Zhao, Xiaoxue Hou, Qusay Bkour, M. Grant Norton, Su Ha
(690d) Homogeneous Reaction Kinetics of Carbohydrates with Viologen Catalysts in Biofuel Cell Applications 267
Randy S. Lewis, Scott Carter, Meisam Bahari, Hilary Bingham, John Harb, Gerald D. Watt
(690e) Hydrogen Fueled Micro-Fuel Cell with Microfluidic Channels on a PFSA Plane 268
Seyed Reza Mahmoodi, Ronald S. Besser
(690f) Design and Analysis of Tri-Generation System Powered By Seaweed Biogas 269
Ivannie Effendi, Peyman Fasahati, J. Jay Liu
(700a) Multi-Functional Mixed Oxide Catalysis in Cascade Chemistry to Convert Ethanol to High Value Oxygenates 270
Mond Guo, Michel Gray, Karthikeyan K. Ramasamy
(700c) Acetaldehyde Condensation-Cyclodehydration to Aromatics over Mg-Al Oxides 271
Marcella Lusardi, Klavs F. Jensen
(700d) Calibration-Free Methods for GC Quantification of Bio-Derived Chemicals 272
Charles S. Spanjers
(700e) A Novel Method of Producing Levulinic Acid at High Concentrations and Yields from Corn Stover Hydrolysate 282
Ravikumar Gogar, Sridhar Viamajala, Patricia Relue, Sasidhar Varanasi
(700f) Hydrodeoxygenation of Acetic Acid As a Model Compound for the Aqueous Phase Catalytic Pyrolysis Oils 283
Hossein Jahromi, Foster Agblevor
(700g) C_5+ Ketones Synthesis from Ethanol over Mixed Metal Oxides 284
Karthikeyan K. Ramasamy, Mond Guo, Michel Gray, Senthil Subramaniam
Author Index

Some abstracts in this book refer to images that were not made available by the publisher

505799 Imaging Single Molecules by Scanning Probe Microscopy

Abstract:

Scanning probe microscopy with functionalized tips has become a powerful analytical method to study a wide range of chemical compounds at the single molecule level. In this talk, I will introduce this novel approach and review recent developments in the field. To this end several studies are discussed where high-resolution scanning probe microscopy has been successfully applied to analyze structural and electronic properties of synthetic model compounds, products formed by on-surface synthesis or atomic manipulation, purified natural samples or complex mixtures.

505819 Non-Contact Atomic Force Microscopy (NC-AFM) Imaging of Model Compound Bispyrene Molecules

Abstract:

Understanding the exact chemical structure of tremendously diverse molecules in hydrocarbon mixtures presents a formidable challenge. State-of-the-art tools such as ultra-high resolution Fourier transform ion cyclotron resonance mass spectrometry (FTICR-MS) yield exact chemical formulae for molecules in complex mixtures but shed little light on the bond structure within molecules. Fortunately, the non-contact atomic force microscopy (NC-AFM) technique recently championed by IBM (Schuler et al., *J. Am. Chem. Soc.*, **2015**, *137*, 9870-9876) demonstrates that exact molecular structure of single organic molecules can be determined for coal and petroleum asphaltenes.

This 2015 asphaltene study observed predominantly the 'island' structure motif. The scarcity of 'archipelago' motifs raised questions on what types of chemical bonds could survive sample preparation by flash heating and sublimating the hydrocarbons onto the imaging substrate. One hypothesis for the apparent absence of 'archipelago' motifs was the susceptibility of weak C-C bond linkages between the multi-core molecules to cracking during the heating/sublimating process. We tested this hypothesis with a series of a,w-substituted bispyrenyl alkane model compounds designed to mimic multi-core asphaltene structures. The successful imaging of these model compounds demonstrates that C-C bonds with bond strengths as low as 65 kcal/mol can survive sample preparation. Further experiments, such as a 50:50 mixture of these bispyrenyl molecules and model compound 'island' molecules, would increase our understanding of potential impediments to definitive observation of 'archipelago' motifs in petroleum asphaltenes.

Chevron

human energy

Molecular Structure of Heavy Oil Revealed – 1
or
Do Asphaltenes Really Exist?

Michael E. Moir
Chevron Technology Company

AIChE Annual Meeting
October 30, 2017

What do we mean when we use the word "Asphaltenes"

Chevron

What do we mean when we use the word "Asphaltenes"

- First defined rigorously *(along with maltenes and carbenes)* by Clifford Richardson in 1905 to help classify pavement feedstocks.

The most accepted definition comes from IP 143/57:

The **Asphaltenes** *content of a petroleum product is the percentage by weight of wax-free material insoluble in n-heptane but soluble in hot benzene.*

Chevron

What do we mean when we use the word "Asphaltenes"

- Unfortunately, dozens of researchers over many years were unhappy with that definition
- They investigated the atomic and molecular composition, molecular weight etc. of asphaltenes in different solvents and conditions, and made assertions about the relative importance of those properties and conditions.
- <u>*This has led to a century of confusion and misunderstanding about the true nature of asphaltenes.*</u>

Chevron

The astonishing work by Schuler, Gross, et al.

- AFM/STEM images of the molecules contained in asphaltenes will profoundly influence our thinking about asphaltenes:

What Do You See?

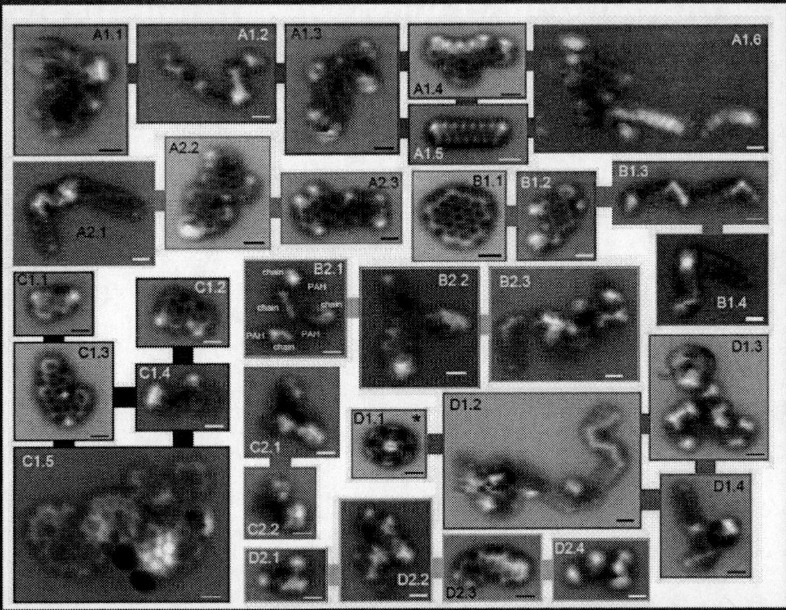

We See an Astonishing Variety and Complexity of Molecules

- In the 200 - 300 molecules imaged in this work, no duplicates were observed, each molecule is unique
- Asphaltenes bear no resemblance to polymers, the language of polymer science, in particular the ideas about the structure of polymers are largely inappropriate
- Notions of average molecule weight and average molecular composition are now seen to be unhelpful if not simply ridiculous

The Canonical View of Asphaltenes

- Widely believed to be hydrogen deficient, contain heteroatoms, consist of a polyaromatic core consisting of six membered rings, polar, and very high molecular weights
- Let's test this belief:

Some examples of molecules found in asphaltenes:

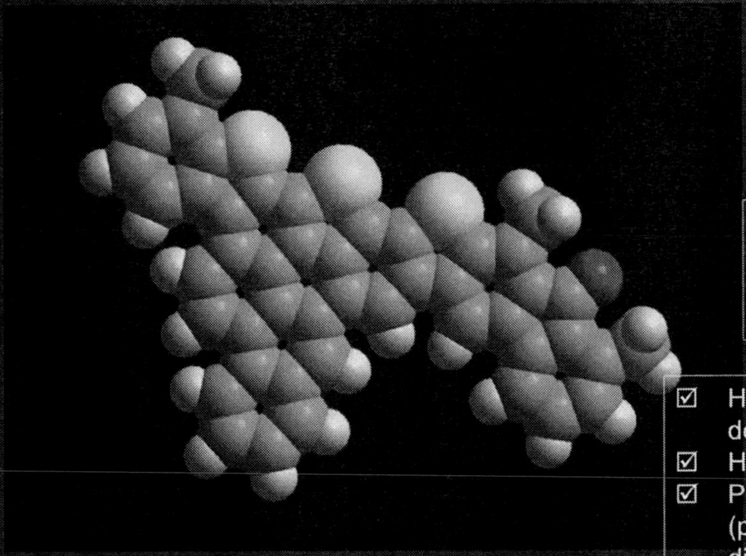

$C_{46}H_{24}OS_3$
Mw = 688
H/C = 0.52
δ = 4.7 D

- ☑ Hydrogen deficient
- ☑ Heteroatoms
- ☑ Polar (permanent dipole)

Calculated from Schuler et al, *Energy Fuels* 2017,31,6856-6861

Chevron

 9

Some molecules are not flat, have no heteroatoms and are not particularly polar

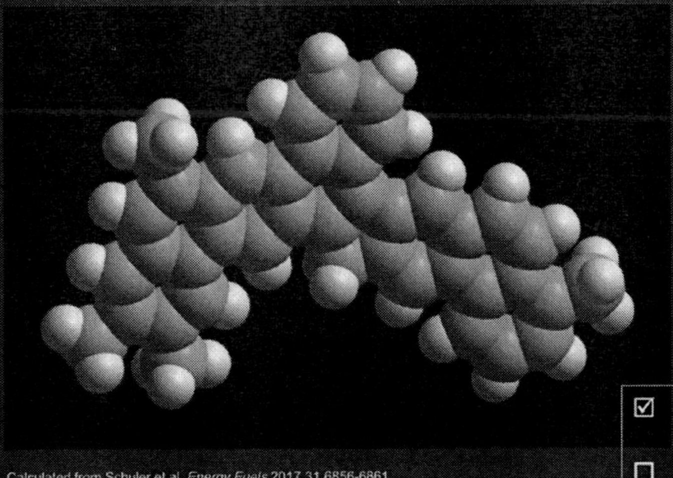

$C_{40}H_{29}$
Mw = 509
H/C = 0.73
δ = 0.78 D

- ☑ Hydrogen deficient
- ☐ Heteroatoms
- ☐ Polar (permanent dipole)

Calculated from Schuler et al, *Energy Fuels* 2017,31,6856-6861

Chevron

 10

Some molecules have very large calculated dipole moments <u>without</u> heteroatoms and contain 5, 7, 8, and 11 membered rings.

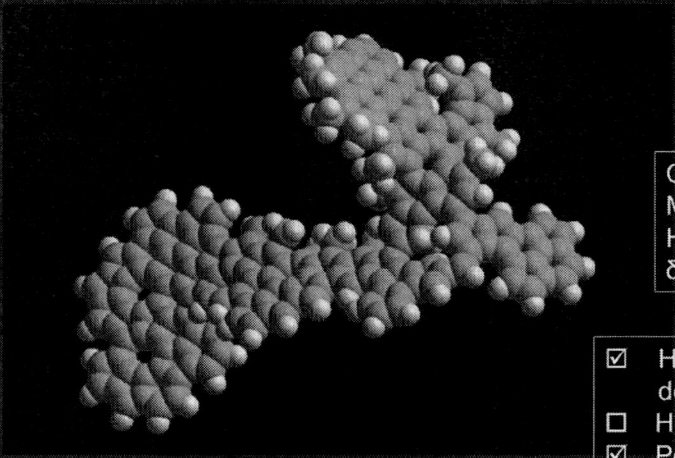

Calculated from Schuler et al, accepted by Energy and Fuels

$C_{196}H_{94}$
Mw = 2446
H/C = 0.48
δ = 19.9 D

- ☑ Hydrogen deficient
- ☐ Heteroatoms
- ☑ Polar (permanent dipole)

Chevron

Some molecules have no heteroatoms and are non-polar

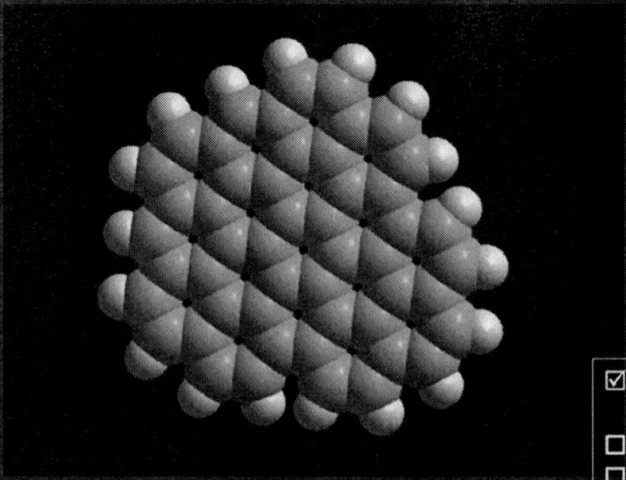

Calculated from Schuler et al, *Energy Fuels* 2017,31,6856-6861

$C_{50}H_{18}$
Mw = 618
H/C = 0.36
δ = 0.05 D

- ☑ Hydrogen deficient
- ☐ Heteroatoms
- ☐ Polar (permanent dipole)

Chevron

The molecules making up "asphaltenes" are not polar, they are polarizable

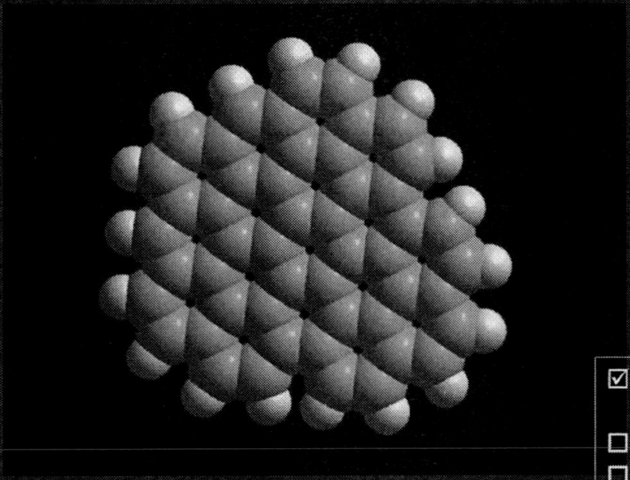

Calculated from Schuler et al, submitted to Energy and Fuels

$C_{50}H_{18}$
Mw = 618
H/C = 0.36
δ = 0.05 D

☑ Hydrogen deficient
☐ Heteroatoms
☐ Polar (permanent dipole)

Chevron

Mass Spectrometry Illuminates the Molecular Weights of molecules present in asphaltenes and deposits

- Rodgers and Marshall were the first to observe the entire astonishing range and complexity of molecules in crude oil.
- The idea of an average molecule is sent to the scrap heap of history
- The molecules in asphaltenes span a wide range of molecular weights but are hydrogen deficient.
- The molecules are part of the continuum of molecular species in crude oil.

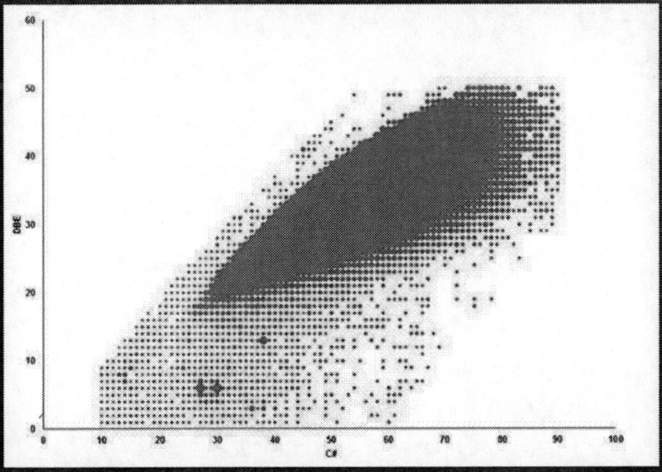

APPI-FTMS of an oil-field deposit.

Chevron

Defining Characteristics of the Molecules in Asphaltenes.

- ~~Contain heteroatoms~~
- ~~Flat hexagonal aromatic rings~~
- ~~Polar (large permanent dipole moment)~~
- 30 to 80 carbon atoms
- Hydrogen deficient core (dense)
  - *the lack of solubility in hydrocarbon solvents, aromatic character, and polarizability are consequences of the hydrogen deficiency*

Chevron

15

Defining characteristics of the molecules in asphaltenes.

- ~~Contain heteroatoms~~
- ~~Flat hexagonal aromatic rings~~
- ~~Polar (large permanent dipole moment)~~
- 30 to 80 carbon atoms
- Hydrogen deficient core (dense)
  - the lack of solubility, aromatic character, and polarizability are consequences of the hydrogen deficiency

But these molecules are not asphaltenes

Chevron

16

Pi-stacking but No π- π Bonding

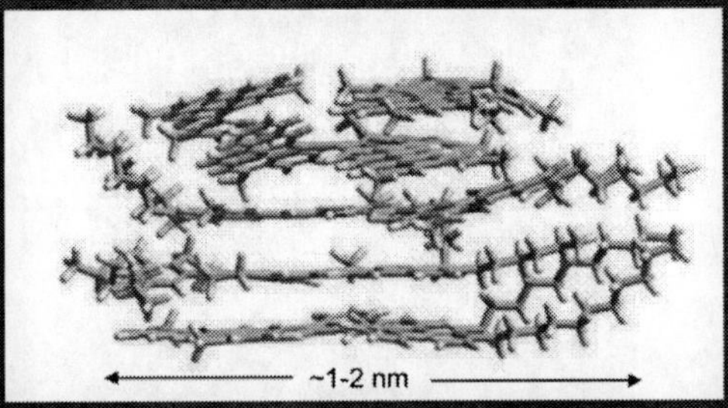

- These molecules while relatively insoluble by themselves associate in hydrocarbon solvent to form stable stacked (but irregular) assemblies .
- The distance between aromatic cores has been known for >20 years to be ~ 3.5 Å, perfectly in line with the predictions from interactions due to London dispersion forces and observations on model systems. There is no pi-pi bonding. They stack to maximize the points of close contact.
- It is London dispersion forces that dominate, all else is secondary.
- But just as the individual molecules are not asphaltenes these assemblies cannot rightly be called asphaltenes either.

17

Pi-stacking but No π- π Bonding

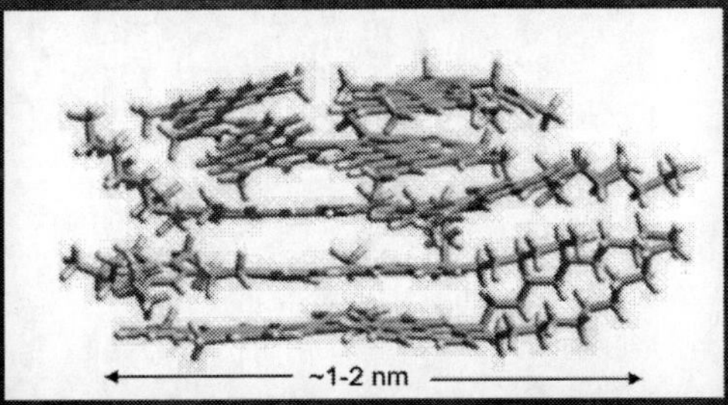

These Molecules are like Fridge Magnets

Energy of Interaction due to London Dispersion Force $w \propto -\sum_{a,b} \left(\frac{\alpha_a \alpha_b}{r6}\right)$

Fridge Magnets $w \propto -\sum_{a,b} \left(\frac{m_a m_b}{r3}\right)$

- These molecules while relatively insoluble by themselves associate in hydrocarbon solvent to form stable stacked assemblies .
- The distance between aromatic cores has been known for >20 years to be ~ 3.5 Å, perfectly in line with the predictions from interactions due to London dispersion forces and observations on model systems. There is no pi-pi bonding. They stack to maximize the points of close contact.
- It is London dispersion forces that dominate, all else is secondary.
- But just as the individual molecules are not asphaltenes these assemblies cannot rightly be called asphaltenes either.

18

Would the Real Asphaltenes Please Stand Up?

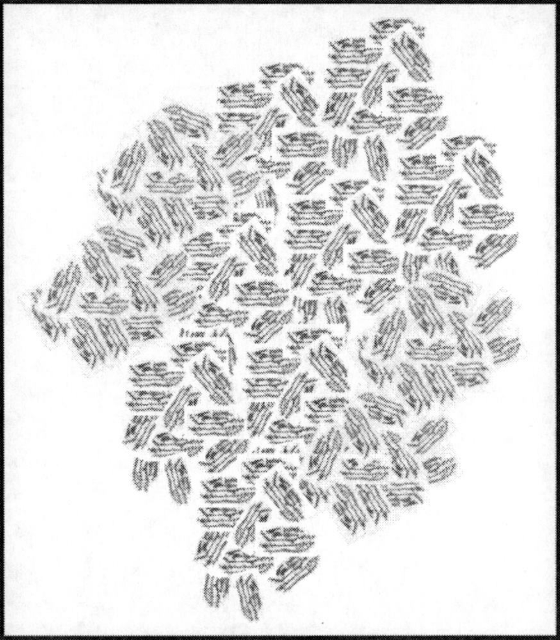

- The stacked assemblies group together in stable colloidal particles containing thousands of individual molecules, each particle being between 40 and 100 nm in diameter.
- It is these colloidal particles that are destabilized by the addition of a poor solvent like n-heptane
- *These are real asphaltenes?*

Chevron

Would the Real Asphaltenes Please Stand Up?

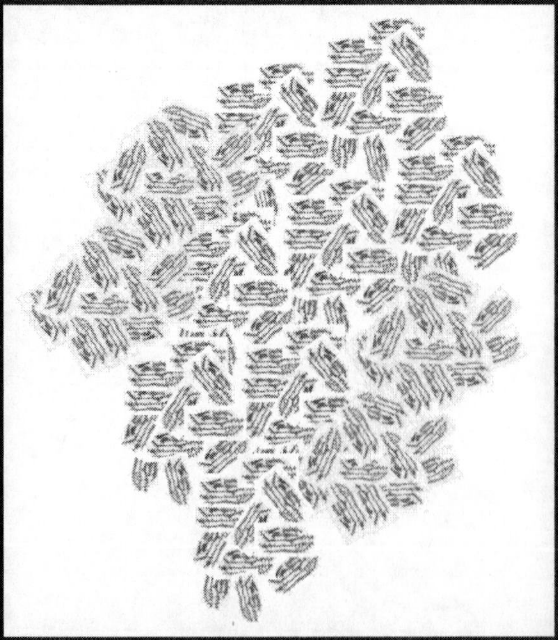

- The stacked assemblies group together in stable colloidal particles containing thousands of

The Unanswered Questions:

- *What is the role of detailed molecular composition in aggregation and disaggregation?*
- *What are the forces that keep these structures stable?*
- *Can we enhance the stability of the system?*

Chevron

So Do Asphaltenes Exist?

- Yes, but perhaps not in the way you might have thought
- Perhaps asphaltenes contain molecules of moderate molecular weight with a hydrogen deficient core in a large colloidal structure

But in reality:

Chevron

So Do Asphaltenes Exist?

- Yes, but perhaps not in the way you might have thought
- Perhaps asphaltenes contain molecules of moderate molecular weight with a hydrogen deficient core in a large colloidal structure

But in reality:

- *The* **Asphaltenes** *content of a petroleum product is the percentage by weight of wax-free material insoluble in n-heptane but soluble in hot benzene.*

Chevron

Petrogenium.

Paper 64e

Molecular Structure of Heavy Oil Revealed 2. The Effect of Severe Hydroprocessing

Dr Frans van den Berg, retired from Shell in October 2016
Current owner of Black Oil Solutions
Principal consultant for Petrogenium BV

October 2017

Samples for this study

B1: Feed asphaltenes. C7 asphaltenes from Athabasca vacuum residue (after partial deasphalting in the field due to paraffinic froth treatment at the mine). Extracted with hot heptane, filtered and Soxhlet extracted until colorless. Vacuum residue contained about 10 %w asphaltenes. Commercial sample from feed to LC-Fining unit.

B2: Product asphaltenes. C7 asphaltenes from the bottom of a mild vacuum column, after severe hydroprocessing of Athabasca vacuum residue (80 % residue conversion) in the LC-Fining unit. Bottoms also contained about 10 %w asphaltenes. Commercial sample from product of LC-Fining unit.

NB: At 80 % conversion, about 4 out of 5 asphaltene molecules are converted as well. Only the most refractory compounds remain. Direct comparison of feed and product molecules is therefore not sensible.

Simplified LC-Fining Reactor Diagram

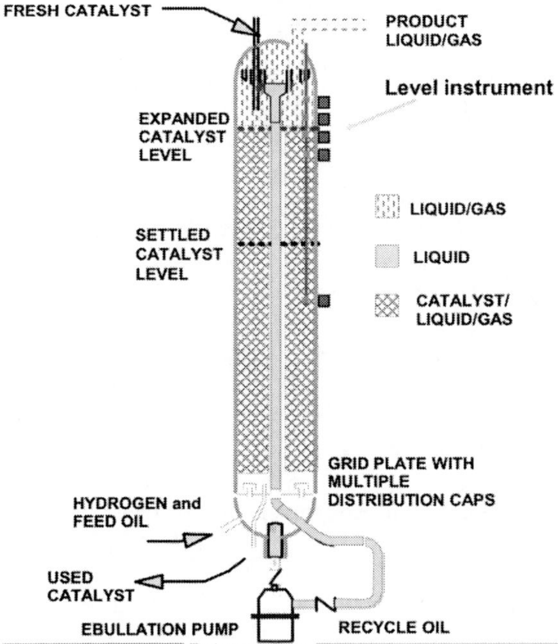

Typical operating conditions:

- **Pressure: 110 – 200 barg** (1600 – 2900 psig)
- **Temperature: 400 – 445 °C** (750 – 830 F)
- **LHSV 0.25 – 1.0 hr^{-1}**
- **Ebullation rate 3-6 x feed rate**
- **Hydrogen rate 3-4 x uptake rate**
- **Bed expansion 130 – 200% of settled level**
- **Expanded bed level 70-80% of tangent**
- **Catalyst Addition Rate 0.4-1.9 kg/mT** (0.15-0.65 lbs/bbl)
- **Catalyst replacement 2-6%**

Capacity: 47000 bbl/d, 7450 tonnes/day

Petrogenium.

CONFIDENTIAL

Properties of feed and product asphaltenes

Property	C7 Feed asphaltenes	C7SR-Product asphaltenes	
Carbon, %w	80,49	87,59	↑
Hydrogen, %w	7,72	5,55	↓
H: C ratio, atomic	1,14	0,76	↓
Sulphur, %w	7,54	1,76	↓
Nitrogen, %w	1,40	1,27	
Oxygen, %w	2,60	0,82	↓
Vanadium, ppmw	1086	595	↓
Nickel, ppmw	406	520	↑
Density, g/ml	1,12	1,27	↑
Apparent Molecular Weight, g/mole (VPO)	2900	1200	↓
Molecular weight from FTICRMS, g/mole	462 (200 – 800)	442 (200 – 700)	
Solubility parameter range, MPa^0,5	19,2 – 21,3	20,6 – 23,0	↑

See also PhD thesis Diana Powers, UofCalgary, 2014

Petrogenium.

CONFIDENTIAL

Typical AFM image of feed asphaltene

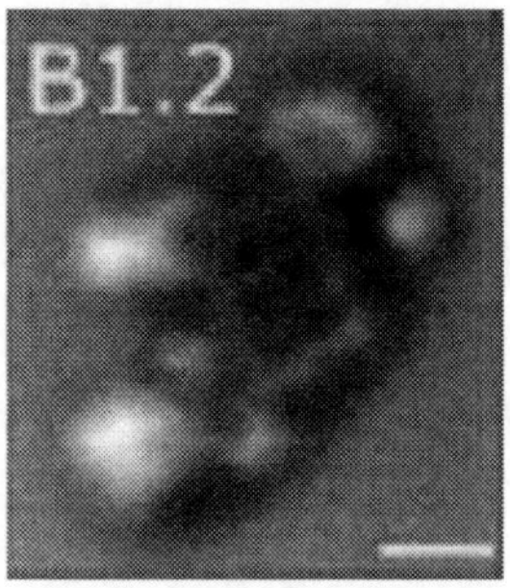

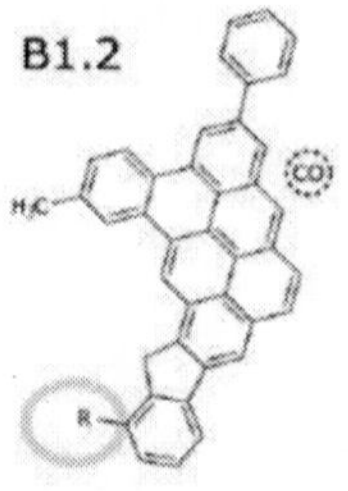

About 70 molecules studied:
- relatively small molecules (no clustering)
- 5 carbon rings a regular feature
- aromatic fragments often connected with shortest possible chain

Unusual molecules in feed asphaltenes

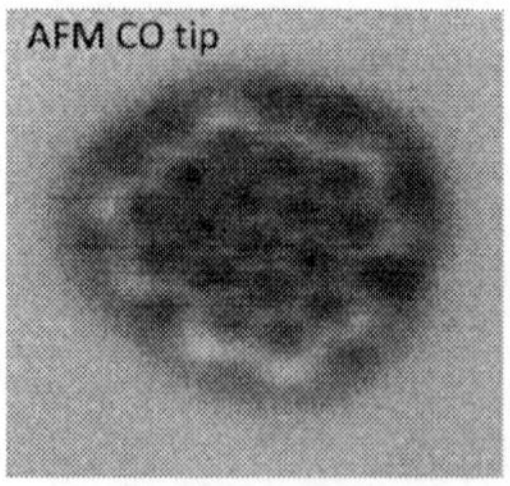

#28

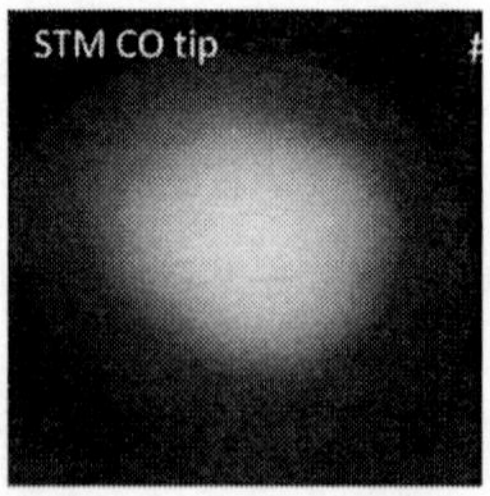

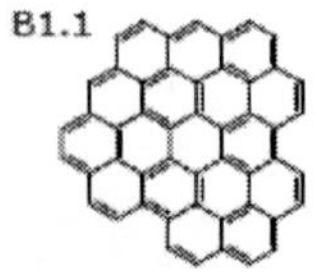

Pure PAH, 17 rings!

on Cu(111)

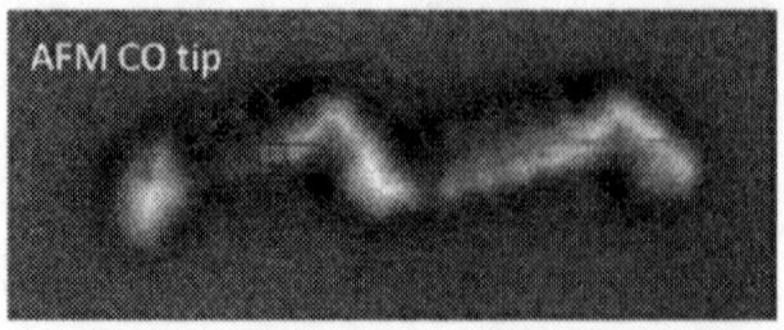

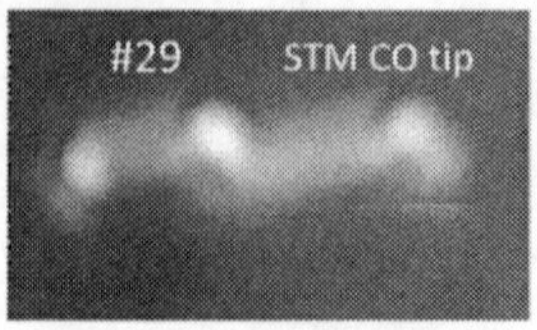

Significant number of paraffins observed in feed
#29 is a linear alkane, $C_{34}H_{70}$
Zig-zag pattern follows surface morphology

Some archipelago molecules present in feed

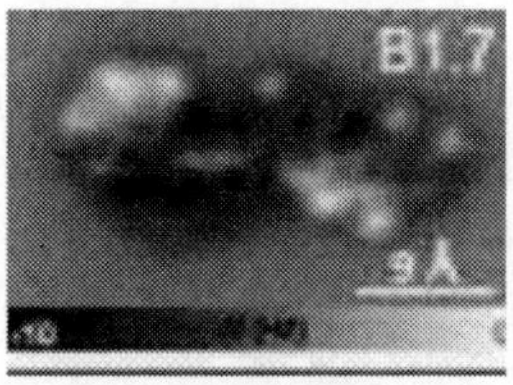

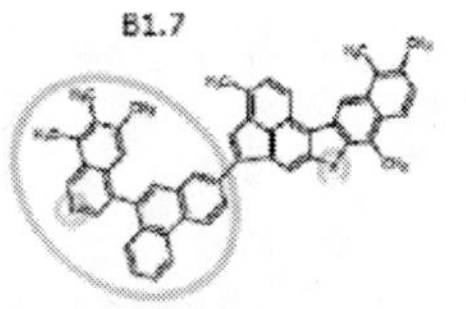

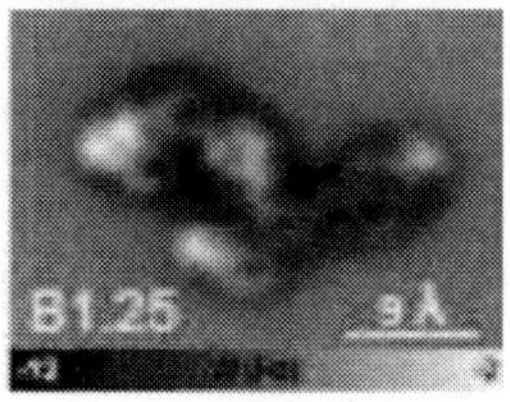

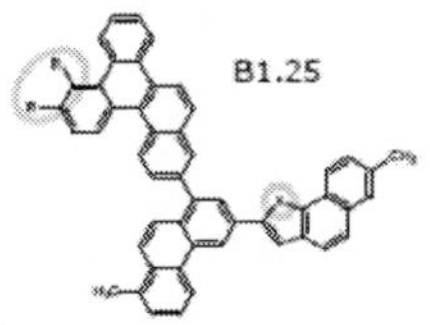

Typical example of product asphaltene

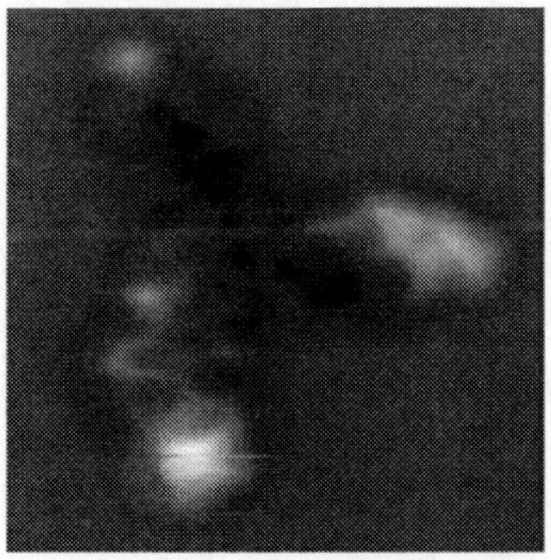

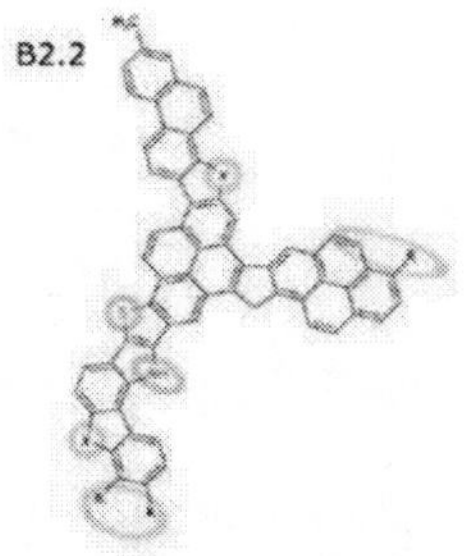

About 50 molecules studied:

- again, 5 atom rings a regular feature
- molecules tend to be more 3D (less planar, more bulky, more difficult to observe)
- no paraffins observed

Unusual molecules in product asphaltenes

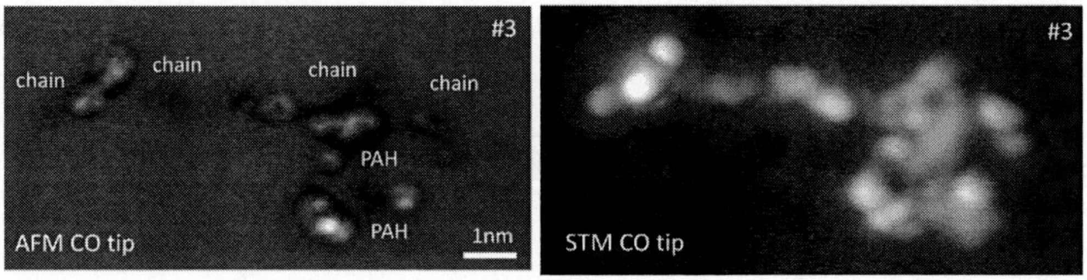

NB: Probing with tip showed this to be a single molecule!

Several archipelago molecules also observed in product. Incomplete cracking???

Product asphaltenes appear more 3 dimensional

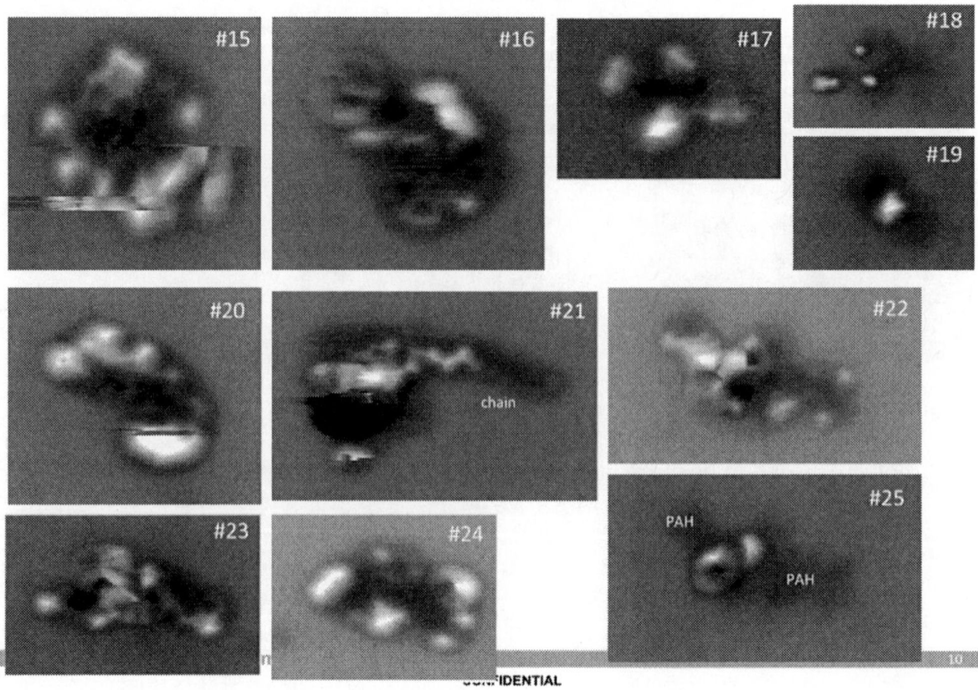

Conclusions

- More questions than answers, but
- 5 atom rings are a regular feature of feed and product asphaltenes
- Molecules observed are mainly island, but archipelago type molecules are also observed, even in products (room for more conversion?)
- Product asphaltenes appear to be 'bulkier' or more 3D than feed asphaltenes.

→ Possible explanation for smaller aggregates in cracked asphaltenes (other studies indicate only monomers and dimers in product asphaltenes)

Petrogenium.

because downstream matters

Back-up slides

Summary of typical product distribution

[redacted]

[redacted]

[redacted]

[redacted]

[redacted]

C7 'typical' examples

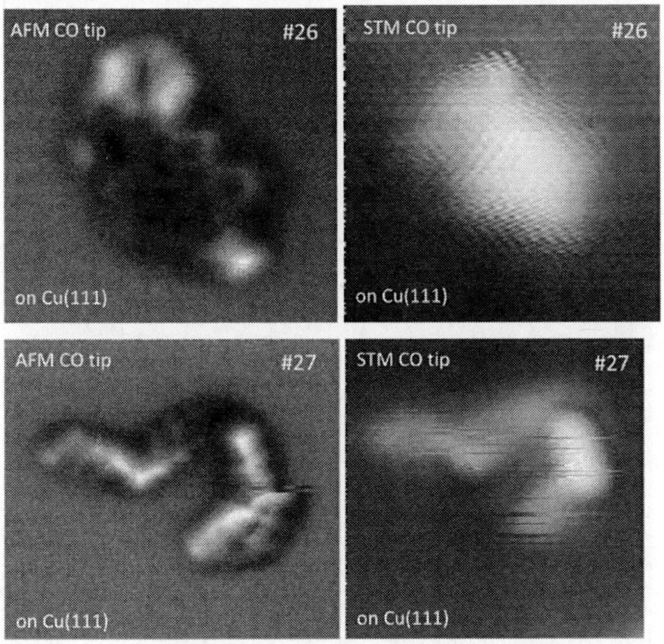

C7 more examples

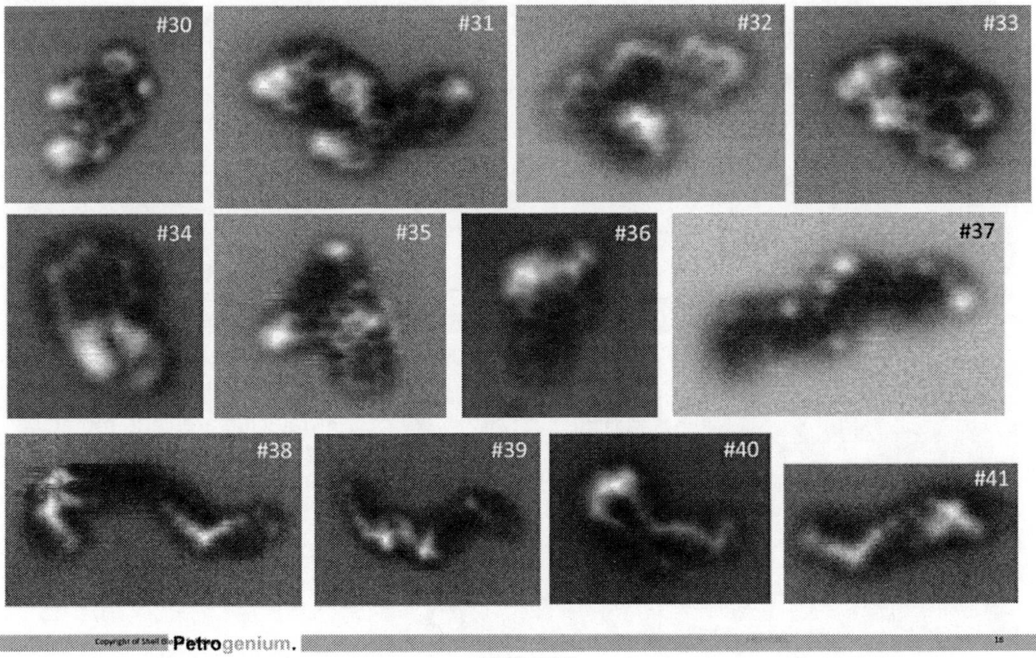

C7 more examples

(this slide: specifically looking for long chains and large molecules → not representative)

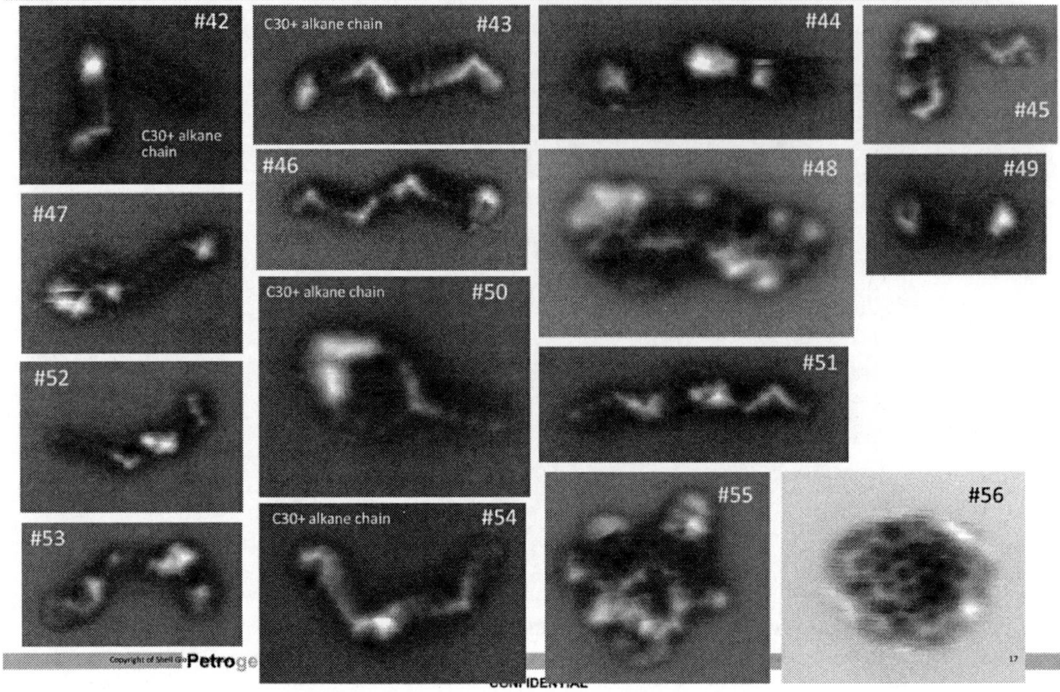

Typical example of product asphaltene

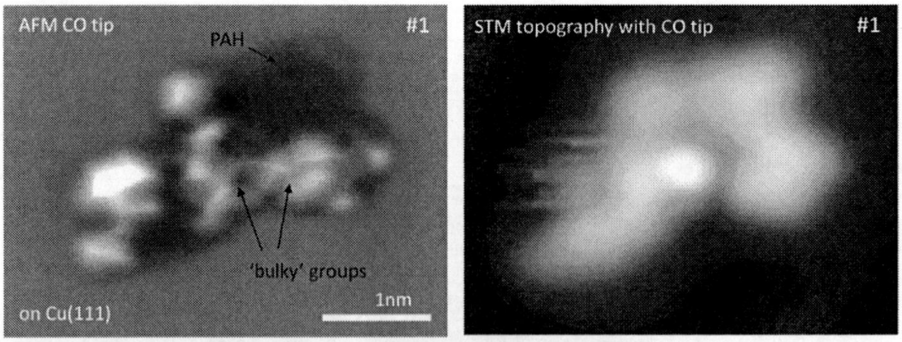

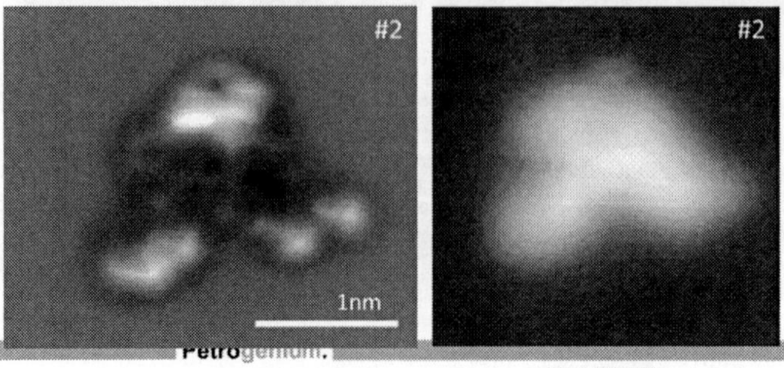

'Typical' examples in terms of:

- Molecule area and molecule types observed
- Resolvable features

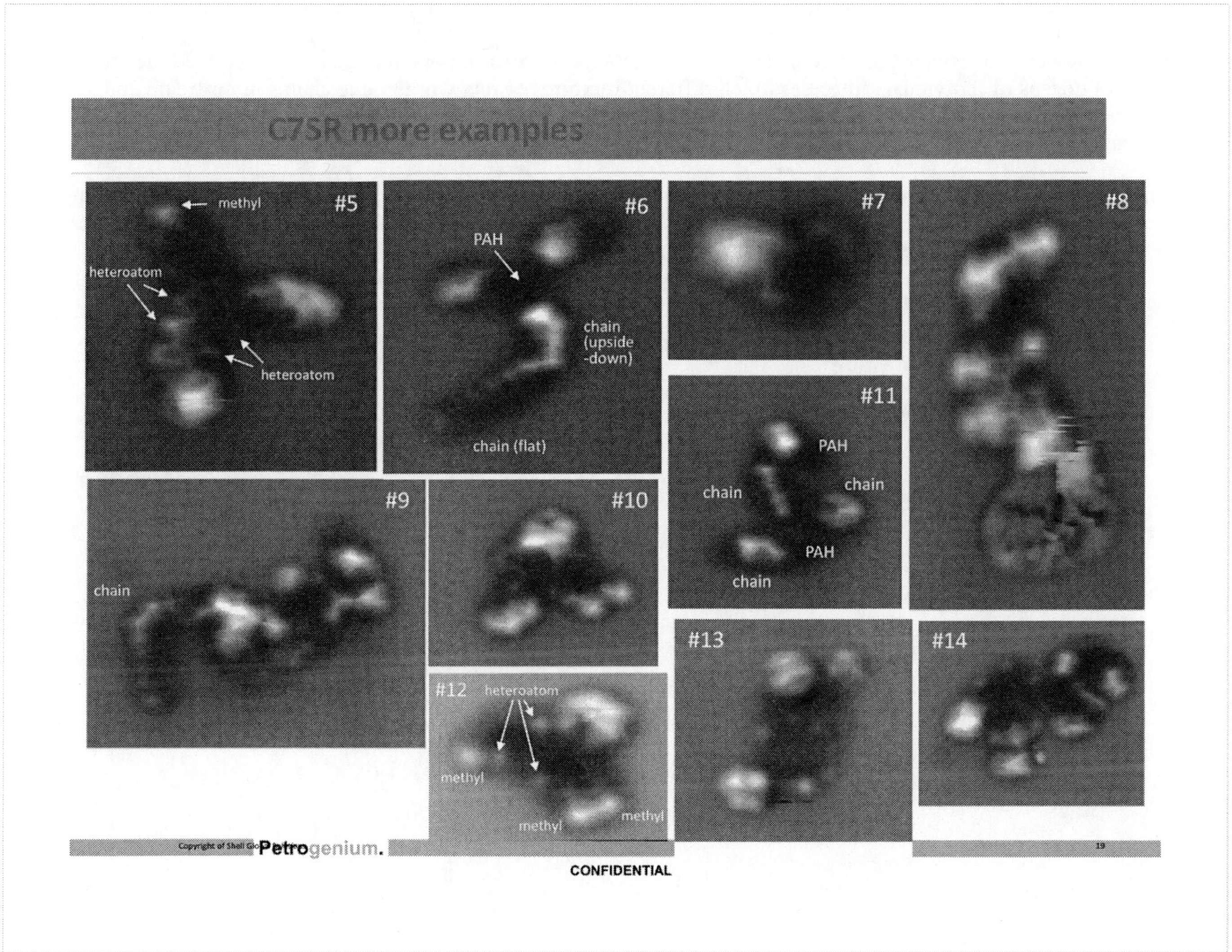
C7SR more examples
#5
methyl
heteroatom
heteroatom
#6
PAH
chain
(upside
-down)
chain (flat)
#7
#8
#9
chain
#10
#11
PAH
chain
chain
PAH
chain
#12
heteroatom
methyl
methyl
methyl
#13
#14
Petrogenium.
CONFIDENTIAL
19

505822 Molecular Structure of Heavy Oil Revealed - 3: High Resolution Atomic Force Microscopy - Perspectives and Examples in Heavy Oil Characterization

Abstract:

Identification of complex molecular mixtures poses a tremendous challenge for petroleum science. In our team at IBM Research, we developed a tool to visualize the structure and other molecular properties of *individual* molecules in such mixtures. In this talk, I will present two recent studies investigating heavy oil mixtures of different origin and treatment that were investigated in a broad consortium of corporate research organizations. We resolved the atomic structure of dozens of individual molecules in ten different mixtures of real world value chains in upstream and downstream oil production. Furthermore, the results are compared to mass spectrometry data to ensure a consistent interpretation.

Additionally, we are establishing a database for AFM identification of molecular substructures. To this end, we studied several model compounds with commonly occurring moieties, among aliphatic chains and rings. These studies already allowed the identification of several motifs in natural samples.

505823 Molecular Structure of Heavy Oil Revealed - 4: Molecular Structure of Heavy Oil Revealed with Non-Contact Atomic Force Microscopy

Abstract:

The structure of heavy oil remains a challenge of significant commercial interest to the petroleum industry and an active area of research. Although techniques such as NMR, XPS, and FTICR-MS can yield detailed chemical information (aromaticity, heteroatom local bonding environment, and exact molecular formulae, respectively), the breathtaking heterogeneity of petroleum samples precludes mapping these analytical results onto unique structures for each molecule.

Schuler et al. (*J. Am. Chem. Soc.*, 2015, 137, 9870-9876) recently imaged individual coal and petroleum asphaltene molecules using non-contact atomic force microscopy (NC-AFM) and found single-core 'island' motifs to predominate over multi-core 'archipelago' motifs. Recent studies with bispyrenyl model compounds (B. Schuler, Y. Zhang and co-workers, *Chem. Sci.*, 2017, **8**, 2315-2320) demonstrated that aromatic cores linked by alkyl spacers can be deposited and imaged successfully. This work extends our study with additional heavy oil petroleum samples:

- A highly aromatic steam cracker tar asphaltene anticipated to have a low degree of methylation and
- A de-asphalted vacuum residue fraction anticipated to have naphthenic rings and alkyl chains

Image analysis of several dozen steam cracker tar asphaltene molecules reveals many unanticipated structures. Specific classes of structural motifs were observed that provide clues for elucidating thermal chemistry pathways. Bulky pendant alkyl groups and non-planar naphthenic rings in the de-asphalted oil compromised their image clarity; further analysis of these images is underway.

505824 Atom Explicit Compositional Models of Heavy Oils: Making and Using State-of-the Art Measurements

Abstract:

The considerable interest in molecule-based models of hydrocarbon structure and reaction is motivated by the need to predict the properties of a given composition. This is because the molecular composition is an optimal starting point for the prediction of mixture properties. The potential advantages of molecule-based modeling are thus clear. Less readily apparent, however, is that the development and operation of molecular models comes with a large requirement for state-of-the-art analytical chemistry measurements and model construction and solution time. This presentation considers current approaches to both the quantitative measurement of molecular compositions and the use of this information in reaction models.

The challenge of building molecule-based models is due to the staggering complexity of the complex reaction mixtures. There will often be thousands of potential molecular and intermediate (e.g., ions or radicals) species. The sheer size of the thus-implied modeling problem creates a conflict between the need for molecular detail and the formulation and solution of the model. Clearly, the use of the computer to not only solve but also formulate the model would be helpful in that it would allow the modeler to focus on the basic chemistry, physics and approximations of the model.

While the fine details vary from one research group to another, all approaches address four general issues. The first is the measurement of the feed molecules. The analytical chemistry will generally be assisted by software tools that transform bulk analytical chemistry into a molecular description of the feedstock. This casts the modeling problem in molecular terms. Reactivity information is then often organized in terms of quantitative structure-reactivity relationships. Network building tools allow for the model equations to be built and coded on the computer. A solution environment allows for reactor simulation, which provides a prediction of the molecular composition that can then be organized into any desired commercially relevant outputs.

This approach is illustrated through the development of molecule-based models for thermal cracking, reforming and hydrotreating.

498855 Perspectives on Catalyst Design for Complex Feedstocks

Abstract:

The past two decades have witnessed significant advances in spectroscopy, characterization, simulation and theory that have allowed for the elucidation of the atomic structure of working catalytic materials and the molecular transformations that occur in simple few â€“component feedstocks. The extension of these advances to fuels and other multicomponent feedstocks has for the most part been non-existent due to complexity and ill-resolution of such feedstocks.

Recent pioneering advances in the characterization of complex hydrocarbon feedstocks including gas oils, heavy oils, biomass and waste stream are beginning to provide detailed insights into the molecules of such feedstocks and have enabled the tracking of these molecules during reaction.Â Such insights will enable unprecedented advances in development of realistic molecule-based kinetic models and the application of computational chemistry to enable the design of optimal catalyst-feedstock pairing.

In this talk, we discuss the how the recent advances in spectroscopy and characterization of complex feedstocks together with ab initio based kinetic Monte Carlo simulations can be used to construct detailed molecular distributions of specific complex feedstocks and follow the transformations of the complex array of molecules over different metal, metal oxide and zeolite catalysts. These simulations are closely coupled with ab initio derive kinetic models that treat the specific detailed atomic and electronic structure of the catalyst (i.e. acidity, alloy composition, adsorbate bond strengths) and their influence on the elementary reaction steps. We show how these models together with advanced characterization will aid in the design of catalysts for specific feedstocks.

505825 New Approaches to Big Chemistry: Rapid Screening and Quantification of Complex Hydrocarbon Mixtures

Abstract:

Advances in the utilization of complex hydrocarbon mixtures require molecular understanding of the chemistry and catalysis of thousands of organic compounds. In the past few decades, transition away from lumped chemistry models to "big chemistry" molecular models has dramatically improved the efficiency of hydrocarbon refining. Parallel experimental efforts in reaction analysis and compositional analysis work in tandem to build and validate large, complex molecular models. Here we present two new experimental techniques for rapid reaction screening and quantification of complex reacting systems containing hundreds of organic compounds.

Mixtures of organic chemicals can comprise thousands of compounds, necessitating expensive and time-consuming separation and detection with conventional calibration techniques. In this first technique, we demonstrate a catalytic microreactor which allows for calibration-free quantification of organic compounds within gas chromatography [[i]]. Analyte compounds eluting a conventional gas chromatograph column flow into the microreactor, where a series of catalytic reactions convert each analyte to methane. Subsequent detection via flame ionization thus results in a common carbon response factor for all compounds. By this approach, mixtures of hundreds of compounds including thiophenes, polyaromatics, and complex hydrocarbons with many heteroatoms can be quantified without calibration [[ii]]. The method is introduced with respect to thermodynamic design constraints, and implementation with conventional gas chromatograph systems is described [[iii]].

Detailed molecular analysis can be combined with advanced chemical reaction screening. Building on the micro-catalytic method first proposed by Kokes and Emmet [[iv]], we have developed a continuous-flow microreactor for rapid evaluation of the reactions of complex molecules. By this method, small quantities of reactant are provided as step pulses to a micro-reaction chamber, with product sampling occurring at steady state. The use of small reactant quantities enables short-time reactor and catalyst studies, which can provide detailed kinetics of rapid catalyst activation/deactivation. When paired with the quantitative carbon detection method, the combined experimental reaction system can rapidly provide automated evaluation of thousands of reaction conditions, leading to multi-dimensional screening and identification of optimal reactor performance [[v]].

[i] **"Quantitative Carbon Detector (QCD) for Calibration-Free, High-Resolution Characterization of Complex Mixtures,"** S. Maduskar, A.R. Teixeira, A.D. Paulsen, C. Krumm, T.J. Mountziaris, W. Fan, P.J. Dauenhauer, *Lab on a Chip*, 2015, 15, 440-447

[ii] **"Quantitative Carbon Detector for Enhanced Detection of Molecules in Foods, Pharmaceuticals, Cosmetics, Flavors, and Fuels"** Connor A. Beach, Christoph Krumm, Charles S. Spanjers, Saurabh Maduskar, Andrew J. Jones, Paul J. Dauenhauer. *Analyst*. 2016, 141, 1627-1632

[iii] **"Increasing Flame Ionization Detector (FID) Sensitivity using Post-Column Combustion-Methanation,"** Charles A. Spanjers, Connor Beach, Andrew Jones, Paul J. Dauenhauer, *RSC Analytical Methods* 2017, 9, 1928-1934.

[iv] R.J. Kokes, H. Tobin, P.H. Emmett, "New microcatalytic-chromatographic technique for studying catalytic reactions," *Journal of the American Chemical Society* **1955**, 5861.

[v] **"Renewable Isoprene by Sequential Hydrogenation of Itaconic Acid and Dehydra-Decyclization of 3-Methyl-Tetrahydrofuran,"** Omar A. Abdelrahman, Dae Sung Park, Katherine P Vinter, Charles S. Spanjers, Limin Ren, Hong Je Cho, Kechun Zhang, Wei Fan, Michael Tsapatsis, Paul J. Dauenhauer, *ACS Catalysis* 2017, 7(2), 1428-1431. DOI: 10.1021/acscatal.6b03335

503670 Digital Alchemy for Assembly Engineeringc

Abstract:

Back in the day, Stuart Churchill was a pioneer in the use of computers for modeling chemical and thermal processes. Since that time, the role of computing in chemical engineering has become indispensable. In this talk, we present new computational methods we call "digital alchemy" that allow for predictive modeling of thermodynamic assembly, and permit the design of building blocks optimized for assembly into target structures with desired properties. In this approach, standard thermodynamic ensembles are generalized to include alchemical variables and their thermodynamic conjugates, allowing for rapid, real-time computational investigation of large, complex design spaces.

503671 Improved Algebraic, Numerical, and Graphical Representations in Fluid Mechanics

Abstract:

The concept of equivalents is applied to fully developed flows through straight channels, and two applications are introduced: the equivalents that produce equal rates of flow through channels of different cross-section, and the diminishment that occurs between the regimes of laminar and turbulent flow. The validity of the eddy diffusivity and the invalidity of the mixing-length as measures of the shear stress due to turbulence have perhaps never before been demonstrated so clearly. The possible choices of primary variables and the consequences are also examined. The objective of these explorations is an improved understanding of the basic fluid mechanics by teachers, students, and industrial practitioners.

503672 The Scaling of Turbulence Near the Wall and the Churchill Turbulent Flux Correlation: Insights with Lagrangian Simulations

Abstract:

The evaluation of mean velocity and mean temperature in wall-bounded turbulent flows has been based on the "classical" theory of turbulence and the law of the wall. The viscous wall parameters, namely the friction velocity, the friction temperature, and the kinematic viscosity of the fluid have been used to scale not only the mean values of velocity, but also other mean turbulent quantities close to the wall, including the Reynolds stresses and the turbulent velocity and temperature fluctuations. Stuart Churchill, as he questioned the validity of the Reynolds analogy between momentum and heat or mass transfer in his works [1], he realized that extending the scaling of momentum to the scaling of scalar quantities in turbulence was not a sound approach. He proposed a scaling of turbulence that depends on the turbulent momentum flux and the turbulent heat flux. In his AIChE Institute Lecture (in Miami, 1998), Churchill presented a summary of his work on this topic, and closed form solutions for the mean velocity and temperature [2]. The turbulent Prandtl number, the most significant parameter for the modeling of turbulent transport not only in theory but also in modern CFD commercial software packages, was also interpreted physically based on these ideas [3]. We have used direct numerical simulation results and results from Lagrangian simulations [4] for the transport of passive scalars in turbulent channel flows to explore Churchill's ideas on turbulent flow scaling and to develop an updated theory for the scaling of heat transport in turbulent flows [5-7]. His passion to remove empiricisms and approximations from the development of predictive correlations in turbulent transport has been an inspiration. In this talk, we will revisit the Churchill scaling in view of our simulation results, and will discuss the counterintuitive finding that wall turbulence can be used to separate rather than mix substances under special circumstances [8].

1. Churchill, S. W., 1996, "A Critique of Predictive and Correlative Models for Turbulent Flow and Convection," Ind. Eng. Chem. Res., 35, pp. 3122–3140.
2. Churchill, S.W., 2000, "Progress in the Thermal Sciences: AICHE Institute Lecture," AIChE Journal, 46(9), pp. 1704-1722.
3. Churchill, S. W., 2002, "A reinterpretation of the Turbulent Prandtl Number," Ind. Eng. Chem. Res., 41, pp. 6393–6401.
4. Papavassiliou, D.V, 2002, "Turbulent transport from continuous sources at the wall of a channel," Int. J. Heat Mass Transfer, 45(17), pp. 3571-3583.
5. Le, P.M., and Papavassiliou, D.V., 2008, "On the scaling of heat transfer using thermal flux gradients for fully developed turbulent channel and Couette flows," Int. Commun. Heat Mass Transfer, 35(4), pp. 404-412.
6. Srinivasan, C., and Papavassiliou, D. V., 2011, "Prediction of Turbulent Prandtl Number in Wall Flows With Lagrangian Simulations," Ind. Eng. Chem. Res., 50(15), pp. 8881–8891.
7. Srinivasan, C., and Papavassiliou, D.V., 2013, "Heat transfer scaling for wall bounded turbulent flows," Applied Mechanics Review 65(3), Art. 031002.
8. Nguyen, Q., Srinivasan, C., and Papavassiliou, D.V., 2015, "Flow induced separation in wall turbulence", Phys Rev E, 91, Art. 033019.

503674 Flow Boiling Using a Piranha Pin Fin Heat Sink

Abstract:

Flow Boiling using a Piranha Pin Fin Heat Sink

Corey Woodcock, Xiangfei Yu, Yoav Peles[1] and Joel Plawsky

Rensselaer Polytechnic Institute; [1]University of Central Florida

The *Piranha Pin Fin* (*PPF*) microdevice is a heat exchanger engineered to dissipate extreme thermal loads using either dielectric fluids such as HFE7000 or water. Fabrication of the device is CMOS compatible and the first generations of the device have been fashioned from silicon substrates. The PPF operates on the principle that heat transfer would be more efficient if heated fluids could be removed nearly as fast as they are created. This process would keep the driving force as high as possible. Thus the heat exchanger is built as two layers; a primary substrate that dissipates the bulk of the heat and shunts the heated fluids to a booster section that continues to heat the fluid and extends the total surface area of the device. Three generations of devices were fabricated. The first two generations were single-layer systems designed to help understand the operation of the primary layer and optimize the shape and location of the pin fins. The third generation was the first, two-layer system to be fabricated and tested.

Figure 1 Representative PPF array

Correlations for the heat transfer coefficient, h_{tp}, based on the Boiling Number and the Weber Number in the PPF design was developed and validated with Gen II. The discrepancy of heat transfer coefficient between predicted and measured values was within 20%. The ratio of two-phase to single-phase pressure drop was correlated with the Boiling Number (Bo) and the Jakob Number (Ja). The discrepancy of the two-phase multiplier between prediction and experimental measurements were within 12%.

The third generation *PPF* heat sinks were the first series to provide both a primary substrate populated with *PPF*s, as well as a booster substrate populated with solid micro pin fins. The substrates are micro-aligned and fusion-bonded to provide a full three dimensional (3D) architecture with bulk silicon properties. Experimental results from the Gen III *PPF* microdevices provide a proof of concept of the true *PPF* heat sink concept with fluid venting into multiple substrates. At low flowrates (G = 670 [kg/s/m^2]) the Gen III *PPF* heat sink has been shown to dissipate heat loads approaching 1.1 [kW/cm^2] at an exit quality of 95%, while at higher flowrates (G = 3,300 [kg/s/m^2]) the thermal limit of the heat sink remains unknown due to exceeding the available power supply at >1.4 [kW/cm^2].

503673 Transport Problems in the Spirit of Stuart Churchill for Teaching and Research at the University of Michigan

Abstract:

Stuart Churchillâ€™s solutions to important heat transport problems have greatly influence both pedagogy and research. The correlation of Churchill and Chu for the natural convection heat transfer to or from a vertical plate, over an astonishing range of Rayleigh numbers is based partly on a similarity solution of a type common in heat, mass, and momentum transfer for problems in chemical engineering. One such problem is asphaltene or wax transport and deposition onto the wall of an oil pipeline, which can be solved using similar methods, as will be discussed. In addition, such problems, and the wealth of their applications make for excellent motivations and examples in the teaching of graduate or undergraduate transport, as will be discussed in the talk.

499961 A Framework for Optimization-Based Design of Heat-Integrated Crude Oil Distillation Units Using a Surrogate Model and Support Vector Machine

Abstract:

Global demand for petroleum-based fuels for industrial, transportation and domestic use has been increasing since inception. These fuels (gasoline, kerosene, diesel, liquefied petroleum gas, fuel oil, etc.) are produced in refineries via crude oil distillation. Crude oil distillation is a complex and capital-intensive process; the crude oil distillation system comprises distillation columns, pump-around loops, side-strippers and a heat recovery system, known as the preheat train. The large number of degrees of freedom (column structure and operating conditions) and complex interactions between individual units makes the design of crude oil distillation unit a highly challenging task. Rigorous tray-by-tray column models are useful for designing crude oil distillation columns, but the approach requires trial and error, or, if optimization is applied, good initial estimates to obtain convergence and significant computational time and poses numerical difficulties.

Surrogate models, including those based on artificial neural networks (ANNs), have enabled significant capabilities in modelling and optimization of various chemical processes, including crude oil distillation. Researchers have shown that artificial neural networks can provide simple, accurate and robust simulation models. Artificial neural network-based distillation models have been implemented in optimization frameworks to perform operational optimization of existing crude oil distillation unit. To date, however, surrogate models for crude oil distillation reported in the literature have not considered design of the distillation column, i.e. optimization-based design, especially considering discrete variables representing the number of trays in each column section.

This paper presents a novel systematic approach for the design of crude oil distillation units that combines artificial neural network-based distillation models and a support vector machine (SVM). The support vector machine aims to ensure that only feasible design alternatives â€“ i.e. those that meet constraints related to product specifications and hydraulic performance â€“ are considered during the search process. This filtering helps to produce an optimal solution that is feasible; it also reduces computational effort by limiting the search to points within the demarcated feasible region.

The proposed framework comprises five main steps. First, an initial feasible design of a crude oil distillation unit is built using a rigorous tray-by-tray model in Aspen HYSYS. In step 2, the most important degrees of freedom are identified and represented using probability distribution functions. Latin hypercube sampling is then applied on these distributions to generate samples (design inputs) that will be used in the calculations. In step 3, all the samples are simulated and the corresponding responses (design outputs) are collected and automatically separated into feasible and infeasible solutions. In step 4, feasible samples are used to build a surrogate (ANN) distillation column model, while the entire set of samples is used to build a support vector machine. In the final step, the resulting models are implemented in an optimization framework, in which a genetic algorithm is used to search for design alternatives that minimize the total annualized cost (the sum of annualized capital cost and annual operating cost). Pinch analysis is used within the design procedure to determine minimum utility targets and estimate the utility costs (i.e. operating costs). The capabilities of the approach are illustrated through its application to an industrially-relevant case study.

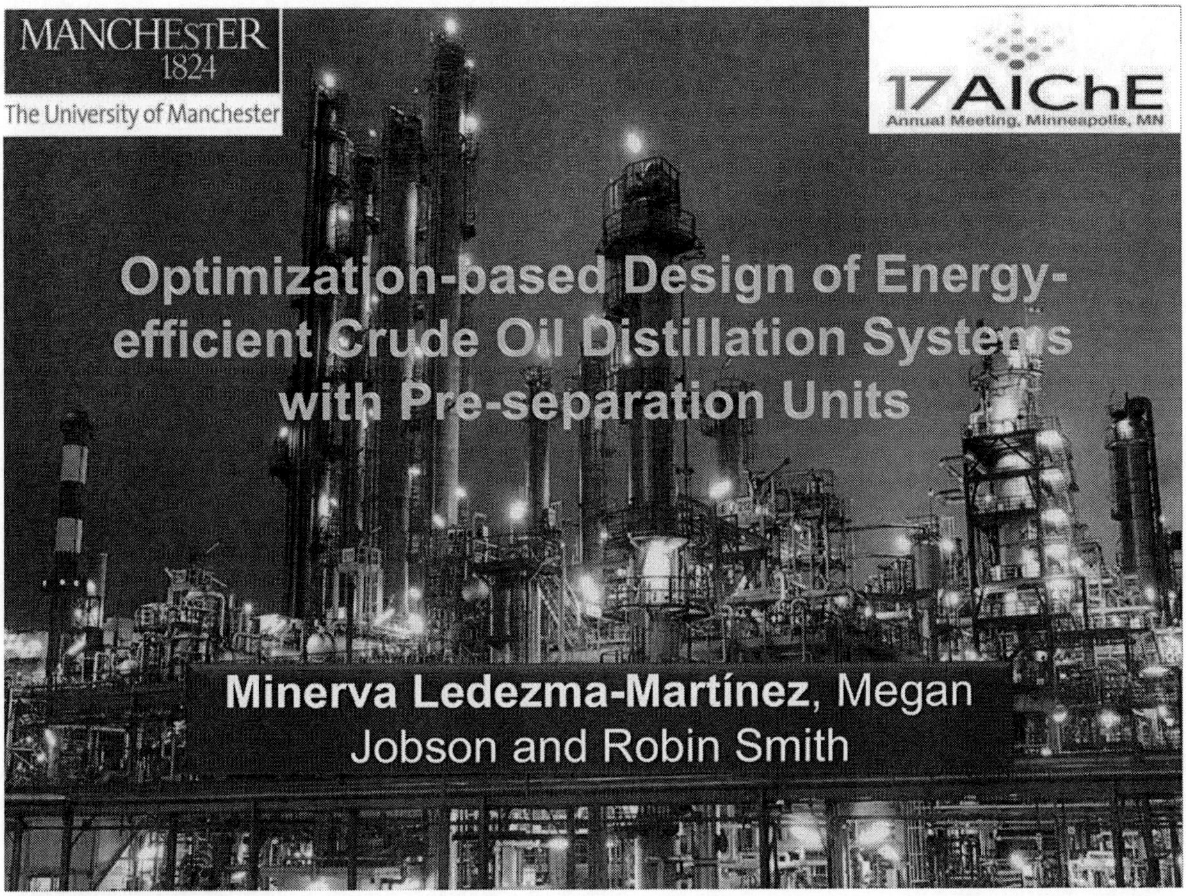

MANCHESTER
1824
The University of Manchester

Contents

1. Introduction
2. Research Objective
3. Optimization-based Design Methodology
4. Case Study
5. Conclusions
6. Future Work

MANCHESTER 1824
The University of Manchester

1. Introduction

MANCHESTER 1824
The University of Manchester

Crude Oil Distillation

- Crude oil distillation systems play a central role in petroleum refining.
- Energy-intensive systems: 2–7% of the crude oil input is consumed for crude oil distillation[1].
- Energy savings can be achieved by exploiting interactions between separation units and the heat recovery system.

Catlettsburg Oil Refinery, Kentucky USA

1. A. Szklo, and R. Schaeffer, 2007, Fuel Specification, Energy Consumption and CO_2 Emission in Oil Refineries, Energy, 32, 7, 1075–1092.

MANCHESTER 1824
The University of Manchester

Crude Oil Distillation System

- Crude oil distillation system:
  1) A preheat train.
  2) Crude oil distillation units with side strippers and pump-arounds.
  3) Pre-separation units.
- **Preflash units** help to enhance energy efficiency (i.e. to reduce fired heat demand).
- Energy efficiency benefits to reduce CO_2 emissions and operating costs.

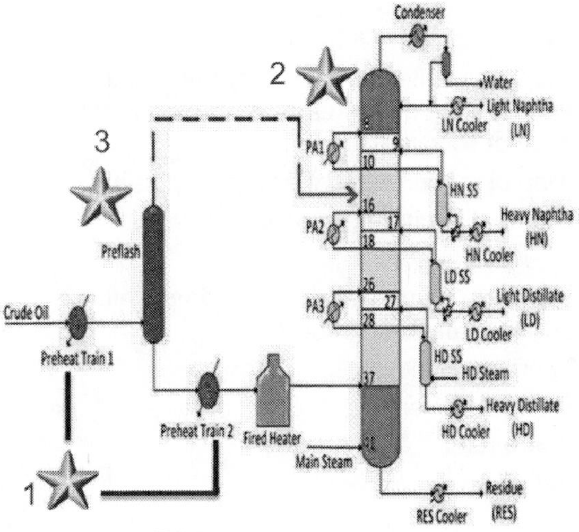

High operating and equipment costs and the complexity of the system motivate the development of systematic approaches for optimal system design.

MANCHESTER 1824
The University of Manchester

Crude Oil Distillation System Design

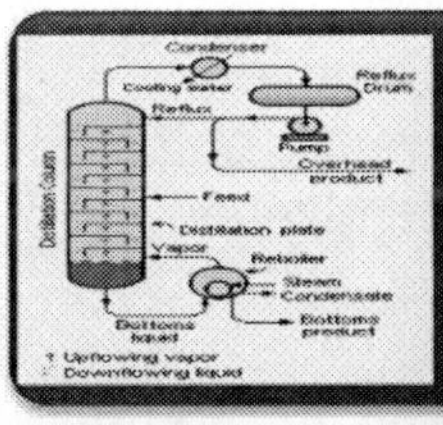

Traditional methods

- Heuristic rules
- Involves trial and error

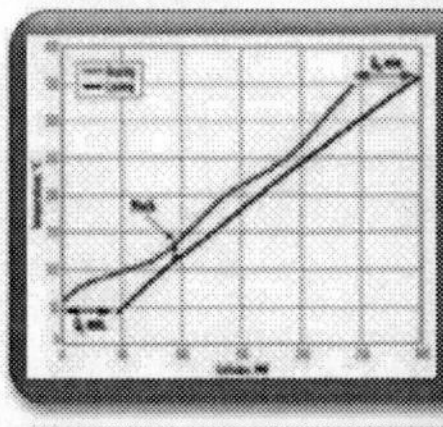

Integrated methods

- Simulations and pinch analysis
- Capital–energy trade offs not considered

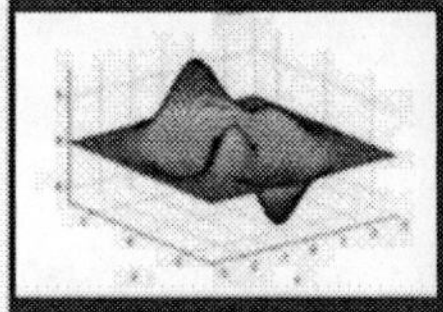

Optimization-based design methods

- Design supported by optimization
- Lack of design methodologies for crude oil distillation systems with pre-separation units

MANCHESTER 1824
The University of Manchester

Pinch Analysis

- Grand Composite Curve shows net heating and cooling needs by temperature interval (after heat recovery).
  - Useful for conceptual design of column.
  - No heat exchanger network details.
- Stream data (supply and return temperatures and duties) are simulation results.
  - Pump-arounds.
  - Product coolers.
  - Condenser.
  - Reboilers.
- High temperature hot utility (**fired heating** = furnace duty) dominates operating costs.

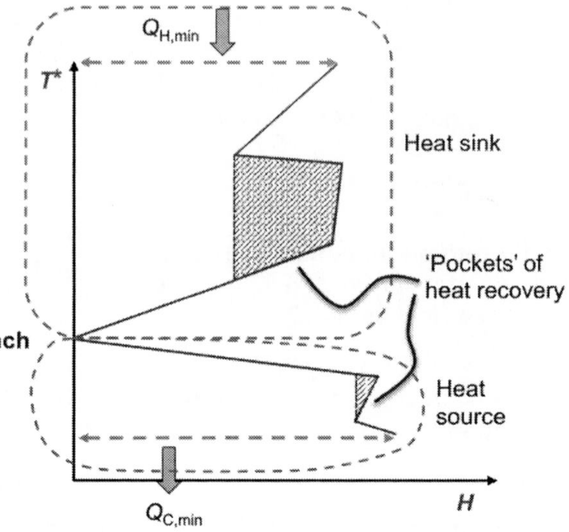

MANCHESTER 1824
The University of Manchester

Optimization (Genetic Algorithm)

- Genetic Algorithms (GA) have found wide acceptance in Chemical Engineering because of their simplicity, and ability to determine optimum solutions for both single and multi-objective, non linear optimization problems[1].

Advantages

- Less prone to getting trapped in local optima than deterministic searches.
- It is possible to handle a variety of decision variables (binary, integer or continuous).

Structure of Genetic Algorithm

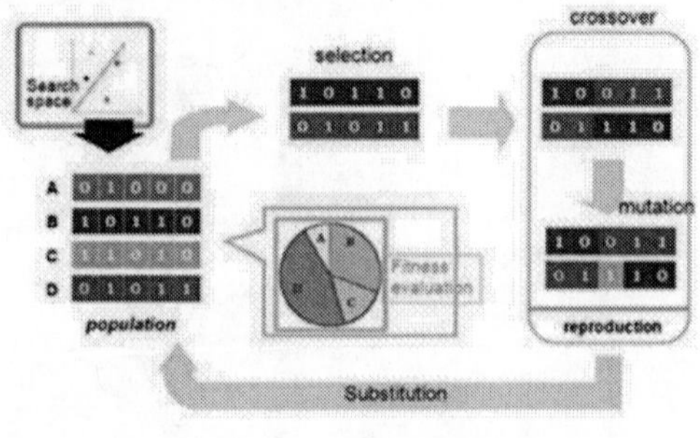

1. Prakash R. Kotecha, Mani Bhushan and Ravindra D. Gudi, 2010, Constraint Programming and Genetic Algorithm in *Stochastic Global Optimization: Techniques and Applications in Chemical Engineering* ,Vol. 2, (ed. G.P. Rangaiah), World Scientific Co., Singapore.

MANCHESTER 1824
The University of Manchester

2. Research Objective

MANCHESTER 1824
The University of Manchester

Research Objective

To develop a systematic approach to design cost-effective, energy-efficient crude oil distillation systems with pre-separation units, accounting for product quality and heat integration.

Preflash/ Prefractionator
- Key opportunity for energy saving.
- Light components bypass fired heater.
- Vapour feed location affects separation within column.

Crude Oil Distillation Unit
- Many degrees of freedom interacting with each other.
- Interactions with Heat Exchanger Network (HEN).
- Product quality specifications must be within allowed range.

Heat Integration
- Reduces energy cost.
- Grand Composite Curve (GCC) sets energy targets.

MANCHESTER 1824
The University of Manchester

3. Optimization-based Design Methodology

MANCHESTER 1824
The University of Manchester

Heat-integrated crude oil distillation unit with a preflash – Degrees of freedom

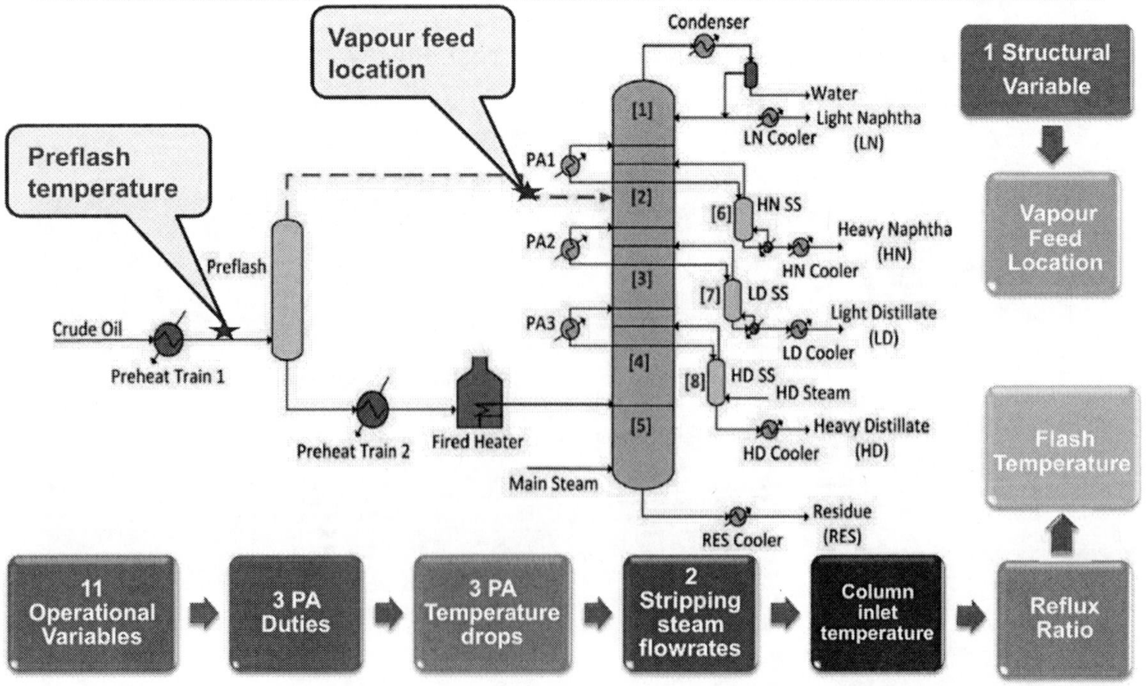

MANCHESTER
1824
The University of Manchester

Simulation-optimization-based design methodology

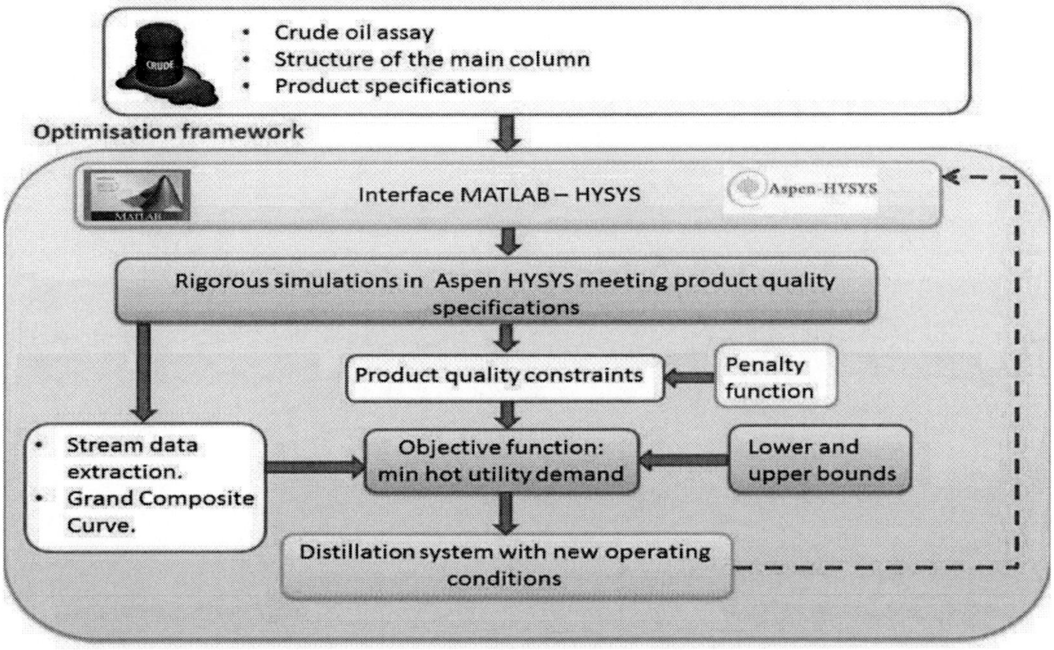

MANCHESTER
1824
The University of Manchester

Direct Optimization with GA

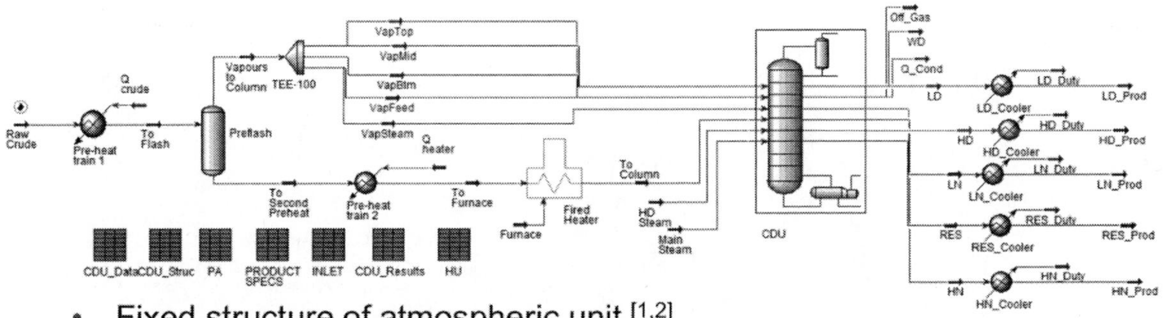

- Fixed structure of atmospheric unit [1,2].
- Optimize preflash temperature and flash vapour feed stage.
  Flash vapour connected to 5 different stages in the main column (1 per section).
- Minimise hot utility demand.
- Constraints: product quality specifications (T5%, T95% ASTM).
- Optimization: Genetic Algorithm.
  Population size = 100; 500 generations.

[1] Watkins R.N., 1979, Petroleum Refinery Distillation, Gulf Pub. Co., Houston, USA.
[2] Chen L., 2008, Heat-integrated Crude Oil Distillation System Design, PhD Thesis, University of Manchester, Manchester, UK.

MANCHESTER 1824
The University of Manchester

4. Case Study

MANCHESTER 1824
The University of Manchester

Crude Oil Distillation Unit with Preflash

Venezuela Tia Juana Light crude oil
100,000 bbl/day (0.184 m^3/s)
25°C, 2.5 bar

Main Steam:
260°C, 4.5 bar
1200 kmol/h

HD Steam:
260°C, 4.5 bar
250 kmol/h

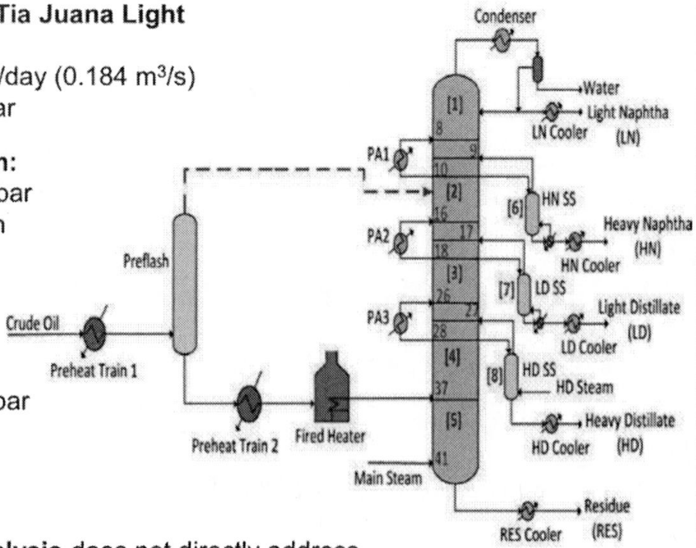

Column section	Stage numbering
Section 1	1–9
Section 2	10–17
Section 3	18–27
Section 4	28–36
Section 5	**37–41**
Section 6 (HN SS)	42–47
Section 7 (LD SS)	48–54
Section 8 (HD SS)	55–59

- **Pinch analysis** does not directly address HEN design.
- Assumption: ΔT_{min} = 30°C.

MANCHESTER 1824

The University of Manchester

Product specifications and flowrates

Product	T5% ASTM D86 (°C)	T95 % ASTM D86 (°C)	Flow rate (m³/h)
LN	26	110*	102
HN	140	196*	87
LD	217*	300*	128
HD	308	354*	54
RES	362	754	292

* Specified in HYSYS.

- For all products, the allowed range was ±10°C.

MANCHESTER 1824

The University of Manchester

Objective function and constraints

- **12 Degrees of freedom**
  - Pump-around duties (3)
  - Pump-around temperature drops (3)
  - Main steam
  - HD steam
  - Column inlet temperature
  - Reflux ratio
  - Flash temperature
  - Vapour feed location (main column)
- **10 Constraints**
  - ✓ Product quality specifications (ASTM D86 T5% and T95% temperatures) for five products

Objective function

$$min\ F(x) = f(x) + \gamma_1 |h(x)| + \sum_{j=1}^{n} \gamma_{1+j}\ max\left(0, g_j(x)\right)$$

$F(x)$ = Unconstrained objective function

$f(x)$ = Constrained objective function

$g_j\ (x) \leq 0$ $h(x) = 0$

γ = Penalty factor

$$T5_i^{lb} \leq T5_i \leq T5_i^{ub}$$

$$T95_i^{lb} \leq T95_i \leq T5_i^{ub}$$

i=1, 2,…..$N_{products}$

Ochoa-Estopier, L. M., Jobson, M., 2015, Optimization of heat-integrated crude oil distillation systems. Part I: The distillation model, *Industrial & Engineering Chemistry Research*, 54 (18) 4988–5000 doi: 10.1021/ie503802j.

MANCHESTER 1824

The University of Manchester

Optimization variables and bounds

Variable	Units	Base Case	Lower bound	Upper bound
1. Main steam flow	kmol/h	**1200**	900	1300
2. HD steam flow	kmol/h	**250**	200	375
3. PA1 duty	MW	**12.8**	14	6
4. PA2 duty	MW	**17.8**	18	6
5. PA3 duty	MW	**11.2**	12	6
6. PA1 ΔT	°C	**30**	20	50
7. PA2 ΔT	°C	**50**	15	60
8. PA3 ΔT	°C	**20**	10	40
9. Column inlet temperature	°C	**365**	350	375
10. Reflux ratio	-----	**4.17**	3.0	4.5
11. Flash temperature	°C	**115**	110	230
12. Vapour feed location (column section)	-----	**5**	1	5

✓ Bounds defined based on sensitivity studies.

MANCHESTER 1824

The University of Manchester

Optimization Results

Variable	Units	Base Case	Optimization Results	
			No Preflash	With Preflash
1. Main steam flow	kmol/h	1200	900	1190
2. HD steam flow	kmol/h	250	200	211
3. PA1 duty	MW	12.8	6.0	6.8
4. PA2 duty	MW	17.8	9.8	8.3
5. PA3 duty	MW	11.2	11.9	8.8
6. PA1 ΔT	°C	30	48.6	25.5
7. PA2 ΔT	°C	50	33.8	23.0
8. PA3 ΔT	°C	20	39.9	39.3
9. Column inlet temperature	°C	365	350	354
10. Reflux ratio	°C	4.17	3.7	3.0
11. Flash temperature		115	----	230
12. Vapour feed location (column section)		**5**	---	3
Minimum hot utility demand	**MW**	**58.3**	**43.4**	**35.9**

MANCHESTER 1824

The University of Manchester

Grand Composite Curves

Optimized base case (no preflash)

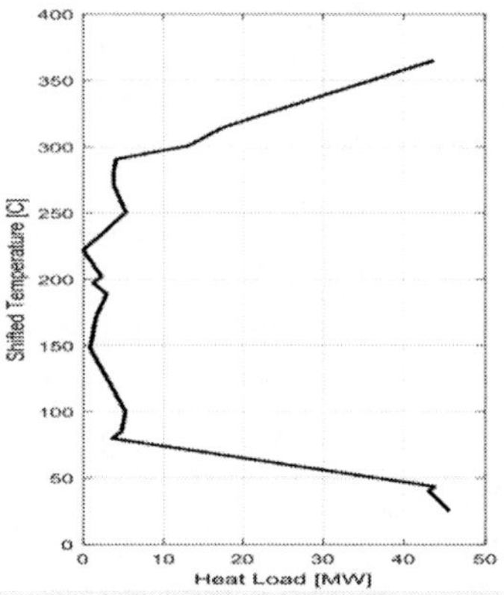

HU = 43.4 MW

CU = 45.5 MW

Optimized design (with preflash)

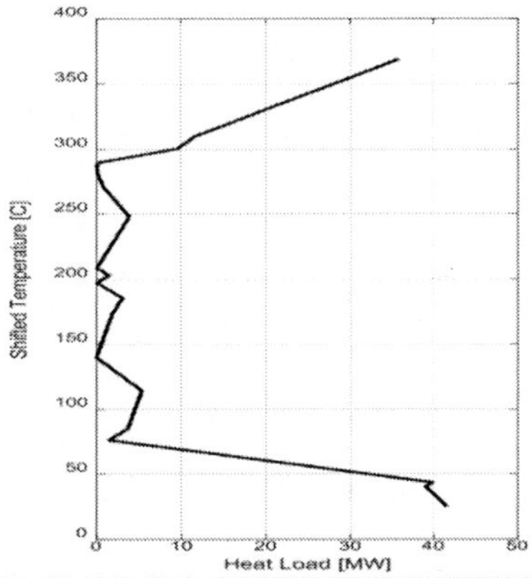

HU = 35.8 MW

CU = 41.5 MW

✓ **Preflash reduced hot utility demand by 21%, compared to optimised base case.**

✓ Energy targets are reduced when implementing a preflash unit within the crude oil distillation system.

➢ $\Delta T_{min} = 30°C$

Significant potential to reduce operating costs

MANCHESTER 1824

The University of Manchester

Product specifications and flowrates

Product	Base Case ASTM (°C)		Optimization Results No Preflash ASTM (°C)		Optimization Results With Preflash ASTM (°C)	
	T5%	T95%	T5%	T95%	T5%	T95%
RES	363	755	354	753	353	753
LN	27	110*	25	110	25	110
HN	143	196*	133	196	133	196
LD	218*	300*	218	300	218	300
HD	308	354*	305	354	302	354

* Specified in HYSYS.

✓ All product specifications are within the allowed tolerance of ± 10°C.

Product Flowrates in m³/h

Product	Base Case	Optimized No Preflash	Optimized With Preflash
RES	292.1	299.4	301.0
LN	102.4	101.1	100.9
HN	86.8	88.3	88.7
LD	127.6	125.5	122.9
HD	53.7	48.4	49.2

MANCHESTER 1824
The University of Manchester

Summary ΔT_{min} analysis

Variable	Units	Base Case	Optimization Results					
			No Preflash			With Preflash		
	°C		ΔT 30	ΔT 20	ΔT 10	ΔT 30	ΔT 20	ΔT 10
1. Main steam flow	kmol/h	1200	900	901	1054	1190	949	1103
2. HD steam flow	kmol/h	250	200	206	284	211	241	365
3. PA1 duty	MW	12.8	6.0	6.0	6.0	6.8	6.0	6.0
4. PA2 duty	MW	17.8	9.8	12.5	11.8	8.3	8.4	8.3
5. PA3 duty	MW	11.2	11.9	11.9	11.7	8.8	11.0	12.0
6. PA1 ΔT	°C	30	48.6	28.6	50.0	25.5	40.1	23.1
7. PA2 ΔT	°C	50	33.8	46.8	48.3	23.0	36.7	41.5
8. PA3 ΔT	°C	20	39.9	40.0	30.0	39.3	22.5	20.2
9. Column inlet temperature	°C	365	350	350	351	354	362	358
10. Reflux ratio		4.17	3.7	3.5	3.7	3.0	3.0	3.0
11. Flash temperature	°C	115	---	----	----	230	230	230
12. Vapour feed location		----	---	----	----	3	3	3
Minimum hot utility demand	**MW**	**58.3**	**43.4**	38.8	**34.7**	**35.9**	31.1	**26.3**
Energy Savings						**21%**	**24%**	**32%**

MANCHESTER 1824
The University of Manchester

Case Study: Summary

- A systematic design optimization approach is demonstrated for a crude oil distillation system with a preflash unit.
- Genetic Algorithm was successfully implemented to optimize a crude oil distillation system with a preflash unit.
  - CPU time: 5.8h using an HP desktop Intel(R) Core(TM) i5-4570 CPU @ 3.20GHz with 16.0 GB in memory RAM.
  - Product quality specifications are all met.
- Minimum hot utility demand was reduced by:
  - 32% for ΔTmin = 10°C.
  - 24% for ΔTmin = 20°C.
  - 21% for ΔTmin = 30°C.

by introducing and optimizing a preflash unit through optimization.

5. Conclusions

Conclusions

1. Methodology has been developed for simulation-optimization-based design of a crude oil distillation system with a preflash unit.
   - Implemented using Aspen HYSYS–MatLab interface with genetic algorithm.
2. An optimization framework accounts for operational and structural variables.
   - Constraints applied to product quality.
   - Heat integration considered simultaneously using pinch analysis.
3. Case study results indicate energy demand can be reduced by using a preflash unit in the crude oil distillation system.
4. The methodology has the merits of using rigorous process models, of being systematic, of using robust optimization techniques and of accounting for configuration as well as operating conditions.

MANCHESTER 1824
The University of Manchester

6. Future Work

MANCHESTER 1824
The University of Manchester

Prefractionation option

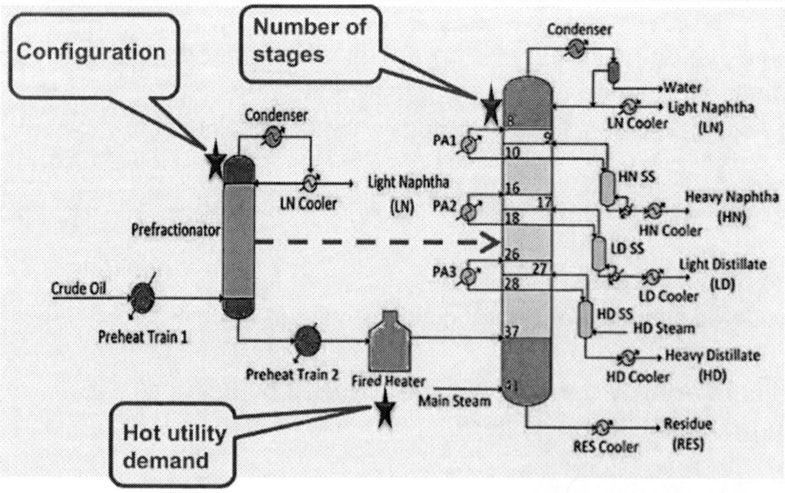

System design

- Main column structure.
- Main column operating conditions.
- Prefractionator design.

Key design parameters

- Number of stages.
- Feed locations.
- Pump-around duties.
- Pump-around temperature drops.
- Stripping steam flow rates

Methodology:

- Rigorous simulation of columns.
- Fuel consumption of the fired heater.

Optimization:

- Continuous and discrete variables.
- Stochastic optimization (GA).

MANCHESTER
1824
The University of Manchester

Thank you

National Council of Science and Technology (Mexico) for the financial support granted for the development of this work.

CONACYT

Contact Details:

Minerva Ledezma-Martínez
minerva.ledezmamartinez@manchester.ac.uk

Paper 175c
(Proceedings 483295)

Vacuum Distillation:

Slop Wax Disposition

Andrew W. Sloley
Advisian (WorleyParsons Group)
Energy Center II
575 N Dairy Ashford St
Houston, TX 77079

Prepared for Presentation at the
American Institute of Chemical Engineers
2017 National Meeting
Minneapolis, MN October 29-November 3, 2017

UNPUBLISHED

1. Summary

Refinery vacuum units include a deentrainment, or wash zone, for removal of heavy material from desired products. A continuous wash oil stream prevents the wash zone (deentrainment zone) from coking. Underneath the wash zone a collector tray gathers the liquid from the wash zone plus any entrainment removed. The total liquid collected on this tray is called slop wax.

Traditionally, slop wax has been drawn from the vacuum tower as a stream and recycled to the feed upstream of the vacuum feed heater. Less common alternate dispositions include sending the slop wax to a stripping section inside the vacuum tower or making the slop wax a product stream.

This review covers:

- Survey of operating units to identify the most common slop wax dispositions
- Evaluation of the major options for towers without stripping sections
- Evaluation of the major options for towers with stripping sections
- Review of the effect of entrainment on slop wax dispositions

2. Refinery Slop Wax

Refinery slop wax is a stream created inside the refinery crude distillation unit. It normally refers to a mixed distillate-entrainment stream from the vacuum distillation tower. However, some plants may use the term slop wax to refer to mixed distillate-entrainment streams from other towers in the crude distillation unit that include de-entrainment sections.

Figure 1 shows the configuration of the de-entrainment section of a refinery vacuum tower wash zone. The two-phase feed enters the vacuum tower below the wash zone. Due to high velocity in the feed line, some liquid droplets entrain up the column with the vapor. These liquid droplets contaminate the distillate products. A wash bed helps de-entrain the droplets. Wash oil irrigates the top of the wash bed to prevent plugging and improve unit reliability. Distillate liquid leaving the bottom of the wash bed (overflash) combines with entrainment. The mixed distillate-entrainment stream is called slop wax. Previous work covered required rates for the wash oil and overflash [1].

The term slop wax comes from lubricants processing. Figure 2 shows an example vacuum tower for manufacturing lubricant base stocks. The lowest lubricant product is "lube base 3". Among other requirements, specifications on this product normally include some combination of viscosity and viscosity index [2] ranges. To make lube base 3 on specification, the rate of production is controlled to reject material with too high a viscosity. The rejected material is left in the liquid inside the vacuum tower.

In lubricant units the vacuum tower bottoms (VTB) often goes to downstream processing to produce "bright stock" lubricants. The bright stock specifications also include some combination of viscosity and viscosity index ranges.

Commonly, material that has too high a viscosity for the lowest lubricant base oil (lube base 3 in this example) has too low a viscosity for bright stock. Without a slop wax stream, any material not drawn in lube base 3 ends up in the vacuum tower bottoms. The unit cannot produce both the highest viscosity lubricant product (lube base 3) and the bright stock on specification at the same time.

The slop wax stream is a drag stream that removes intermediate viscosity material from the process. With the slop wax stream as a product, both the lowest lubricant product and the bright stock meet specifications. Units with bright stock production usually do not recycle the slop wax. Recycled slop wax in these units builds up an inventory of off-specification material inside the unit. The slop wax rate is set by the amount of undesirable material in the crude oil feed that falls into the gap between the product specifications. The slop wax is a product from the crude distillation unit. Conventional dispositions include blending into asphalt, sending it to a fuels crude unit, or delayed coking.

Fuels vacuum units produce feeds for downstream catalytic processing to produce transportation fuels. Fuels vacuum units do not have a gap between product specifications that prevents recycling the slop wax. While the recycle problem does not exist, the term slop wax has been retained for the stream name.

Since the slop wax can be recycled in fuels vacuum units, two immediate question are raised:

- When is slop wax recycle desirable?
- If desirable, what is the best way to recycle slop wax for distillate recovery?

3. Survey

Basic configurations for slop wax recycle in the refining industry include

- no recycle, slop wax is a product (Figure 3);
- no recycle, external piping with gravity flow (Figure 4);
- no recycle, external piping with a pump (Figure 4);
- no recycle, internal flow (Figure 5, Figure 8);
- recycle to upstream of fired heater (Figure 6); or
- recycle to an upstream tower boot or other upstream location (Figure 7).

Table 1 summarizes a recent survey of vacuum towers and shows the percent use of major options.

Table 1 Summary of Recycle Dispositions and Configurations

Slop Wax Disposition	Occurrence, %	
Product	14	
Recycled, total	36	
Recycled, vacuum tower has no stripping section		28
Recycled, vacuum tower has stripping section		8
Not recycled, total	50	
Not recycled, vacuum tower has no stripping section		20
Not recycled, vacuum tower has stripping section		30
Total	100	
Recycled, Destination (Always pumped external configuration)		
To vacuum heater feed	67	
To upstream tower boot	33	
Total	100	
Not Recycled, Destination		
To boot or tower bottoms	40	
To stripping section	60	
Total	100	
Not Recycled, Configuration		
Gravity flow, external	32	
Gravity flow, internal	52	
Pumped, external	16	
Total	100	

Figure 3 to Figure 7 show the different configurations in use. All the figures show vacuum towers without stripping sections. However, all configurations shown can be used with or without a stripping section. Figure 8 shows an example with internal flow and a stripping section in the tower. When a stripping section is present, systems without recycle return the slop wax to the top of the stripping tray. The slop wax then mixes and becomes part of the feed to the stripping section.

4. Slop Wax Recycle

Slop wax is the liquid collected on the slop wax collector. Slop wax is the combination of two components. The first component is entrainment carried by the vapor rising from the feed zone. The second is overflash, distillate liquid leaving the bottom of the wash bed. The overflash is material that was originally flashed as vapor in the feed. All vapor in the feed is potential distillate product from the vacuum tower. Overflash is vapor in the feed, but ends up in the slop wax instead of the distillate product.

Since the slop wax contains entrained liquid (residue), it is not suitable for downstream use as a gas oil product in most plants. The concept behind recycling the slop wax is that reprocessing the slop wax allows for recovery of the distillate contained in the slop wax. Based on a recent survey of slop wax disposition, 36% of the surveyed units recycled slop wax (Table 1).

Figure 9 illustrates the possible benefits of slop wax recycle in an example unit. The example is for a realistic commercial unit [3] with a feed very close to the average currently processed in U.S. refineries (Tia Juana Light, 31.6°API, SG=0.8676 [4]).

At a best-practice liquid irrigation rate of 0.15 gpm/ft^2 (6.11 lpm/m^2) in a tower without a stripping section, the slop wax recycle increases yield from a cutpoint[a] of 1,026°F to 1,038°F (552°C to 559°C). This is equivalent to a liquid distillate yield increase of 0.62 volume percent. The relative benefit is nearly the same for a tower with a stripping section. Slop wax recycle improves unit yields at fixed tower inlet conditions and fixed tower operating pressure.

Given the possible benefits, the survey results surprise. They show that only 36% of units have slop wax recycle compared to 50% of units that do not (the 14% of units that have slop wax as a product lie outside the scope of the comparison). For every 2 units with recycle, 3 units do not have recycle.

5. Alternatives to Slop Wax Recycle

To understand why slop wax recycle is not more widely used, alternatives to slop wax recycle must be examined. Refiners have three straightforward alternatives to maintain yield.

- Operate the vacuum tower at lower pressure.
- Operate the vacuum tower at higher temperature.
- Add a steam stripping section (or make it bigger).

5.1. Operate The Vacuum Tower at Lower Pressure

Reducing the vacuum tower pressure increases feed vaporization. Table 2 outlines the main effects of lower pressure operation.

---

a Cutpoint is a measure of distillate yield. It is the atmospheric equivalent temperature on an ASTM D2892 test, or equivalent, that gives the same liquid distillate yield as seen on the operating unit.

Table 2 Effects of Lower Pressure Operation (For Same Yield)

Effect	Cost or Benefit
Larger tower diameter for same capacity	Cost
Larger vacuum system	Cost
Higher vacuum system utility demand	Cost
Lower temperature streams for heat integration	Cost

Compared to slop wax recycle, the costs are higher. Both more capital is required and operating costs rise for the same capacity. Economically, lower operating pressure instead of slop wax recycle is not attractive.

5.2. Operate the Vacuum Tower at Higher Temperature

Many units operate at as high a temperature as possible within the constraint of a target operating duration between shutdowns to decoke the furnace. Crude oil thermal stability can limit operating temperatures. However, some units can raise temperature or heaters can be modified to allow operation at higher temperatures.

Vacuum tower effects of higher operating temperatures versus slop wax recycle are more complex. The higher temperature increases the vapor load by heating up the entire vapor stream but allows for reducing the vapor load by eliminating the slop wax recycle. Figure 10 shows the relative vacuum tower cross-section required in the wash section for a slop wax recycle system versus a higher heater outlet temperature. In Figure 10 the vapor load is defined by a maximum C-factor of 0.45 ft/sec at the top of the wash section [5]. Overall, the vacuum tower is slightly larger with higher heater outlet temperatures.

Table 3 summarizes the effects of using higher heater temperature to get the same yield as slop wax recycle. Compared to slop wax recycle, the costs are higher. Both more capital is required and operating costs rise slightly for the same capacity. Economically, higher heater outlet temperature instead of slop wax recycle is not attractive.

Table 3 Effects of Higher Temperature Operations (For Same Yield)

Effect	Cost or Benefit
Larger tower diameter for same capacity	Cost
Larger vacuum system (more cracked gas from heater)	Cost
Higher vacuum system utility demand	Cost
Higher temperature streams for heat integration	Benefit
Higher coking rates in the fired heater	Cost

5.3. Add a Steam Stripping Section

Steam stripping the residue recovers light material from the residue [6]. Stripping adds steam and removes recycle vapor from the tower but adds stripping steam load. Figure 10 includes a comparison of using stripping steam to get the same yield as slop wax recycle. The basis used is a one-stage stripping section with enough steam to match the slop wax recycle yield at the same tower inlet conditions. The net effect of steam stripping is that a larger tower diameter is required for the same capacity and yield.

Table 4 Effects of Stripping Steam (For Same Yield)

Effect	Cost or Benefit
Larger tower diameter for same capacity, Entire Tower	Cost
Taller tower to add stripping trays	Cost
Larger vacuum system (steam load)	Cost
Higher vacuum system utility demand	Cost

Table 4 summarizes the effects of using a stripping section to get the same yield as slop wax recycle. Compared to slop wax recycle, both the capital costs and the operating costs are significantly higher. Economically, stripping is less attractive than slop wax recycle.

6. Problems with Slop Wax Recycle

Evidently, if most plants to not use slop wax recycle, it must have other problems. Most of the problems with slop wax recycle come from two areas, thermal cracking and system reliability.

6.1. Thermal Cracking

Slop wax recycle takes hot liquid from the slop wax collector and recycles it to upstream of the heater feed. This liquid gets reprocessed back to heater outlet temperatures. The initial feed material may be subject to some thermal cracking at high heater outlet temperatures. The thermal cracking results in products that are less thermally stable. Heating them a second time can result in aggressive coking.

The coking may take place in either the fired heater or the vacuum tower. Several plants have eliminated slop wax recycle to extend plant run length between shutdowns. Native crudes with high paraffin and asphaltic content tend to have more coking problems with slop wax recycle. Syncrude blends also have more problems as they usually contain reactive material. Highly aromatic crudes tend to be more thermally stable and have fewer coking problems.

Coking is the result of a time-temperature-reactivity profile. The same crude may have significantly different behavior depending on the unit layout and operating point. Coking problems will often be found only by experience with specific crudes on a given unit. Several plants suspend slop wax recycle when running specific crudes.

6.2. Pump and Mechanical Issues

The major problem with slop wax recycle comes from the recycle pump, its control, and associated equipment. The slop wax recycle pump has the following characteristics:

- High operating temperatures
- Low flow rate in all except very large units
- High discharge head
- Unstable feed conditions
- Fluid with a high freezing temperature

- Difficult level control

These combine to make pump reliability in this service difficult. Many refiners find the cost and potential safety problems of maintaining the pump outweigh the benefits.

Operating temperatures routinely exceed 700°F (371°C). High temperatures create potential problems with seals, piping stress, and nozzle loads. Proper pump selection and installation may be extremely expensive.

Nearly all pumps in this service are built to API-610 [7] standards to maximize reliability and safety. API-610 pump specifications set criteria that make building low-flow pumps challenging. Additionally, if the recycle is returned immediately upstream of the fired heater (Figure 6), the pump will have a high discharge head requirement. The combination of low flow plus high discharge head can make selection of a reliable pump difficult [8].

To reduce discharge head requirements, the recycle may return to the boot of an upstream tower (Figure 7). This reduces the head requirement and makes the pump more reliable. However, the recycle now has to go through two pumps and the bottoms pump on the upstream tower needs to be larger.

The pump feed conditions tend to be unstable in both state and flow rate. Solutions to these problems tend to create other problems.

The entrainment and the overflash mix on the collector tray. The entrainment is hotter than the overflash. The mixing can cause vaporization. If the outlet nozzle of the tower is too small or the liquid level on the collector is too low, vapor can be drawn into the pump suction. Vapor in the feed line to the slop wax pump creates cavitation, impeller damage, and seal problems.

Entrainment rates may vary dramatically from the average value. If liquid hold-up on the slop wax collector is low, the level may vary dramatically. Large level changes can result in needing to change the recycle pump rate quickly.

One solution might be to hold a substantial liquid level on the slop wax collector. This allows for degassing the mixture and for control to avoid flow rate changes. However, large liquid inventory in the column increases thermal cracking. The liquid is at a high temperature. Excessive residence time results in coke formation on the tray, eventually plugging it. Thermal cracking also produces light gas that increases the load on the vacuum system. High vacuum system load increases tower operating pressure.

Economics strongly favor keeping the tower pressure low. In nearly all cases, the refinery attempts to minimize slop wax inventory. This makes control and reliability problems difficult to solve.

Inevitably, pumps in this service require more-than-normal amounts of maintenance. With a pump out of service, slop wax in the system can cool off. Slop wax has a high freezing temperature. Even a little bit of excessive cooling can lead to a cascade of colder slop wax-higher viscosity-lower flow velocity-colder slop wax until the system solidifies. Returning systems to service after repairs or maintenance can be difficult.

Reliable level control is also difficult in this service. Many instruments fail. Poor level control can lead to pump problems. The most reliable equipment uses gamma ray level detection on the pump suction pipe or on the slop wax collector.

All these problems can be solved. The solutions cost money and may require considerable effort and attention to stay in service. For these reasons, the majority of plants do not use slop wax recycle.

7. Alternatives

If the slop wax is not recycled and cannot be used as a product, it must still go somewhere. The alternative is to put the slop wax into the bottoms product. Figure 4 shows two options for external flow from the slop wax collector. Figure 5 shows internal flow inside the vacuum tower. Both the external flow and internal flow configurations can be used in units with or without a stripping section. In vacuum towers without stripping sections, the external flow may go to the bottoms product line from the vacuum tower.

7.1. External Flow to Stripping Section, Boot, or Bottoms

Table 1 shows that approximately ½ of the units with the slop wax going to the tower stripping section, boot, or bottoms, have external flow. Of the units with slop wax with external flow, 2/3 use gravity flow and 1/3 use pumped flow. Gravity flow is shown on the right side of Figure 4 and pumped flow is shown on the left side of Figure 4.

The major purpose of external flow is to measure the flow rate of the slop wax. An accurate flow measurement makes optimization to minimize wash oil more straightforward and accurate. Based on field review of flow measurement installations, most flow meters in this service are incorrectly installed. In addition, if special measures are not taken, the slop wax may contain vapors that make the measurement inaccurate. With properly installed equipment [9], flow measurements can be made in this system.

Full optimization of wash oil requires analyzing the slop wax stream as well as measuring its flow rate. With an analysis, the amount of entrainment versus the amount of overflash can be estimated. The analysis method uses the amount of a non-volatile material, such as sodium, in the sample to estimate the entrainment rate. The purpose of the pumped systems is to have enough pressure to take a sample for analysis. With both the slop wax rate and the slop wax entrainment known, the wash oil rate can be fully optimized.

However, only about half the towers with no slop wax recycle have either a flow meter or a sampling point. The same problems that plague operability and maintenance of slop wax recycle systems also create problems with external flow meters and sampling stations.

If present, the pump required in this service is a low-head pump so has fewer problems. Nevertheless, all the problems with pump suction level control persist, as do problems with slop wax freezing.

In addition, taking and cooling the sample is difficult. The sample stream is very hot and handling requires extensive safety precautions. The sample systems tend to be unreliable (they plug up) and expensive. Overall, many plants have decided the problems and costs of the sample outweigh the potential benefit.

7.2. Internal Flow to Stripping Section or Boot

Internal flow is the simplest method of running the system. It has few maintenance or safety issues. Special piping and collector tray arrangements can make it highly reliable. The cost of internal flow is that neither the flow rate nor the composition are known. The most effective control method with internal flow is to use temperature profiles in the wash bed to estimate the wash oil rate [1]. Internal flow is recommended for most units.

8. Anecdotes

Based on interviews with plant operating personnel, two points merit further discussion.

- Based on operating experience, units with functioning stripping trays show dramatic performance changes when the stripping trays are damaged. This is evidence that the stripping trays function as expected.
- Based on operating experience, units with slop wax recycle often show little to no performance changes when the slop wax pumps are out of service. This is evidence that slop wax recycle systems often under-perform expectations.

The major reason that slop wax systems might underperform is that the slop wax contains little overflash. In some units slop wax is nearly 100% entrainment. In these units the slop wax is material that did not vaporize at feed conditions. That material will not vaporize the second time it is heated and sent to the tower either.

9. Conclusions

Slop wax is a combined distillate-entrainment stream from the de-entrainment section of a refinery vacuum tower. Slop wax recycle should be an effective way to increase distillate yields in vacuum units. Nevertheless, just over a third of vacuum units recycle the slop wax. A small number take the slop wax as a product. However, half of the surveyed vacuum units send the slop wax to a stripping section, the vacuum tower boot, or directly to the tower bottoms stream. The reason for this is mechanical difficulty and expense in keeping slop wax systems running. Maintenance cost and effort often exceed benefits.

Additionally, many units show little to no performance change when the slop wax system is out of service. Under these conditions, the survey results showing that many plants use systems without slop wax recycle are unsurprising.

More expensive alternatives to increase unit yield include higher heater outlet temperatures and enhanced stripping sections. However, these have limits and are more expensive in most cases. They are more reliable within their appropriate limits.

External flow systems to improve control of the vacuum tower also suffer from many of the problems associated with slop wax recycle. The systems are expensive and difficult to keep in service. The alternative of sending slop wax to the tower bottoms with internal piping has the highest reliability and lowest capital cost, even though it may have slightly lower yields. Many plants find the reliability improvement and lower capital cost worthwhile.

10. References

[1] A. W. Sloley, "Vacuum Distillation: Wash Zone Configuration," presented at the AIChE Spring Meeting, San Antonio, TX, 2017, p. Paper 34a, Proceedings 479206.

[2] ASTM, "D2270-93, Standard Practice for Calculating Viscosity Index from Kinematic Viscosity at 40 and 100 C (Reapproved 1998)," ASTM International, West Conshohocken, PA, 1998.

[3] A. W. Sloley, "Crude Distillation Sequences," presented at the AIChE National Meeting, San Francisco, CA, 2013.

[4] R. N. Watkins, "Crude Assay: Tia Juana Light," in *Petroleum Refinery Distillation*, Houston, TX: Gulf Publishing Company, 1973, pp. 251–283.

[5] A. W. Sloley, "Refinery Vacuum Towers: State of the Art," presented at the AIChE National Meeting, San Francisco, CA, 2013.

[6] A. W. Sloley, "Crude Unit Stripping Sections," presented at the AIChE Southwest Process Technology Conference, Galveston, TX, 2017.
[7] API, "STD 610, Centrifugal Pumps for the Petroleum, Petrochemical and Natural Gas Industries," American Petroleum Institute, Washington, D. C.
[8] F. Korkowski and L. Glassburn, "Low Flow / High Head Centrifugal Pump Options," presented at the Calgary Pump Symposium, Calgary, ALB, Canada, 2005.
[9] A. W. Sloley, "Get Some Inside Information," *Chem Proc*, vol. 76, no. 8, pp. 43–44, Aug. 2014.

11. Figures

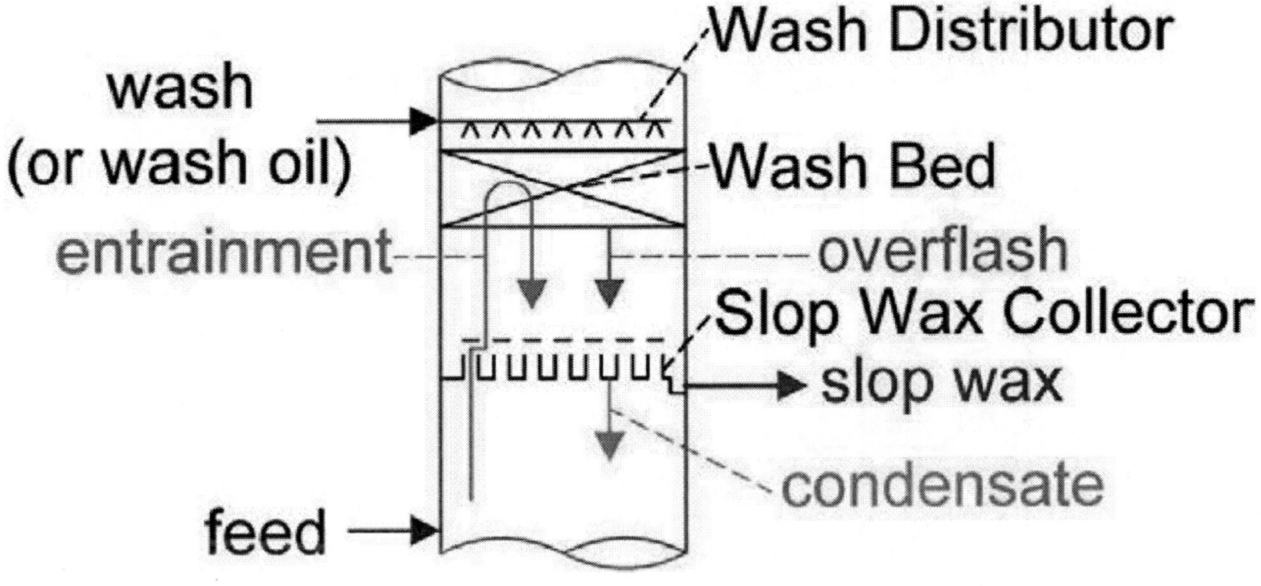

Figure 1 Wash Zone Terminology

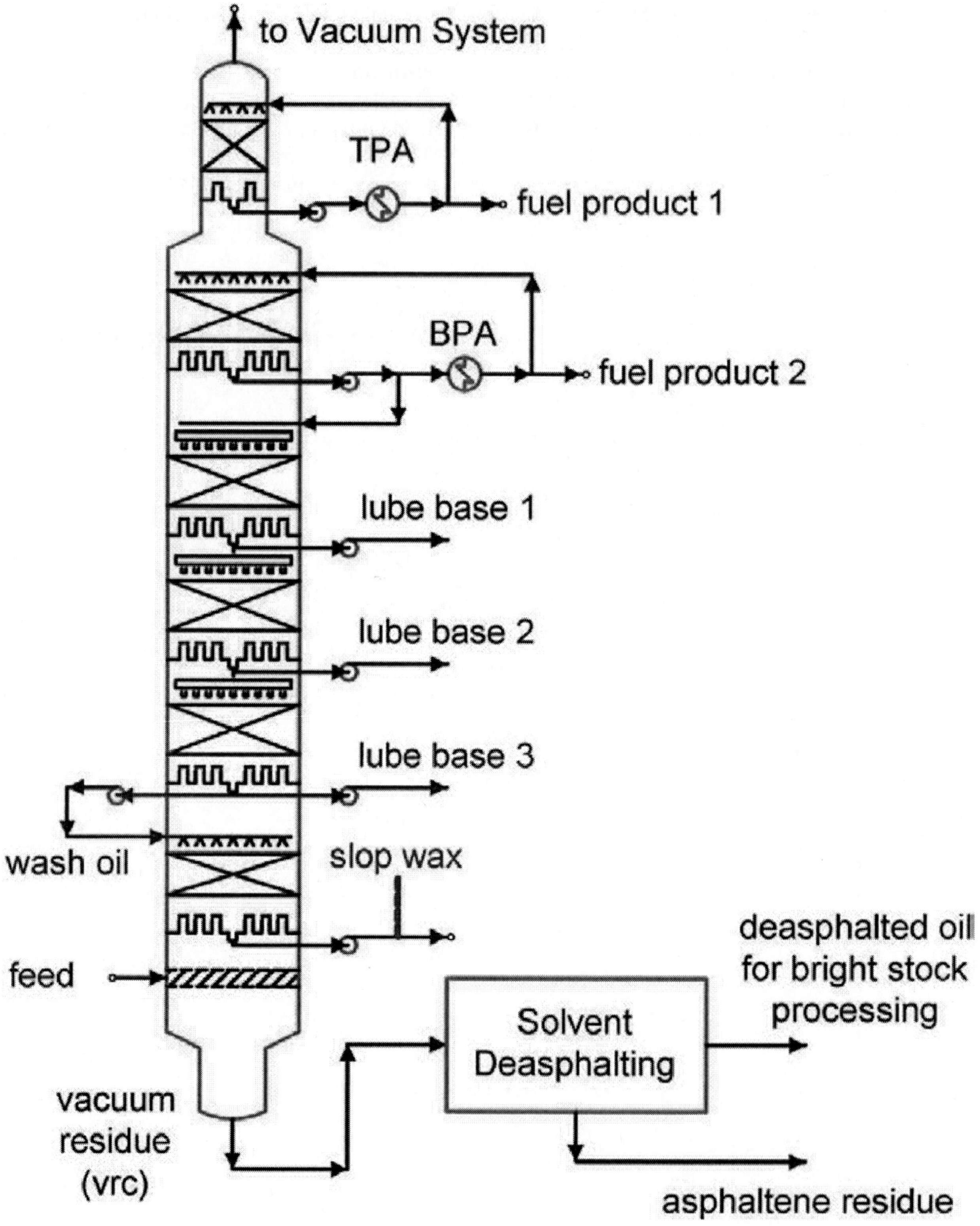

Figure 2 Vacuum Tower and Downstream Lubricant Processing

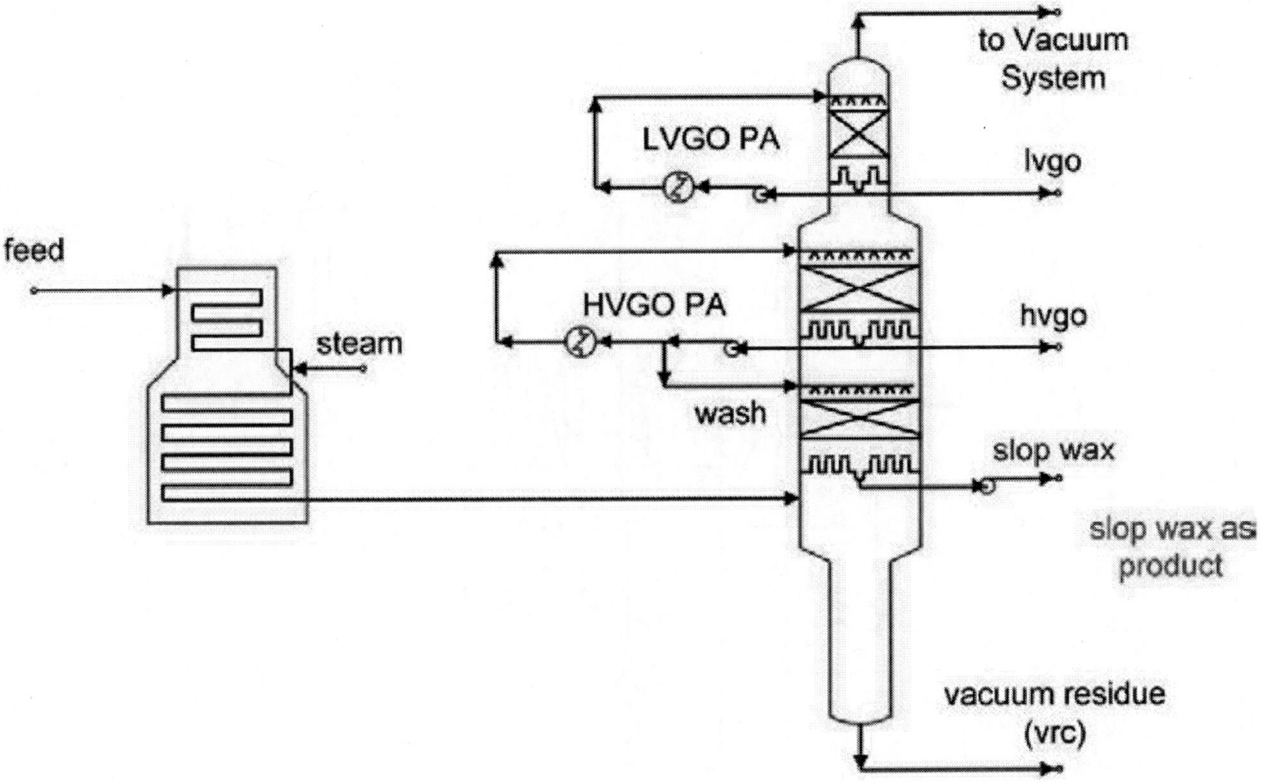

Figure 3 Slop Wax as Product

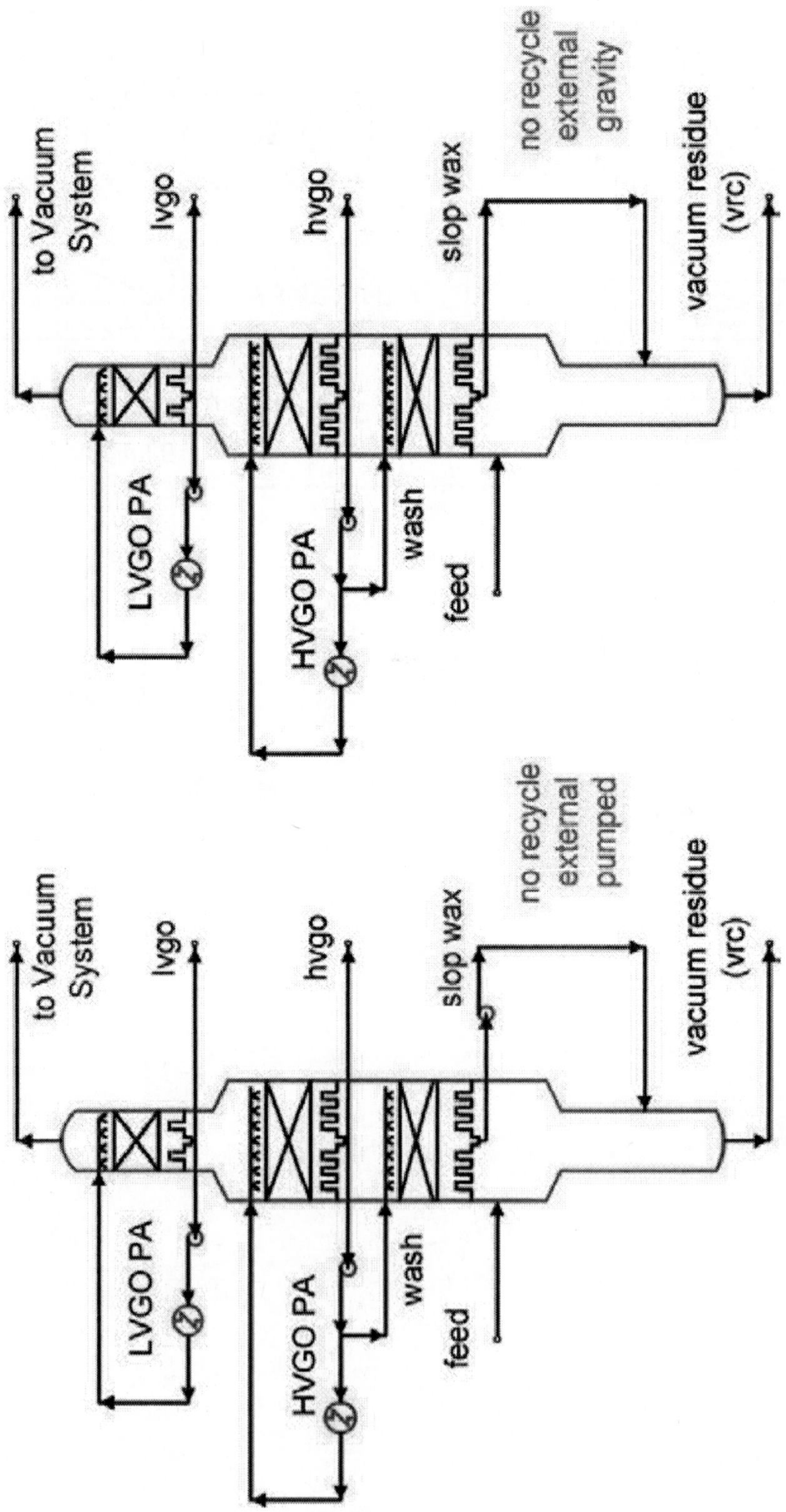

Figure 4 External Slop Wax Recycle (Pumped and Gravity Flow)

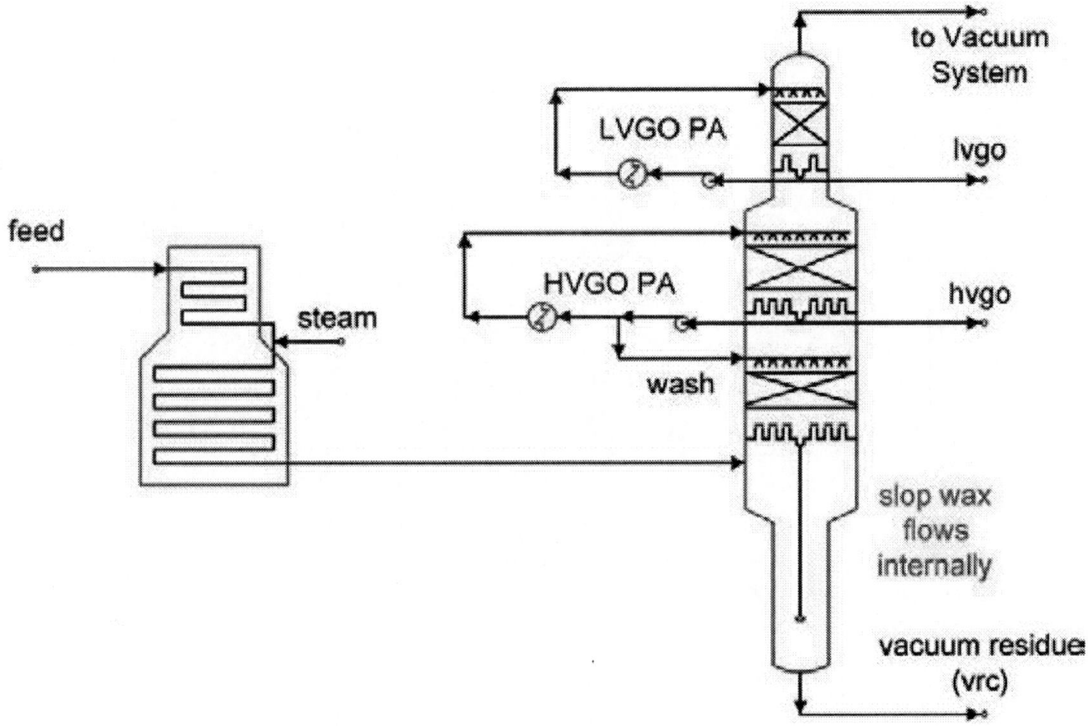

Figure 5 Slop Wax Internal Flow

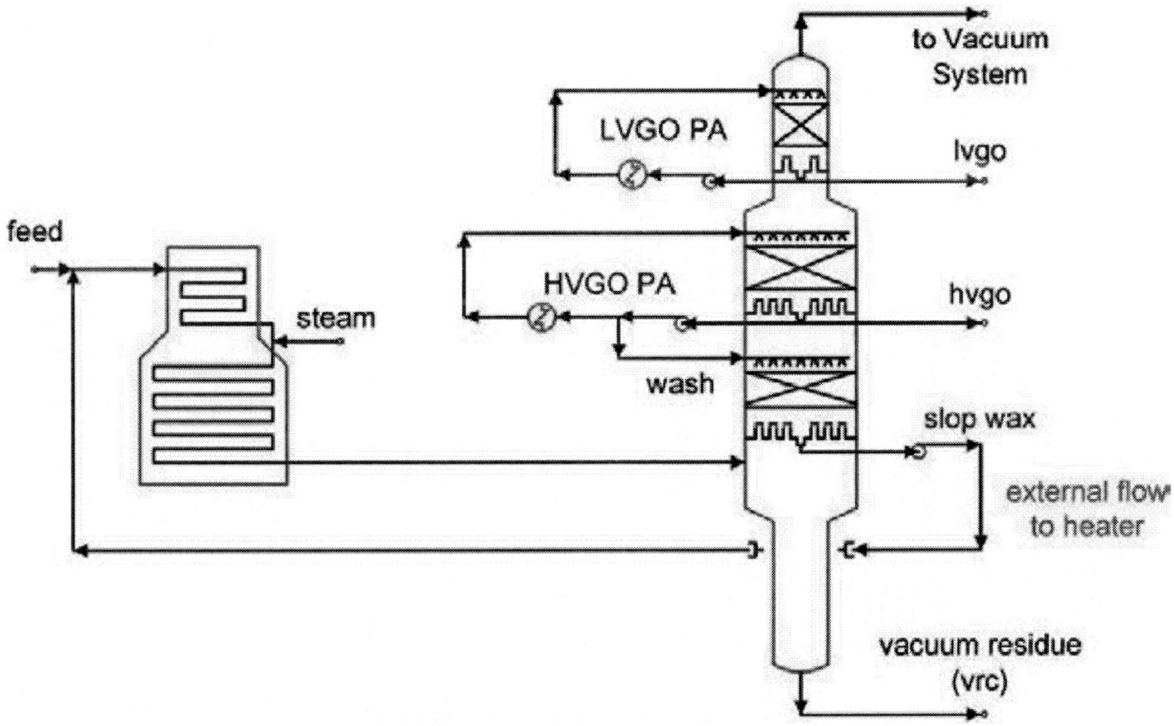

Figure 6 Slop Wax Recycle to Heater Feed

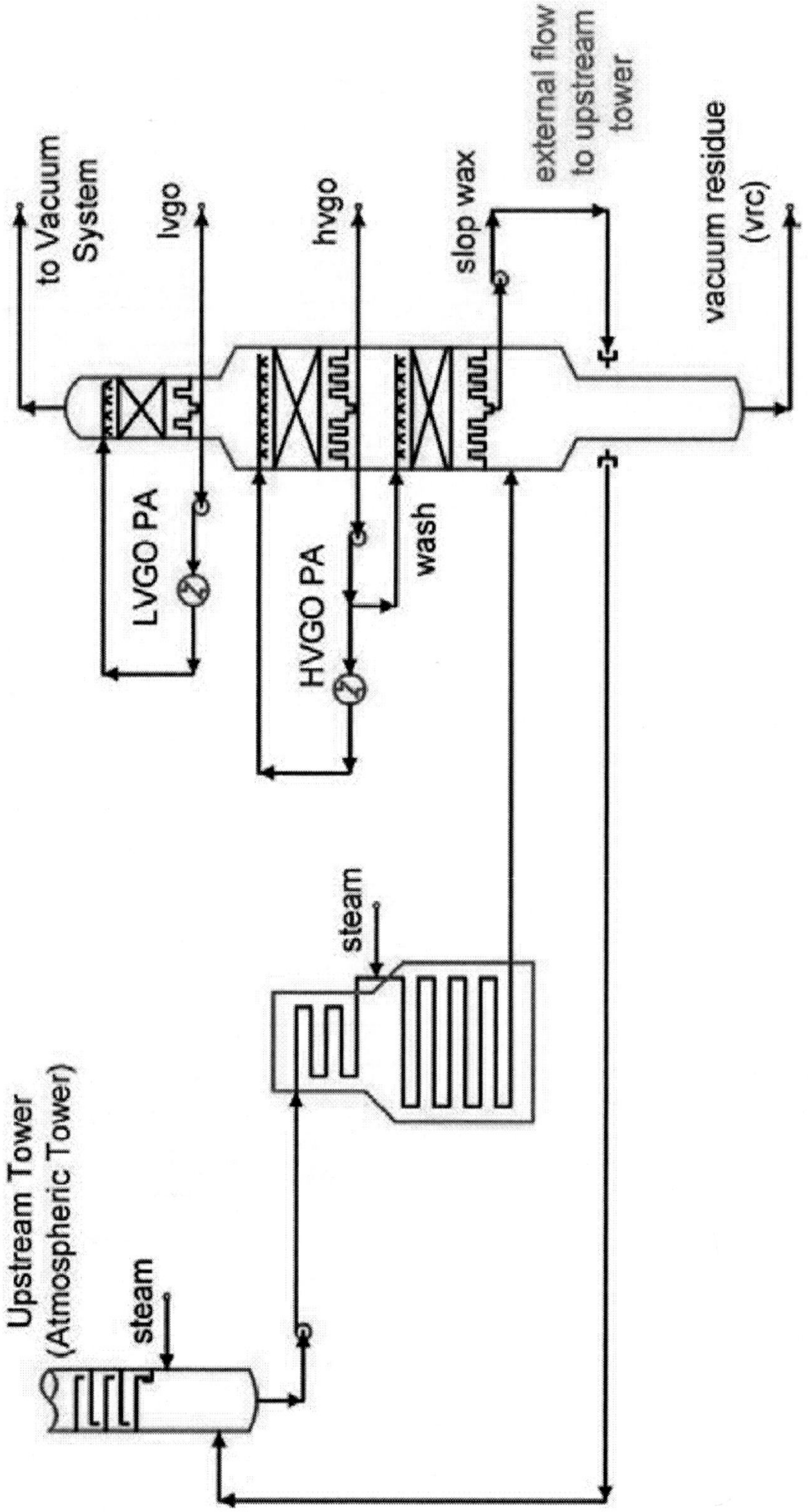

Figure 7 Slop Wax Recycle to Upstream Tower

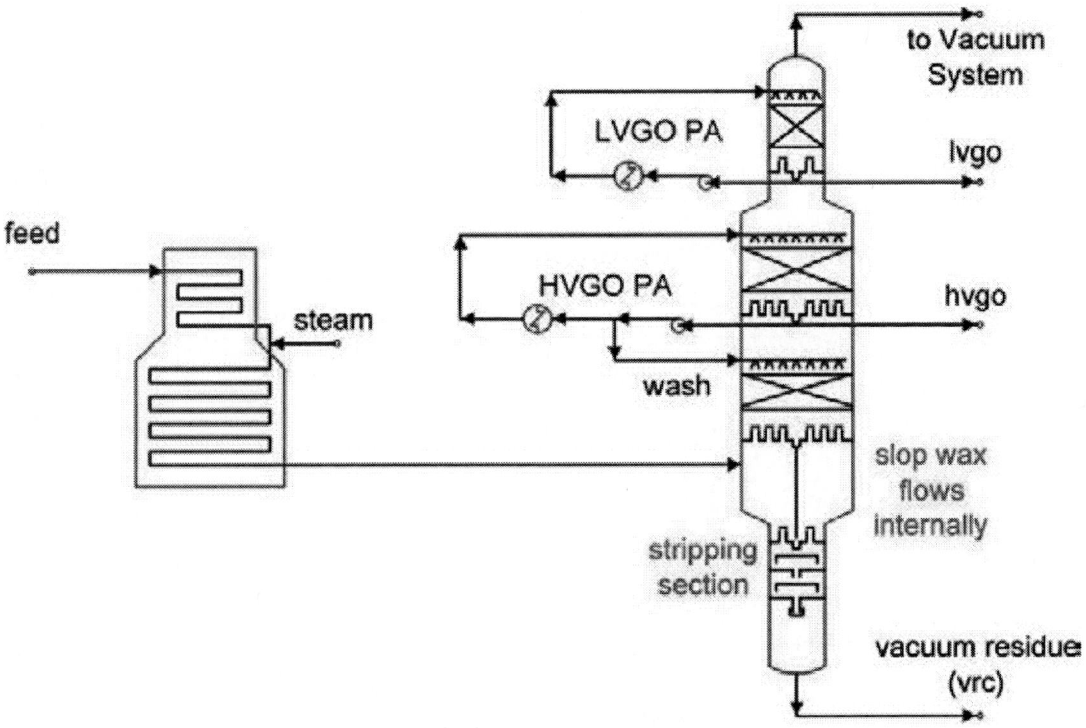

Figure 8 Slop Wax, No Recycle, Internal Flow to Stripping

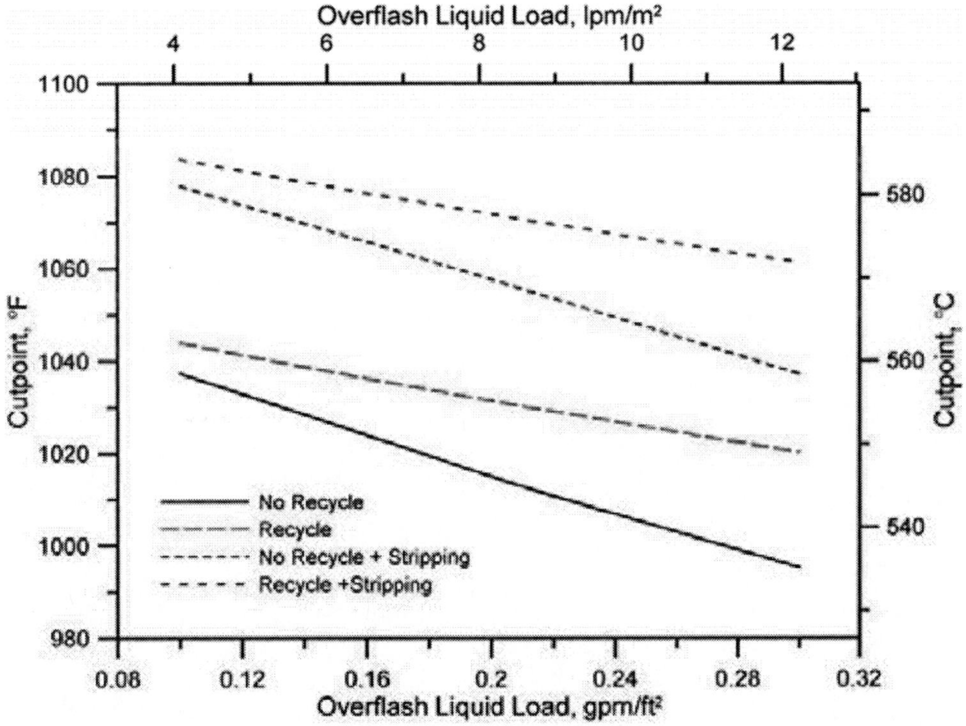

Figure 9 Slop Wax Recycle Yield Benefits

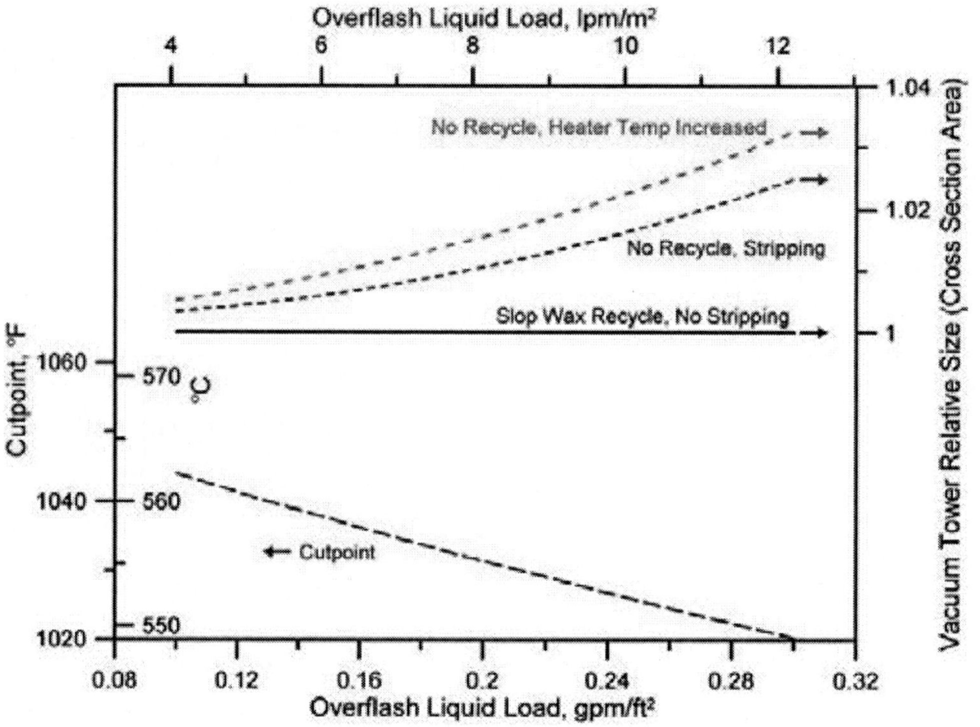

Figure 10 Slop Wax Recycle versus Higher Heater Outlet Temperatures or Stripping, Fixed Yields

489909 Acid Regenerator Optimization for Increased Unit Throughput

Abstract:

This discussion presents steps taken to increase unit throughput by pushing existing equipment beyond historical maximum rates using an acid regenerator tower case study in an HF Alkylation unit. Process changes at the acid regenerator contributed to record breaking alkylate production in March of 2017.

Acid regenerator optimization included process step tests, tower scans, and evaluation of adjacent equipment capacity. Regenerator optimization allowed for over 15% tower feed rate increase. This supports demonstrated acid strength recovery at increased unit throughputs for record breaking monthly average alkylate production.

Vacuum Distillation: Fractionating Wash Zones

Paper 175e

Proceedings 483296

Andrew W. Sloley, Principal Consultant, Advisian

Prepared for Presentation at the
American Institute of Chemical Engineers
2017 National Meeting
Minneapolis, MN, October 29 – November 3, 2017

www.**advisian**.com

Advisian
WorleyParsons Group

Overview

Wash Section

- Mainly considered a de-entrainment device
- Limited literature discussion on wash section as fractionating bed for controlling content of volatile metal-containing compounds:
  - — Importance of vacuum system type
  - — Importance and costs vs. benefits of reflux vs. product quality
- Nickel and vanadium porphyrins most important commercially
- Nickel and vanadium affect downstream fluid catalytic cracking and hydrotreating catalysts

Advisian / 3

Wash Section

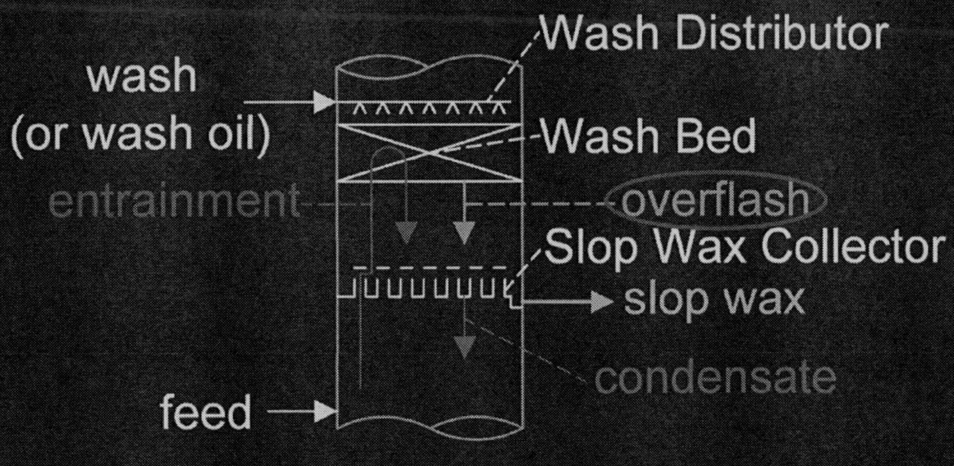

Advisian / 4

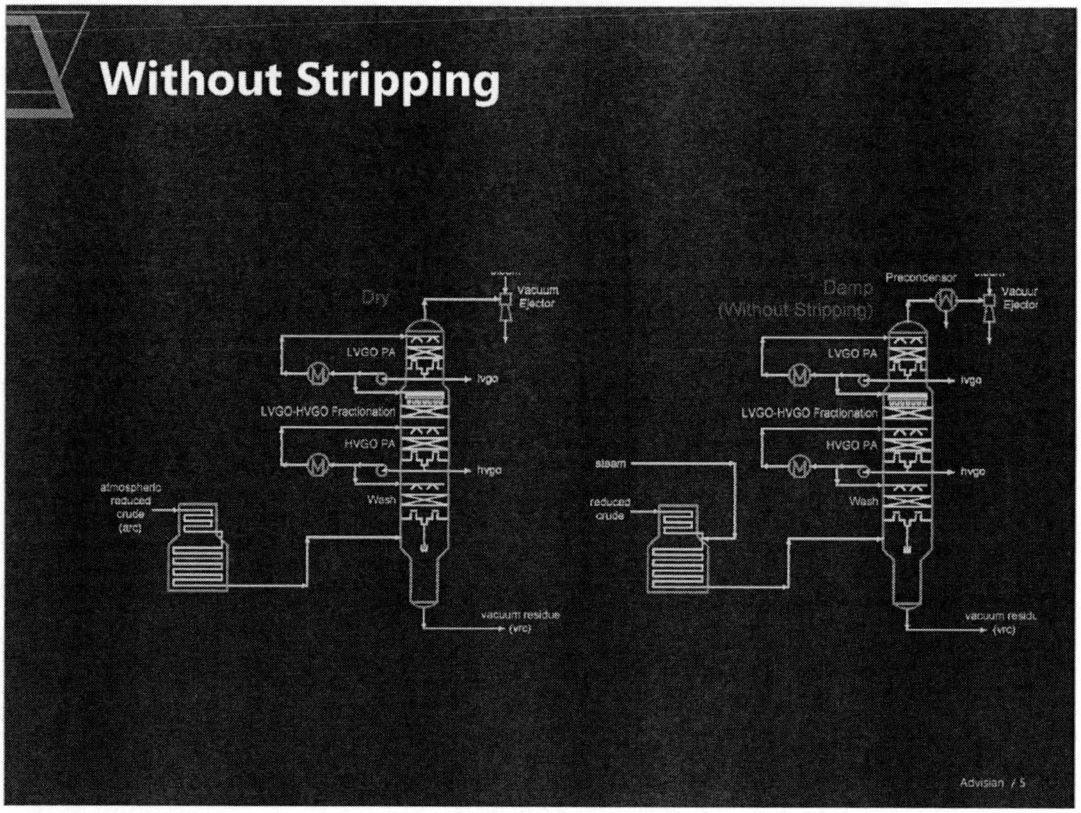
Without Stripping
Dry
Vacuum
Ejector
LVGO PA
lvgo
LVGO-HVGO Fractionation
HVGO PA
hvgo
atmospheric
reduced
crude
(arc)
Wash
vacuum residue
(vrc)
Damp
(Without Stripping)
Precondensor
steam
reduced
crude
Advisian / 5

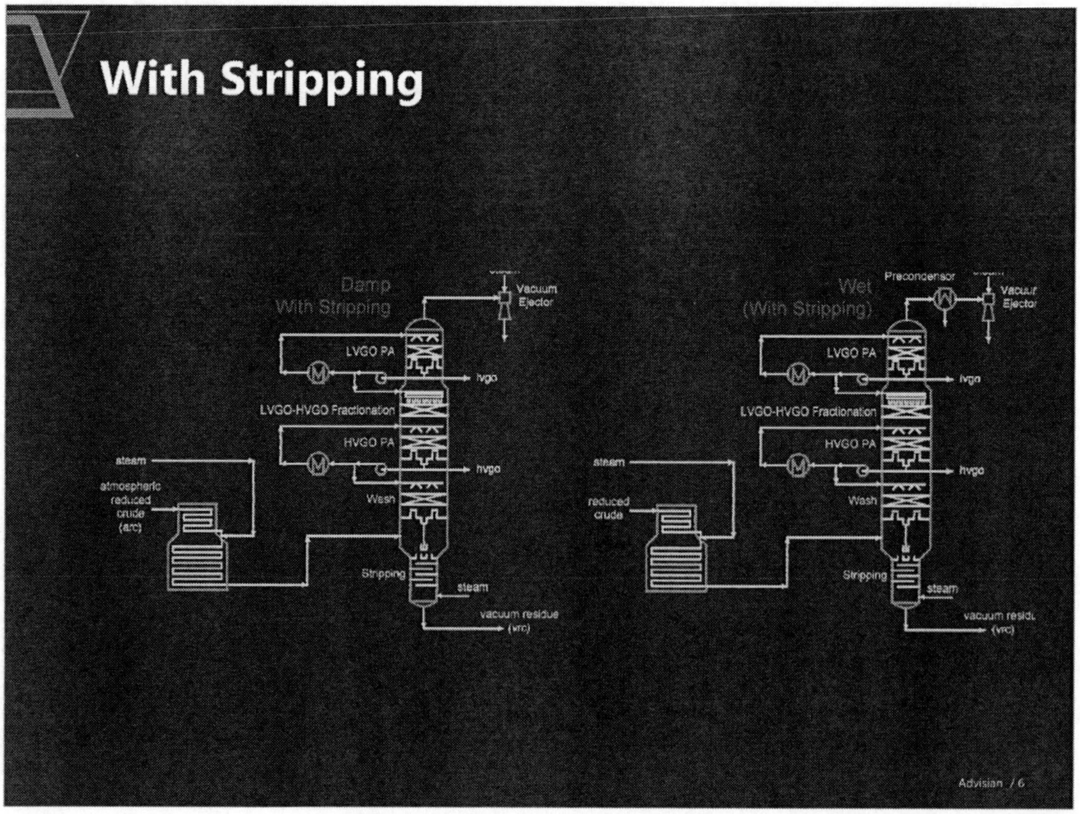
With Stripping
Damp
With Stripping
Vacuum
Ejector
LVGO PA
lvgo
LVGO-HVGO Fractionation
HVGO PA
hvgo
steam
atmospheric
reduced
crude
(arc)
Wash
Stripping
steam
vacuum residue
(vrc)
Wet
(With Stripping)
Precondensor
reduced
crude
Advisian / 6

Basis

- Volatile metal data extremely difficult to obtain
- Comparison against previous work
- Reverse-calculate metals distributions
- Duplicated 1991 publication results to within round-off error
- Looking at:
  - Nickel (Ni)
  - Vanadium (Va)
  - Conradson Carbon Residue (CCR)

Advisian / 7

Feed TBP (Arab Atmospheric Residue)

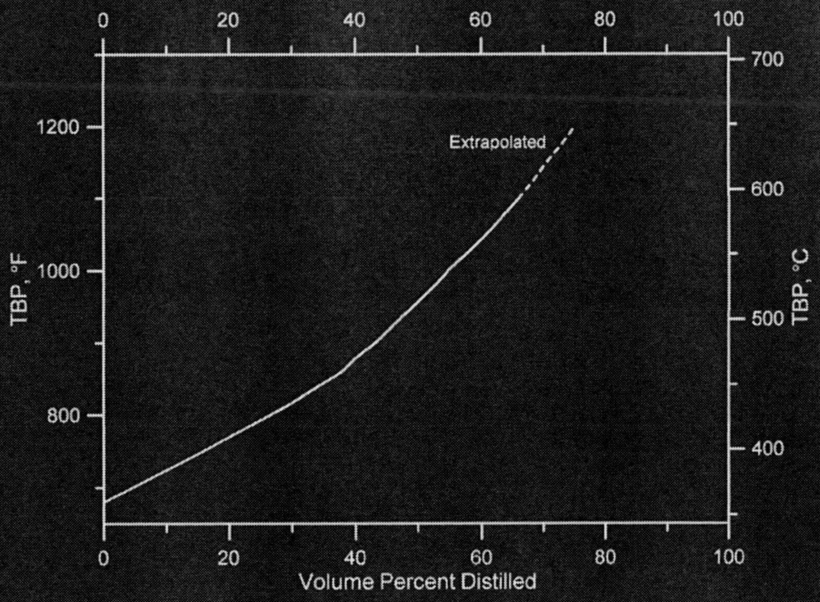

Advisian / 8

Feed Properties

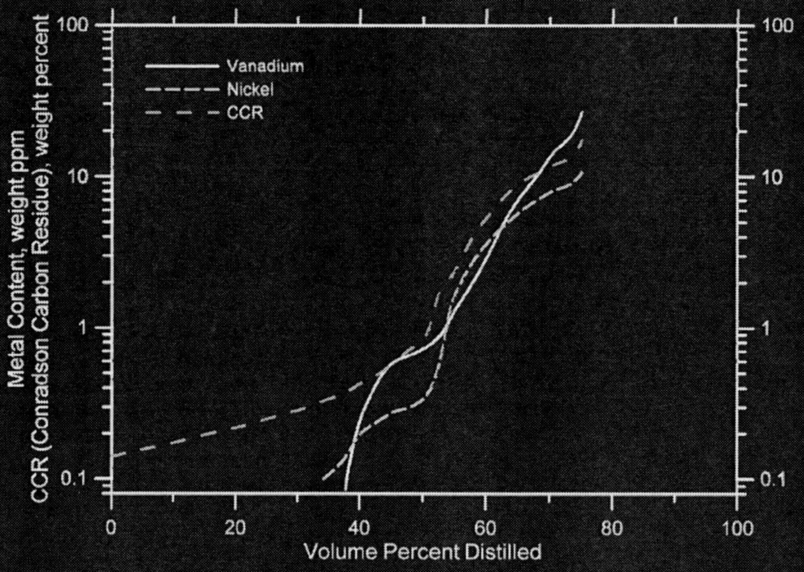

Type Comparison (Duplicates 1991)

Operation	Dry	Damp 1	Damp 2	Wet
Heater steam, lb/hr		5918	2948	2948
Stripping steam, lb/hr			4004	4004
Upstream cutpoint, °F	680	680	680	680
Vacuum tower cutpoint, °F	1,063	1,063	1,063	1,063
Overflash rate, vol % on feed	1	1	1	1
Flash zone pressure, mmHg	18	35	35	66
Wash stages		1	1	1
Stripping stages			1	1
Ni, wppm	1.93	2.10	1.56	1.79
Va, wppm	3.10	3.40	2.43	2.84
CCR, wt percent	2.09	2.16	1.84	1.94

Old Conclusion and Follow-Up

- Stripping sections inherently improve product quality (lower Va, Ni, CCR)
- No real mechanism proposed
- Always thought partial pressure and relative volatility was the answer
- Time to test conclusion
- Keep operating types constant and vary partial pressures
- Vary damp partial pressures by varying both system pressure and steam rates

Advisian / 11

Partial Pressure Determines Separation

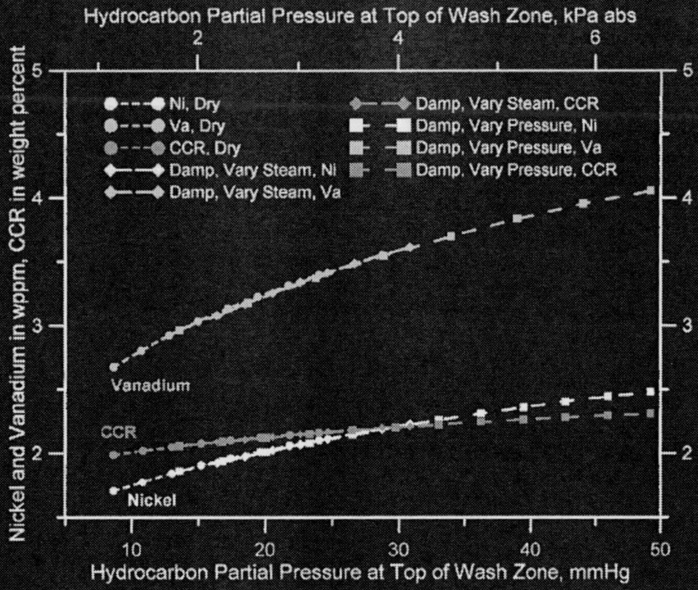

1063°F (573°C) Cutpoint

Advisian / 12

New Conclusion

- Partial pressure determines product quality improvement
- At lower partial pressures hydrocarbon compounds have higher relatively volatilities
- Low partial pressure comes from either:
  - — Steam injection
  - — Low operating pressures

Advisian / 13

CCR vs. Overflash Rate

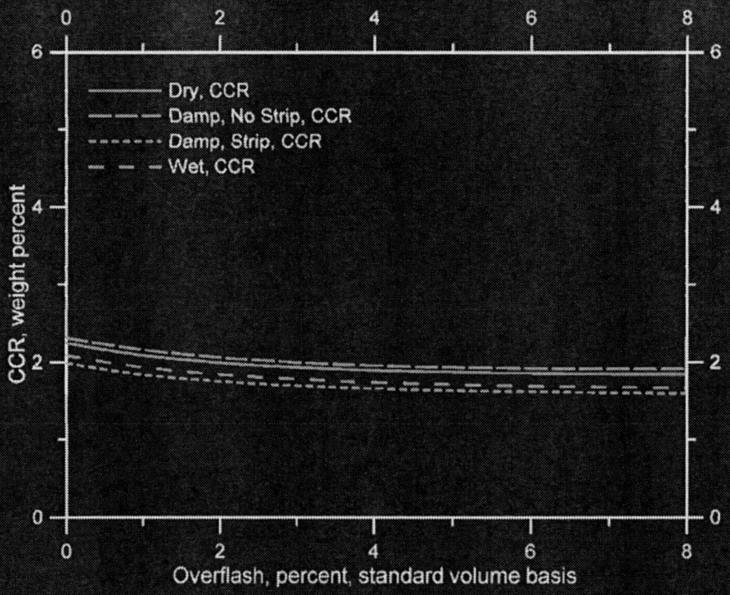

1063°F (573°C) Cutpoint

Advisian / 14

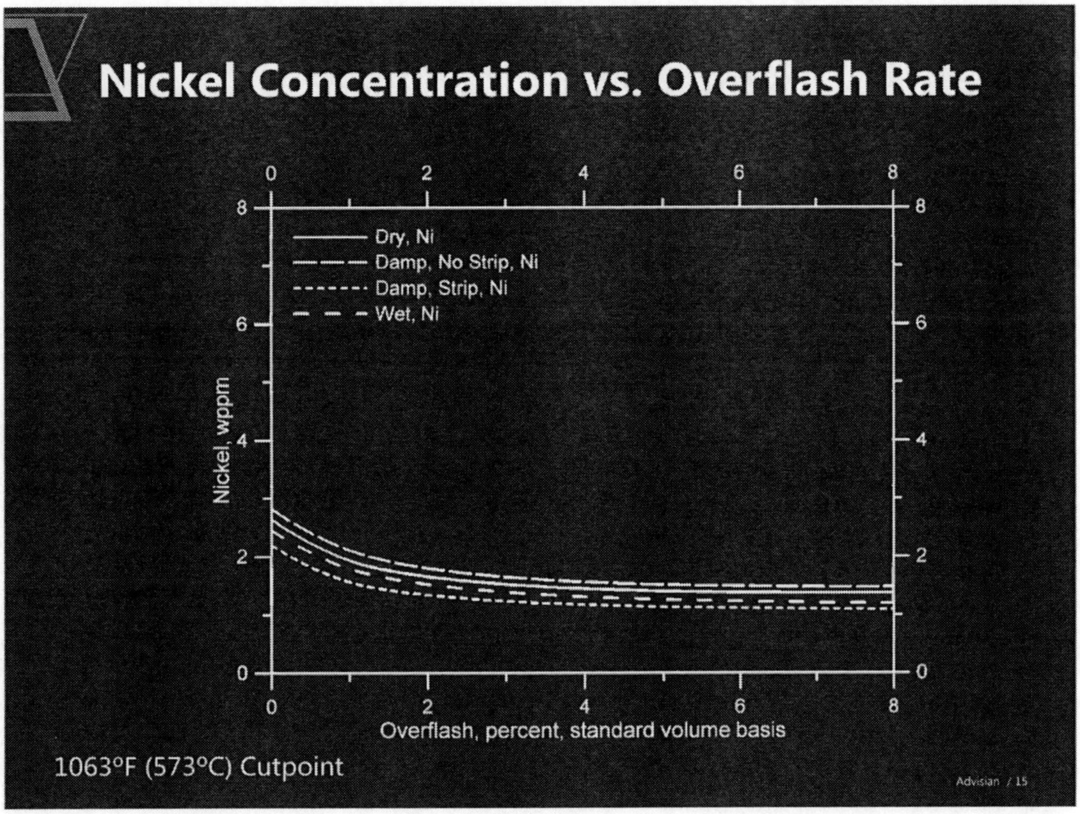

Nickel Concentration vs. Overflash Rate
0
2
4
6
8
Dry, Ni
Damp, No Strip, Ni
Damp, Strip, Ni
Wet, Ni
Nickel, wppm
Overflash, percent, standard volume basis
1063°F (573°C) Cutpoint
Advisian / 15

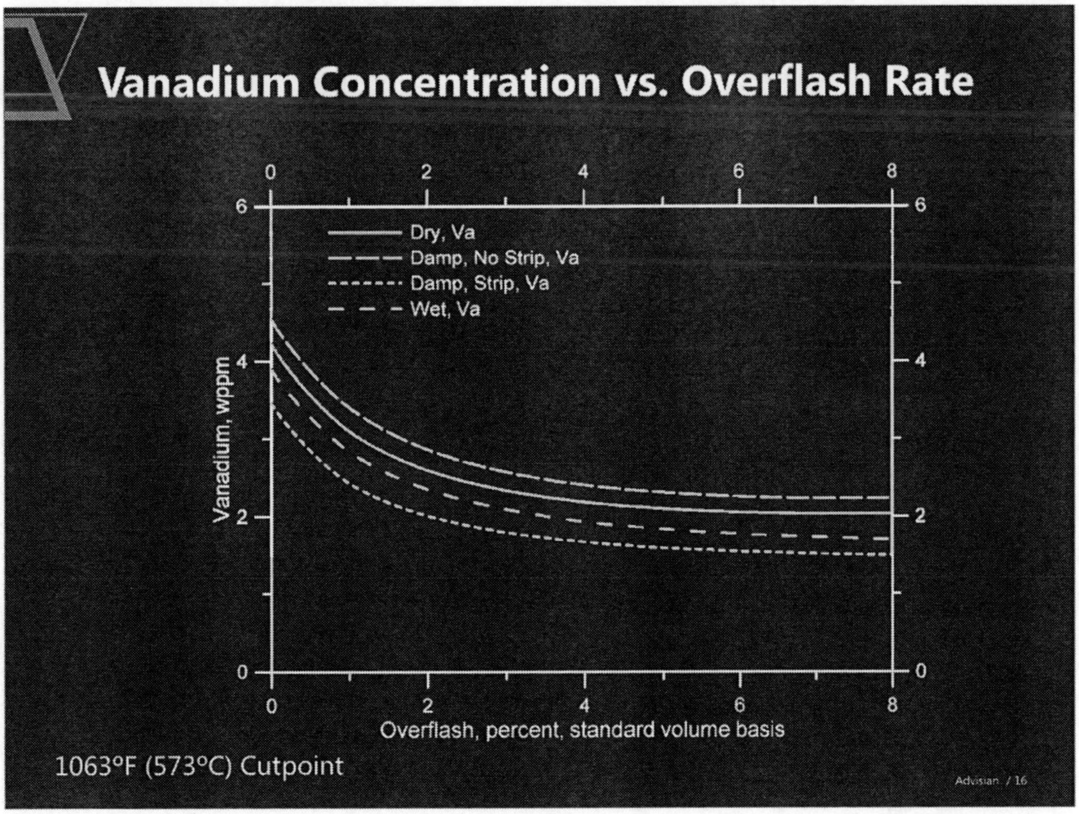

Vanadium Concentration vs. Overflash Rate
0
2
4
6
8
Dry, Va
Damp, No Strip, Va
Damp, Strip, Va
Wet, Va
Vanadium, wppm
Overflash, percent, standard volume basis
1063°F (573°C) Cutpoint
Advisian / 16

Other Factors Limiting Overflash

- Irrigation required for reliable operations (reduces coking rate):
  - ~3 to 4% overflash required for reliable commercial operation
- This pushes operation into flat part of performance curve
- Little benefit for overflash vs. fractionation on most commercial units: agrees with plant observation
- Impurity concentration becomes highly correlated with yield (cutpoint)
- Other observations:
  - Vanadium has largest response for this crude
  - Vanadium is "best case" for justifying process changes
  - Only show vanadium results for other options

Alternatives

Alternatives to Reduce Volatile Metals

- More wash stages
- More stripping stages
- Slop wax recycle
- Lower pressure operation – large equipment, high capital

Advisian / 19

Wash Stages

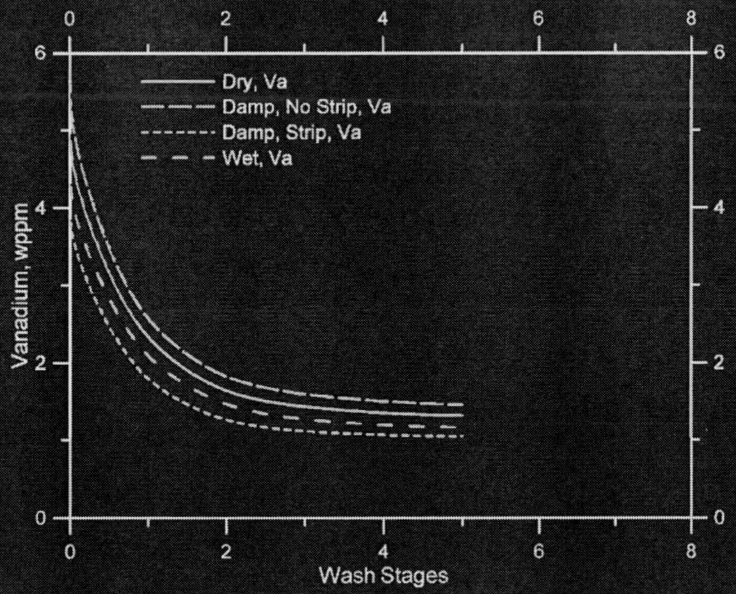

1063°F (573°C) Cutpoint, 3% Overflash

Advisian / 20

Stripping Stages

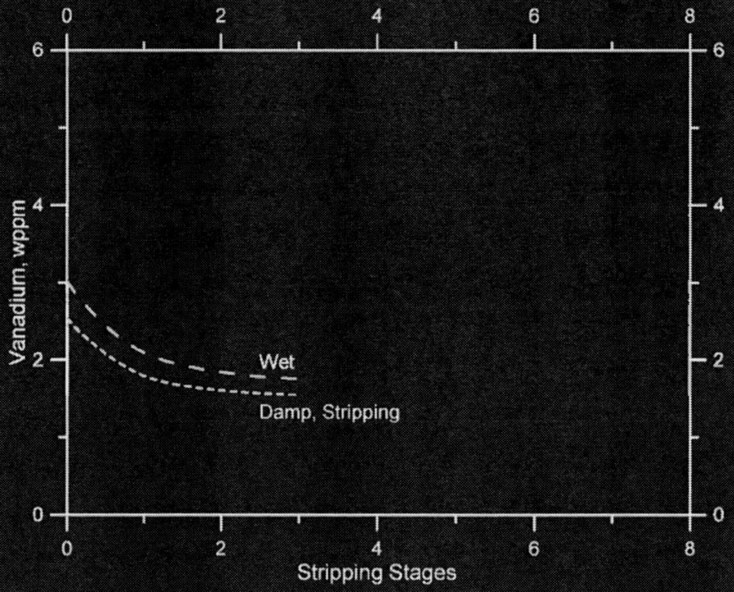

1063°F (573°C) Cutpoint, 3% Overflash

Slop Wax Recycle

Operation, without recycle	Dry	Damp 1	Damp 2	Wet
Ni, wppm	1.52	1.65	1.24	1.39
Va, wppm	2.33	2.57	1.79	2.09
CCR, wt percent	1.94	2.00	1.70	1.78

Operation, with recycle	Dry	Damp 1	Damp 2	Wet
Ni, wppm	1.44	1.56	1.19	1.34
Va, wppm	2.17	2.40	1.70	1.99
CCR, wt percent	1.89	1.96	1.68	1.76

1063°F (573°C) Cutpoint, 3% Overflash

Reducing Volatile Metals

- More wash stages: modest improvement
  - — Very little benefit past 2 stages
- More stripping stages: small improvement
  - — Very difficult to do
- Slop wax recycle: very small improvement
  - — Not worth the cost

Conclusions

Reducing Volatile Metals

- Commercial units have little benefit for increasing overflash rate:
  - — Reliability sets a minimum rate high enough that extra overflash has little value
- More wash stages produce a modest, but possibly useful benefit
- More stripping stages and slop wax recycle produce small benefits – usually not worth the cost

Advisian / 25

Questions?

Advisian / 26

Disclaimers

DISCLAIMER

This presentation has been prepared by a representative of Advisian.

The presentation contains the professional and personal opinions of the presenter, which are given in good faith. As such, opinions presented herein may not always necessarily reflect the position of Advisian as a whole, its officers or executive.

Any forward-looking statements included in this presentation will involve subjective judgment and analysis and are subject to uncertainties, risks and contingencies—many of which are outside the control of, and may be unknown to, Advisian.

Advisian and all associated entities and representatives make no representation or warranty as to the accuracy, reliability or completeness of information in this document and do not take responsibility for updating any information or correcting any error or omission that may become apparent after this document has been issued.

To the extent permitted by law, Advisian and its officers, employees, related bodies and agents disclaim all liability—direct, indirect or consequential (and whether or not arising out of the negligence, default or lack of care of Advisian and/or any of its agents)—for any loss or damage suffered by a recipient or other persons arising out of, or in connection with, any use or reliance on this presentation or information.

Advisian / 27

About Us

Advisian
WorleyParsons Group

Consulting and Engineering in

Refinery and Petrochemicals

- Molecule Management
- Economics and Planning
- Technology Selection
- Refinery-Petrochemical Integration

Logistics

- Hydrocarbon Logistics
- Bulk liquid terminals
- Railway corridors
- Market analysis
- Risk Analysis
- Safety and Environment

Sulfur and Gas Treating

- Oxygen enrichment
- Gas treating solvents

Infrastructure and Environmental

- Water
- Materials handling

EcoNomics

Advisian / 28

Advisian

WorleyParsons Group

Backup Information

Basis Crude Performance

(Tia Juana Light)

Advisian / 30

CCR vs. Cutpoint

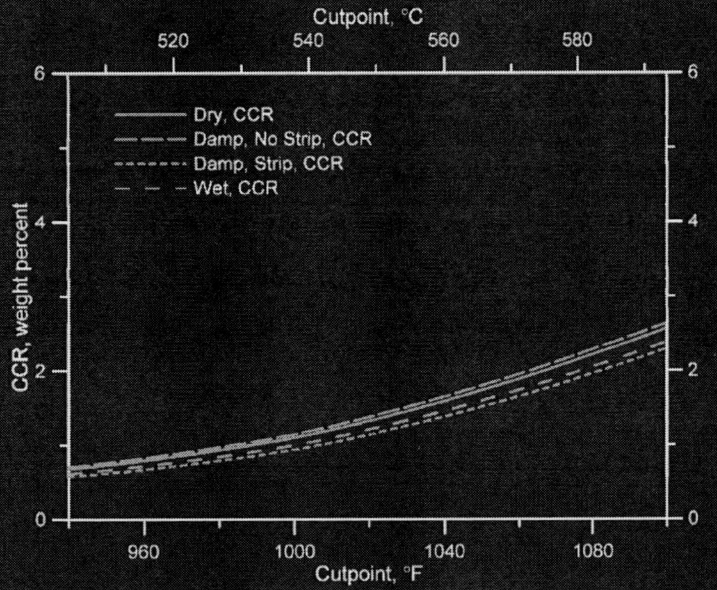

Advisian / 31

Nickel vs. Cutpoint

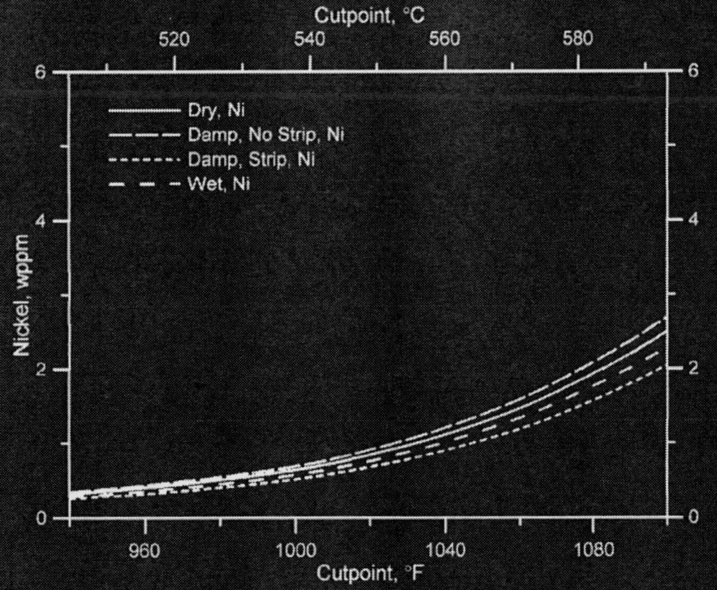

Advisian / 32

Vanadium vs. Cutpoint

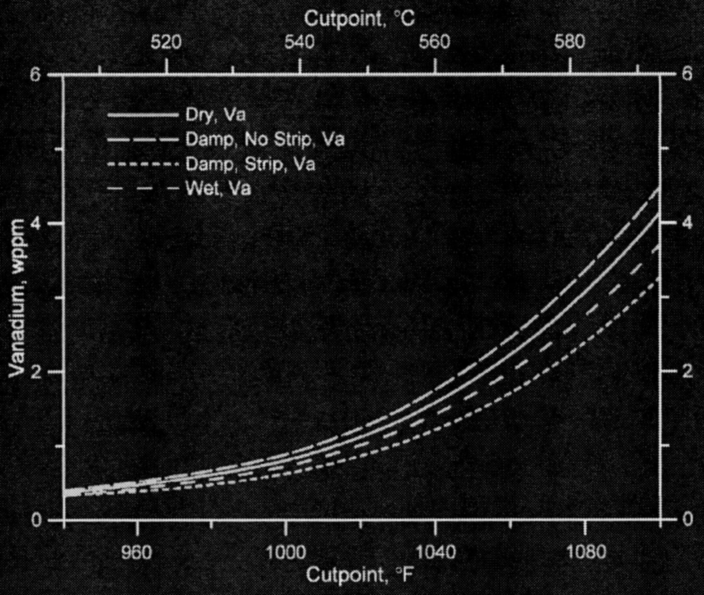

Advisian / 33

Backup Information
Crude Metals Contents

Advisian / 34

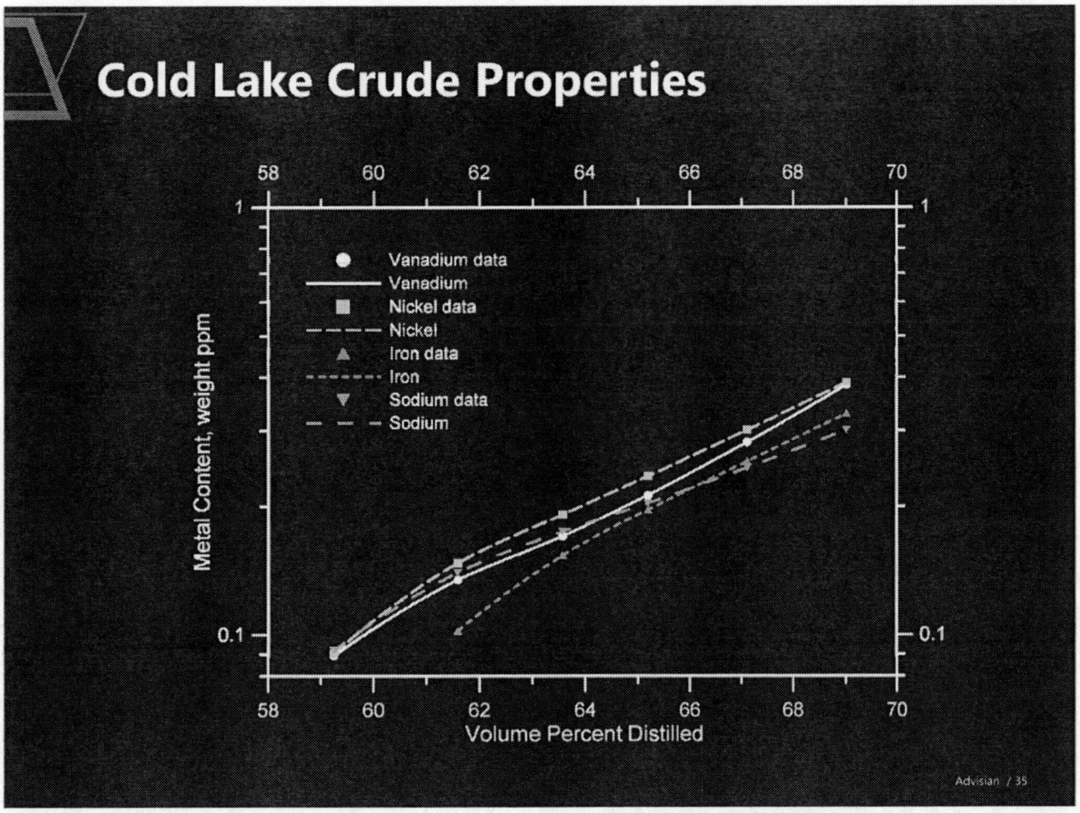
Cold Lake Crude Properties
Vanadium data
Vanadium
Nickel data
Nickel
Iron data
Iron
Sodium data
Sodium
Metal Content, weight ppm
Volume Percent Distilled
58
60
62
64
66
68
70
1
0.1
Advisian / 35

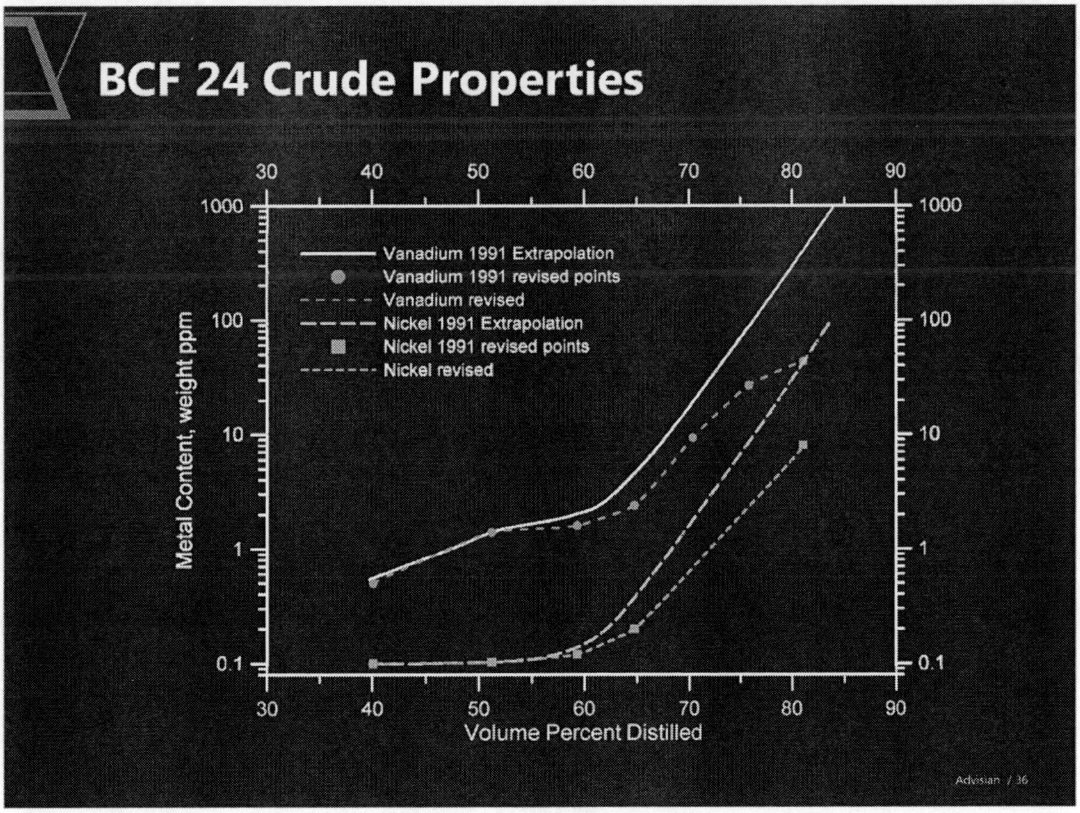
BCF 24 Crude Properties
Vanadium 1991 Extrapolation
Vanadium 1991 revised points
Vanadium revised
Nickel 1991 Extrapolation
Nickel 1991 revised points
Nickel revised
Metal Content, weight ppm
Volume Percent Distilled
30
40
50
60
70
80
90
1000
100
10
1
0.1
Advisian / 36

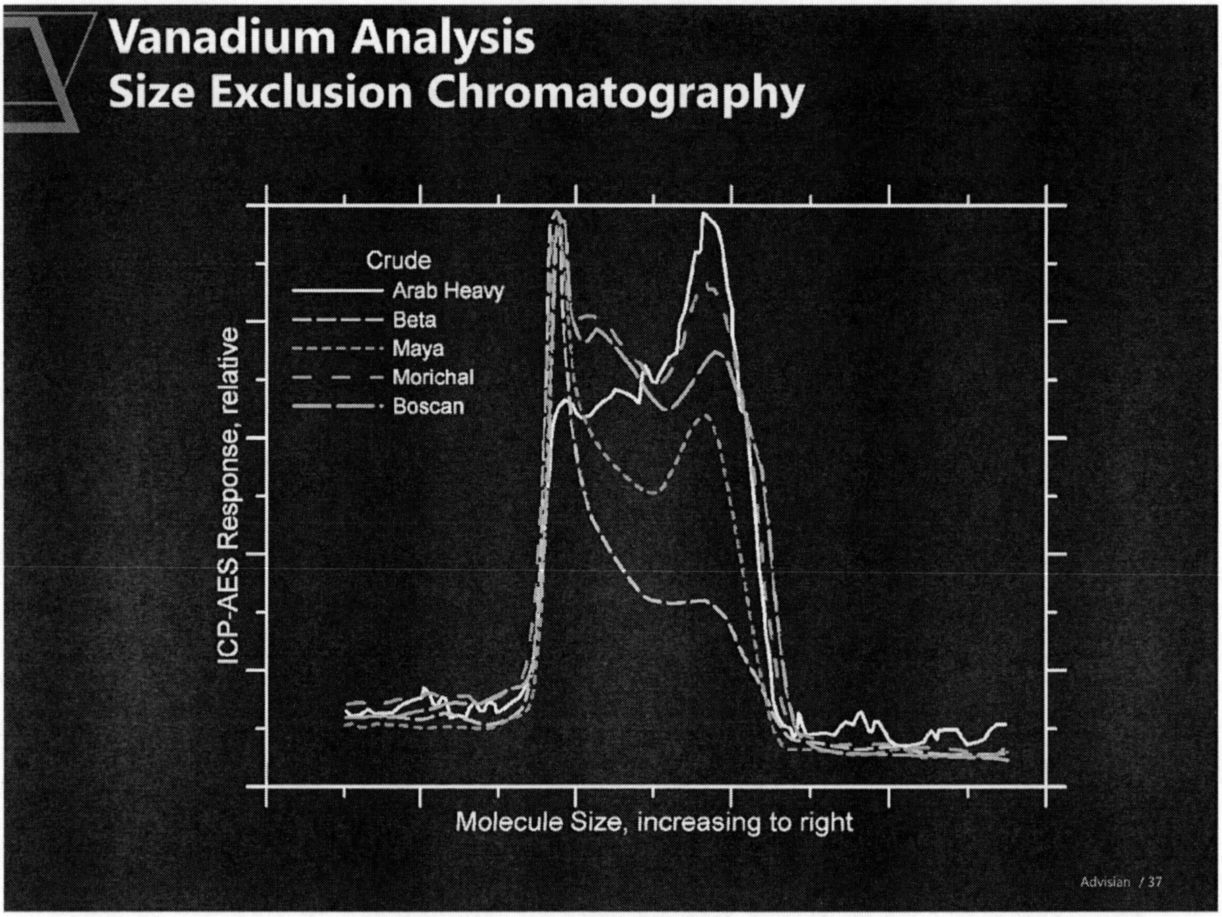
Vanadium Analysis
Size Exclusion Chromatography
Crude
Arab Heavy
Beta
Maya
Morichal
Boscan
ICP-AES Response, relative
Molecule Size, increasing to right
Advisian / 37

Advances in Salt Point Management Practices: Mitigating Overhead Corrosion with Big Data

Collin W. Cross, Ph.D.; Wan Zhaoyang, Ph.D.; Keyur Patel, Ph.D.
9-27-2017

Abstract

In the world of overhead corrosion, it has long been recognized colloquially that ninety percent of metal loss occurs during ten percent of the time in operation. In many types of fractionating tower overheads, a confluence of fluctuating dynamical factors lead to the formation of amine hydrochloride salts and ongoing associated corrosion. Because of the constantly changing tendency to form corrosive salts in an operating unit, it has remained difficult using traditional approaches to systematically identify the proverbial ten percent of time where damage-causing episodes occur so that effective intervention and mitigation are possible.

The challenges above have caused the fundamental principles of overhead corrosion mitigation to remain largely unchanged in the last decades. While advances in metallurgy, inhibitors, injection practices, and monitoring techniques have occurred, overhead corrosion measurement and predictive practices have not. The result is that crude unit overhead systems continue to experience corrosion that leads to unexpected failures, fouling, outages, all contributing significantly to lost profit opportunities.

Salt point management practices continue to rely largely on series of unconnected salt point calculations over time using sparse data sets. Here we will present a new computational framework that allows continuous and quantitative sensitivity computations for amine hydrochloride salt formation in an operating crude unit. This capability then leads to the rapid identification of seemingly random corrosion events and their precise causes. During an emerging corrosion event, the onset can be rapidly flagged and alerted while the driving forces of salt formation and their sensitivities are quantitatively revealed. The demonstrated method takes near real-time data from operating crude units and uses it to compute how changing factors, such as crude diet, processing objectives, operations, control schema, physical constraints, set points, and randomly varying amine and chloride levels, cause corrosion events. The practical objective of the methodology is to drive timely, precise, and proactive mitigation options in a granular and systematic fashion across a wide variety of event types. Real-world examples taken from operating refineries will be used to demonstrate the technique.

Introduction

In the world of overhead corrosion, it has long been recognized colloquially that ninety percent of metal loss occurs during ten percent of the time in operation. In many types of fractionating tower overheads, a confluence of fluctuating dynamical factors leads to the formation of amine hydrochloride salts and ongoing associated corrosion. Because of the constantly changing tendency to form corrosive salts in an operating unit, it has remained difficult using traditional approaches to systematically identify the proverbial ten percent of time where damage-causing episodes occur so that effective intervention and mitigation are possible.

The challenges above have caused the fundamental principles of overhead corrosion mitigation to remain largely unchanged in the last decades. While advances in metallurgy, inhibitors, injection practices, and monitoring techniques have occurred, overhead corrosion measurement and predictive practices have not. The result is that crude unit overhead systems continue to experience corrosion that leads to unexpected failures, fouling, outages, all contributing significantly to lost profit opportunities. Bundle washes continue to yield unequivocal evidence of amine hydrochloride salt-induced corrosion that went un-detected and, therefore, un-mitigated.

Salt point management practices surrounding overhead systems continue to largely rely on series of unconnected salt point calculations over time using sparse data sets around key thermodynamic drivers. Salt point deposition temperature computations continue to be largely performed manually, meaning users must provide and manage over time all data necessary for calculating results. The result of this situation is that overhead corrosion mitigation practices fall into a spectrum of sophistication surrounding mitigation options. At the most basic level of the spectrum, corrosion control depends primarily upon injecting corrosion inhibitors and maintaining traditional service and monitoring metrics, such as pH, chlorides, iron throw, ER data, coupons, inspection, etc.

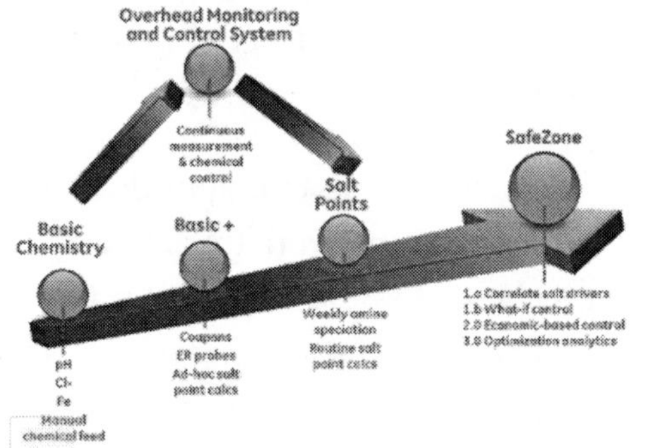

Figure 1*: Continuum of corrosion mitigation and monitoring practices, shown as moving from more simple to more complex.*

Figure 1 shows a way to conceptualize the progression of corrosion management practices from more basic to more advanced. Currently, the bulk of refineries globally do not use high frequency, wide spectrum amine speciation as a regular part of their corrosion control program. Salt point deposition temperature computations are often used, but are not granular enough, predictive enough, reliable enough, nor do they provide enough operational or economical information to be practically useful towards optimizing a crude unit in a quantitative way. In cases where amine speciation analyses are routinely performed at sufficient frequency, often the number of amines sampled are insufficient. Alternatively, if enough amine species are sampled, either routinely or for specific events, the labor-intensive traditional method of performing calculations with such large data sets largely prevent effective salt point analysis from being carried out daily as a routine procedure.

Here, we will present a new computational framework and supporting set of infrastructure and processes that allows rapid, continuous, and quantitative computations for amine hydrochloride salt formation in an operating crude unit to be easily performed. From this basis, the dynamic sensitivities, or gradient, of the salt point to its input variables is available and can be used towards the realization of many practical capabilities. Furthermore, in addition to rendering the salt point and its gradient in close to real-time, the dynamic variabilities of each input parameter, as well as their impact on the output, are additionally calculated. This capability then leads to automatic detection of emerging corrosion events, enabling their rapid and dynamic identification, as well as their precise causes and sensitivities. During an emerging corrosion event, the onset of the conditions making salt deposition likely are flagged and alerts are sent digitally to concerned parties. At the same time, the driving forces of salt formation and

*Trademark of SUEZ; may be registered in one or more countries

their sensitivities are quantitatively revealed and deconstructed graphically and tabularly into a group of indicators and actionable options towards mitigation. The demonstrated method automatically takes multiple sources of asynchronous, high-frequency data from operating crude units and uses it to compute how changing factors, such as crude diet, processing objectives, operations, control schema, physical constraints, set points, randomly varying amine and chloride levels, cause corrosion events, and other factors affect salt formation. Sophisticated signal processing and analytics algorithms are then used to make a wide array of valuable capabilities become practical, seamless, and potentially fully automated. All of these computations are done subject to the actual physical and operational constraints on the unit during its operation, which are handled quantitatively. The practical objective of the methodology and its supporting infrastructure is to drive timely, precise, and proactive mitigation options in a granular and systematic fashion across a wide variety of event types to improve reliability and productivity of the asset being monitored.

Computational Framework for Automated Salt Point Calculation

Salt point calculations require operational data and chemistry data from the crude distillation tower overhead system. Operational data required for salt deposition temperature calculations are temperature, pressure, reflux flow rate, product flowrate, and feed flow rate, etc., whereas required chemistry data are a regularly repeating wide list of individual amine concentrations and the chloride concentrations. Usually, high frequency operational and process data are readily available from controllers installed in a crude distillation column. Chloride levels in accumulator boot water can be cost effectively and reliably obtained at high frequency by a variety of methods. Additionally, accumulator boot water pH data are also easily measured and tabulated as a time series. These data are all then retrieved into a central data storage platform, which is a key part of the computational pipeline. After being received, data is then trended and analyzed to identify and monitor specific corrosion events as they emerge. This enables proactive actions to be continuously evaluated towards optimal mitigation by means which are in line with processing objectives.

Accumulator boot water amine concentrations are normally the rate limiting factor in the deployment of high frequency salt deposition computations. The lack of a suitable processing schema has also made the routine manual calculation for salt points of all amines widely known to affect corrosion seem cost prohibitive. The value and usage of such a data set to provide tangible operational predictive power has remained elusive due to both the expense and complexity of amine speciation as well as the inability to effectively convert a series of broad-based, high-frequency amine data into reliable and actionable information. Here, we will show that using such data is very worthwhile. The combination of broad-based, high-frequency amine data and a high speed automated processing pipeline can allow previously large, unseen, and/or randomly occurring corrosion events to be systematically identified and revealed for mitigation in a proactive rather than reactive way. The ability to identify, understand, and prioritize a practical list of weighted options towards the mitigation of large and previously unseen corrosion events on a regular daily basis can have a dramatic impact on unit reliability over the length of a production run. This is especially true during the processing of challenging crudes.

To complete the computational process, all the high-frequency telemetry discussed above is extracted hourly, daily, or semi-weekly, in an automatic fashion, and is used in a new platform called **SafeZone*** – a proprietary, automated, salt-point analytics framework, offered exclusively by **SUEZ - Water & Solutions Technologies**. This is done via a suite of integrated hardware, data warehousing, high performance computing, analytics, expert analysis, and reporting. The platform uses the output of the computations to provide a set of tools which enable the ability for a team of people to graphically optimize salting risk against a refinery's production objectives on an ongoing basis with little training or effort.

The suite of tools and the data analysis pipeline for the SafeZone tool suite is designed with digital, team-oriented communication and decision-making support as key objectives. Because of this system, the tools are rigorous, systematic, and automatic, but distill the data into a simple set of streamlined information widgets that are capable of being displayed using a diverse set of reports built around them. This system is then able to provide a large array of powerful and standardized information to a dispersed group of people operating in a busy environment. The information is provided in a very flexible, yet structured and quantitative way. This provides a platform for rapid, streamlined, and quantitative decision making by the team over a wide variety of dynamic situations as they emerge during a production run.

At the heart of the **SafeZone*** process is a sophisticated set of proprietary algorithms developed by **SUEZ** scientists and engineers that use a state-of-the-art high-performance computing infrastructure to implement a stochastic, feed-forward control schema in a crude fractionators multi-dimensional state space and subject to all its constraints, which can also be modeled for evaluation. The results of the complex mathematical analysis are used to provide graphical output as a set of indicators in a variety of expert tools, visualization dashboards, and structured reports. This platform is then used to iteratively understand the drivers of both systematic and episodic corrosion and relate them to the objectives of the crude unit as well as its constraints and control scheme. These elements are then used to iteratively optimize an operating crude unit on a day-to-day basis (or faster) using a team-oriented stochastic feed-forward decision loop facilitated by the **SafeZone*** framework.

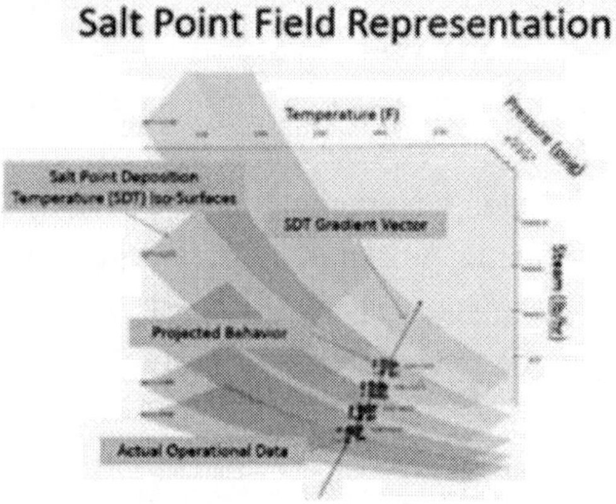

Figure 2: *Cross section of multidimensional state space containing the scalar and vector fields computed via the* ***SafeZone**** *high performance computing platform.*

Figure 2 shows a cross section of the multidimensional state space containing the scalar and vector fields for the margin differential between salt point deposition temperature (SDT) and tower top temperature and the SDT's gradient. **Figure 3** shows a graphical schematic of the telemetry gathering process and computational framework as well as a snapshot of an example dashboard.

For ensuring reliable mitigation of corrosion for arbitrary emerging events, it must be ensured that the key telemetry has sufficient granularity to enable a sufficient numerical estimate of salt deposition temperatures as well as their sensitivities and variabilities to both the controlled and

*Trademark of SUEZ; may be registered in one or more countries

stochastic aspects of the multidimensional input space. The algorithms are then used to predict the probability of a salting event occurring daily given the state of the system, its immediate history, and its path-dependent trajectory through the local input space. The probability of salting computed at the core of the process is called the "SPI," or Salting Probability Index. This index represents the probability that salting will occur during a current or historical time, given the location of the system in state space and the variability of the input data and confidence intervals around the output data. The dynamic rendering of the SPI is what drives the core of the **SafeZone*** process. **Figure 4** shows a two-dimensional, graphical representation of the SPI within a manifold of its input space.

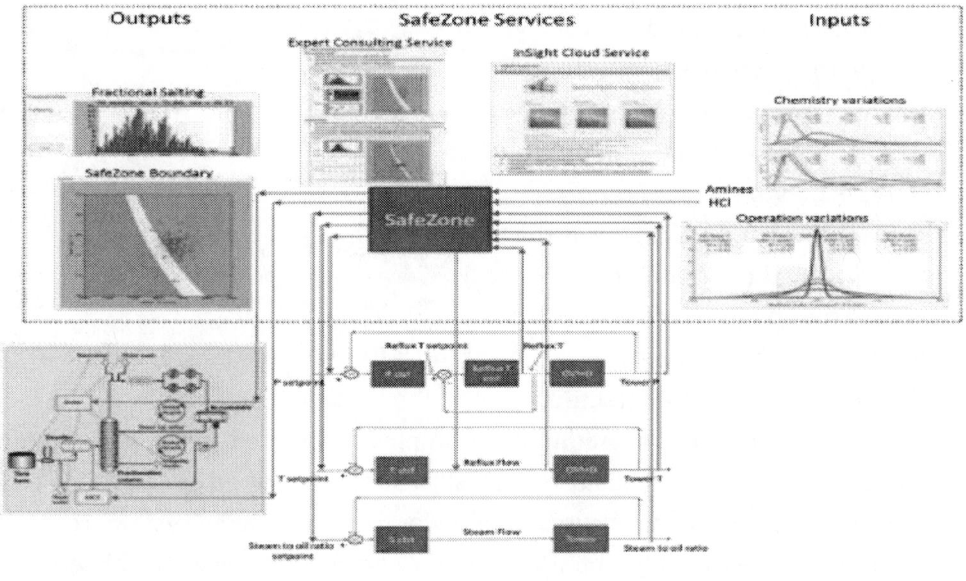

Figure 3: *Computational Framework for Automatic Salt Point Calculations*

Amine Testing Requirements

Traditionally, to eliminate corrosion risk from amine hydrochloride salt deposition, a typical margin, or differential, of 15-25 degrees is assumed and then maintained between either the mixed exit temperature (MET) or tower top temperature (TT), respectively, and the salt deposition temperature (SDT). However, because amine and HCl measurement frequencies are often limited by sparse data for a variety of reasons, using these various margins as indicators has not been found to be a reliable means to predict corrosion risk. To address this challenge, the salt point calculation algorithms used here consider also the random variations of operational and chemistry variables over a given timeframe. These are then used to compute a more representative indicator of salt formation potential, called the salting probability index (SPI). Over a specified timeframe, and for several random variations of operational and chemistry variables, the SPI indicates the number of times salt formation occurs, divided by the total number of operational points being sampled. The SPI represents

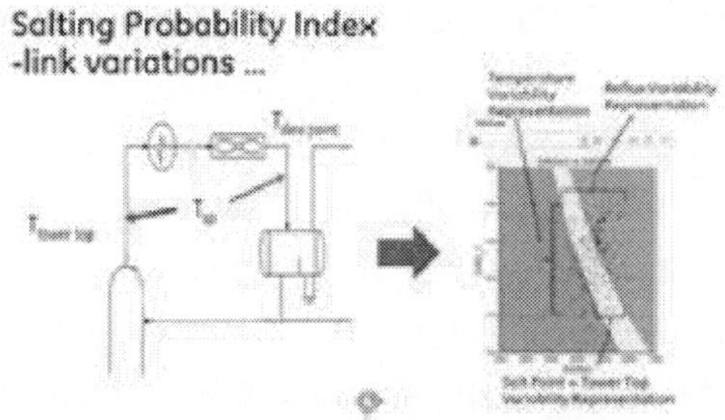

Figure 4: *Graphical representation of the integral SPI and its operational ensemble within a two-dimensional cross section of the fields input space.*

*Trademark of SUEZ; may be registered in one or more countries

the dynamical probability of salting based on neutralizer, tramp amine, chloride concentrations, and the current operational situation.

After automatically gathering the data and performing the calculations above, the platform then reduces the information to graphical and numerical output and delivers high-frequency, advanced reporting via a variety of possible channels. The executive summary of the daily report shown below gives visual indications of salting in both the overhead condensing system and the tower top by showing both respective SPIs independently. **Figure 5** shows a snapshot of the executive summary section of the daily report. While the report itself is beyond scope, this is the first of many pages designed to offer unique insight into dynamic aspects of corrosion and their relationships to both operations and chemistry so emerging events can be detected, identified, categorized, and acted upon. Additional examples of routine weekly, monthly, and strategic reports are not shown here. Within the daily report, SPI values are plotted over a specified period and color coded to indicate whether salting occurs in the tower top or overhead system, respectively. Not only are SPIs presented, but sensitivities are also output for both operational and chemistry variables. This allows prioritization and ranking of multiple potential mitigation options to best bring the tower top or overhead system out of the salting region if desired. A large range of competing mitigation options can easily and systemically be evaluated and all stakeholders can have a similar viewpoint and a reference system by which to make quantitative comparisons over time.

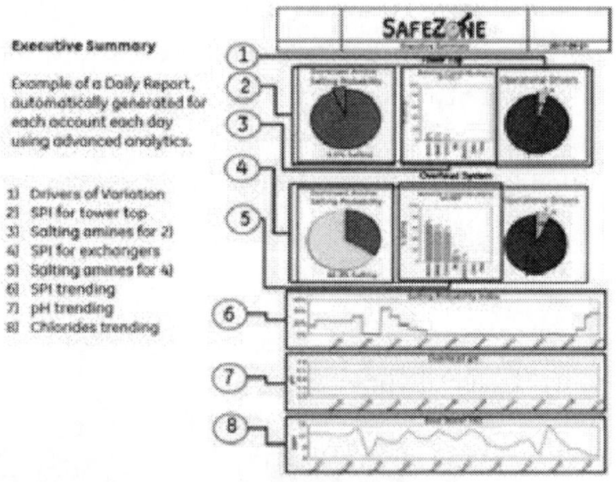

***Figure 5**: SafeZone* daily report executive summary page. Detailed explanation is beyond scope. Executive summary answers 3 questions each morning. Q1) Is there a problem? Q2) If so, what is causing it? Q3) How can it best be addressed? The answer to Q3) extends onto the reports other 10 pages.*

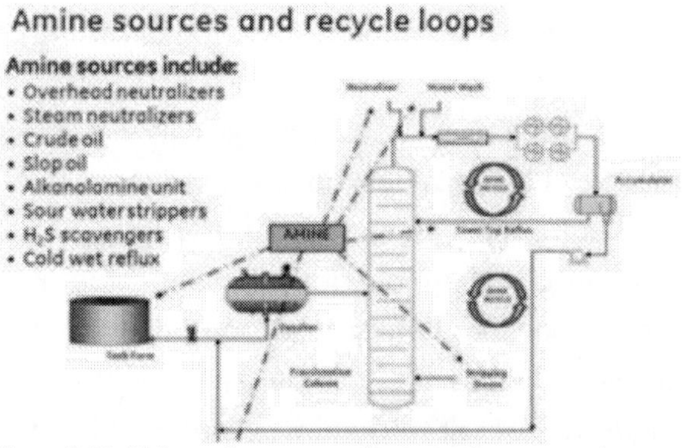

***Figure 6**: Multiple varying sources of entry to the system, along with multiple complex recycle mechanisms having different initiation criteria and time scales, along with the different physical properties of amines with respect to pH, boiling point, and relative oil/water solubility, cause the overhead of a typical refinery to experience a randomly fluctuating tendency (stochastic) for all the various amines to impact the amine hydrochloride deposition process and dependent ongoing corrosion.*

The system also calculates many other useful parameters such as HCL margins. This is the dynamical amount of HCl reduction needed to move the system from a salting region to a non-salting region, or vice-versa. The HCl margins are one of many parameters automatically calculated and trended daily and used towards weighting and evaluating multiple potential mitigation options when needed. This parameter will become important for the case study discussed below. Additionally, along with salt deposition temperatures, mixed exit temperatures, and dew point temperatures, many other metrics are reported for easy comparison with

*Trademark of SUEZ; may be registered in one or more countries

SPI charts to help unravel the driving causes of ongoing events and their relative sensitivities towards mitigating corrosion upon manipulation.

Amine Dynamics

As discussed above, the lack of an adequate computational platform as well as the expense and complexity of broad-based high-frequency amine speciation, have made it difficult to observe, document, and understand the complex, random, dynamical aspects of amine hydrochloride salt deposition that have been shown to underpin ongoing corrosion. While detailed discussion is beyond scope of this document, **Figures 6, 7, and 8** show various data representations that detail how different amines participate in the corrosion process over time and that MEA, while a predominate actor, is only one of many problematical amines that affect the corrosion process over time. Often, common suspect amines are completely misidentified or focused on as a primary driver of corrosion while other hidden causes remain unknown. This lack of granular visibility during operations often leads to the misidentification of more significant drivers, and thus often prevent ongoing problems from being effectively revealed, identified, and eliminated, eventually leading to subsequent unplanned impacts on reliability and production.

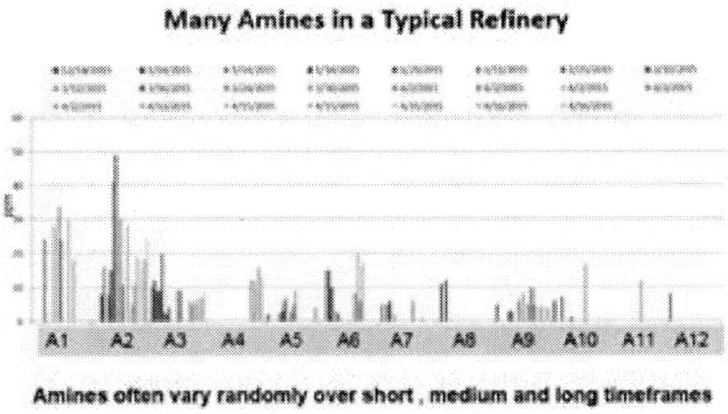

***Figure 7**: Overhead receiver water amine concentrations sampled over a 4-month timeframe. Continuous daily sampling will reveal large random variations over time.*

Chloride	DMAE	Morph	MDEA	MA	DMA	TMA	MEA	NH3
7	171	249		4	11		10	16
10	129	178		3		4		17
10	82	122		8	4	10	22	10
12	92	138		4		12	14	25
14	40	54	34	3	3	10	24	11
15	76	124		3			13	14
16	132	196		4		6		15
20	198	270		4	4	14		15
60	145	241		2		10	23	12

Crude Unit Neutralizer
Steam Neutralizer
Plant Amines
Scavenger, Byproducts or Other

amine:chl ratio above SPBP @ 260°F

***Figure 8**: Multiple samples of Salt Point Deposition (SDT) temperatures for a series of receiver samples taken over a 2-week timeframe. Color coded boxes are examples where the SDT is problematical from a corrosion perspective. Some of the randomly varying tendencies of the corrosion potential for multiple amines can be seen here.*

To effectively identify, understand, de-construct, categorize, analyze, down select, and prioritize effective and dynamic corrosion mitigation actions that have a meaningful affect upon the trajectory of overhead corrosion during an ongoing run, it is necessary to use high-frequency, broad-based amine speciation. Along with a comprehensive feed of amine speciation data, as well as the other telemetry, a high-performance processing and signal analysis platform makes it possible to gain visibility, understanding, and higher orders of control over the fluctuating and stochastic tendencies of amine hydrochloride deposition potential leading to active corrosion in the overhead of refinery fractionators. This, in-turn, has an ongoing and direct impact on operational reliability.

Example Study

To see how the **SafeZone*** technique differs from traditional methods of overhead corrosion control an example case is shown that illustrates how the **SafeZone*** technique was used to smoothly, reliably, and

*Trademark of SUEZ; may be registered in one or more countries

profitably, handle a very difficult opportunity crude with a bad reputation and a deep discount. The opportunity crude referenced here is corrosive to the overhead because it contains very high levels of volatile organic acids and the refinery had no previous experience running the crude. It is important to realize that there are three physical realities largely defining the data granularity requirements needed to achieve the given objectives. These three physicalities are:

1) Salt points cannot be computed unless there is a valid set of amine partial pressure estimates available.

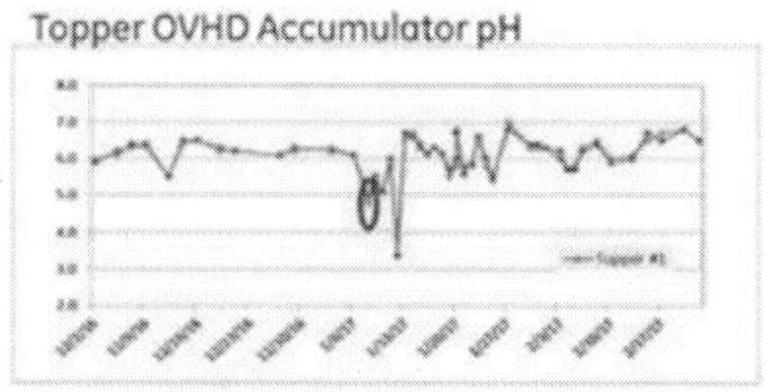

Figure 9: pH trace of overhead receiver during the introduction of a corrosive opportunity crude. pH quickly drops to near 3 until it is stabilized with an increase of the neutralizing amine DMAE.

2) The set of amine partial pressures driving overhead corrosion in most refineries is extremely variable and highly random.

3) The specific individual natures and patterns of the amine variations driving corrosion are often specific to a given situation, but fluctuate widely due to operations and other factors.

Because of these physical constraints, any salt point calculation methodologies attempting to predict salt points by use of recycle models of partial pressure and neutralizer feed rates are not sufficient. Furthermore, techniques seeking to base calculations on sets of assumptions wherein past amine values continue to be used longer than relevant (a few days usually) cannot reliably detect episodic corrosion events. This is true for both short and medium timeframes, which are normally the most important timeframes of interest from a proactive refinery overhead corrosion mitigation perspective.

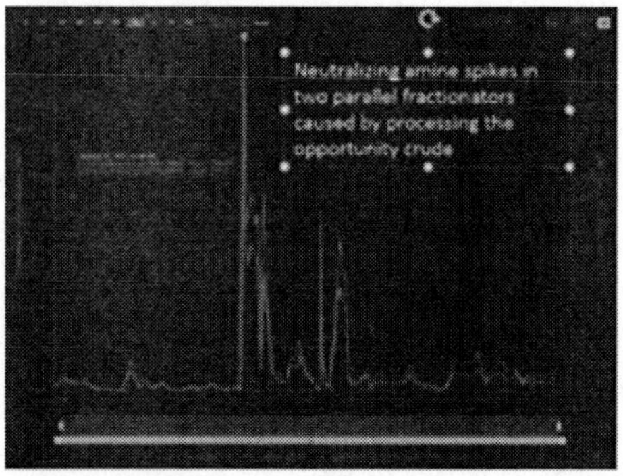

Figure 10**: **Insight data system which feeds the **SafeZone*** engine. The first DMAE spike on the left is the one covered by this case study.*

In this example, as the opportunity crude was introduced, the pH of the overhead receiver rapidly dropped to near 3, as shown in **Figure 9**. As a response to this observed change in the corrosion landscape, the neutralizer was quickly and manually increased to bring pH back into the proper control range of 5.5-6.5. Continuing throughout the run, which lasted several months, chlorides remained well controlled and always less than 15 ppm, with less than 10 ppm being the norm. Furthermore, no other indicators of corrosion had manifested until after the run was well in progress. Caustic was used initially at less than 0.5 pounds per thousand barrels to control chlorides and keep then less than 10 ppm and there was also an upper sodium spec on the residua streams of 10 ppm.

*Trademark of SUEZ; may be registered in one or more countries

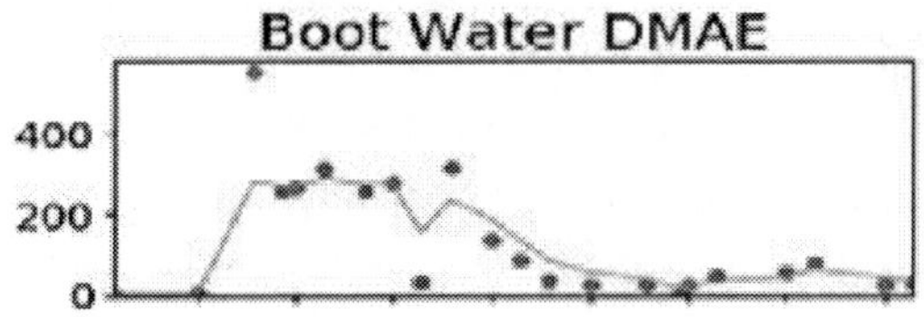

*Figure 11: Chart showing a trace of the high frequency amine data used to detect the DMAE spike, whose corrosion impact is being mitigated using **SafeZone***. The timescale of the trace is 2 months and begins just before the introduction of the increased neutralizer injection used to combat the drop in overhead pH.*

Normally, after the depressed pH was brought into range upon introducing the opportunity crude, no further actions would have appeared necessary using traditional corrosion mitigation techniques. This is because pH was in range, chlorides and caustic were low and controlled, and no iron throw or ER probe readings yet indicated a problem. Traditionally corrective actions would have ceased at this point because there was no reliable indication of active corrosion. Since organic acids and amines are not routinely speciated at high frequencies as part of normal service, the DMAE increase and other drivers of this problem, as well as many other potential dynamical contributions, would not likely have been detected or triggered any further corrective actions.

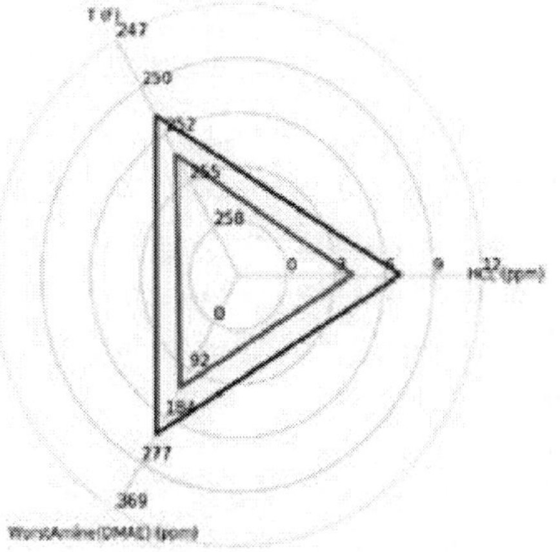

*Figure 12: Constrained optimization envelope chart. If the blue triangle is outside the red triangle the system is in the salting zone. The diagram also shows three geometrically equal ways to bring the system back into the **SafeZone***. Since the system is now in the salting zone, it would theoretically require approximately 3 degrees of temperature, vs. 3 ppm of chloride reduction, vs. nearly 100 ppm of DMAE reduction to achieve the necessary lowering of salt point. Combinations of the above changes can also be utilized if desired.*

In this case, however, when the neutralizer was increased to bring pH into range, the **SafeZone*** SPI indicator immediately alerted a high probability of induced salting and forced a response from the **SafeZone*** team. The salting event was quickly found to be due to the neutralizing amine itself, and its salt point information was dynamically computed based on the actual low levels of amines and chlorides in the system and the actual operational state of the unit at that particular time.

Figure 10 shows a data trace of two DMAE spikes occurring during the opportunity run, while **Figure 11** shows only the first spike in expanded scale. This is the spike of interest and is the one being primarily covered in this case. The first spike occurred between January 13th and February 14th and caused significant corrosion. The second event was better controlled and corrosion was mitigated effectively. Because the **SafeZone*** program detected a high likelihood of salting directly at the event's onset, the chloride margins defined above were evaluated. The team determined from the daily report and offline "what-if" analysis tools that if chlorides could be lowered to 4 ppm then salting would no longer be favorable at the current conditions. **Figure 12** shows how a constrained envelope chart from the **SafeZone*** report can be used to quantitatively evaluate the impact of several options. Many such charts are available for a wide variety of situations as the complete gradient vector is always available.

*Trademark of SUEZ; may be registered in one or more countries

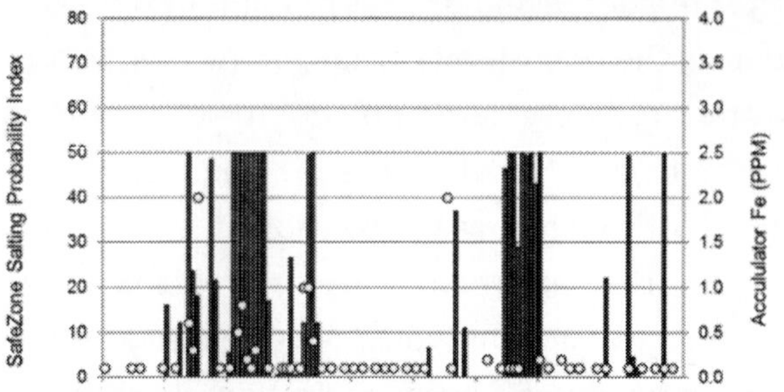

***Figure 13**: Chart of iron throw vs SPI. Only the initiation of positive SPI and the initiation of iron throw should correlate. Not the relative magnitudes of the two responses. On the far right, chlorides could now be driven lower than 2 ppm using allowed caustic. Thus, even though the SPI was sometimes positive and roughly equal to the responses on the left, there were now many fewer chloride ions available to deposit as a salt. Therefore, actual corrosion was dramatically attenuated relative to the SPI signal. This is like a forecast for 100% probability of rain. Such a forecast equally describes both a light mist and a torrential downpour. Both are given a 100% probability of precipitation. The outcomes are very different however, on one case a person gets very wet, in the other they will not. Further information must be invoked in both cases to detail the degree of impact after its initiation.*

Around this point in the chronology of the case study, Iron throw and ER probe signals were also positively detected as a delayed response to the corrosive conditions, thus validating the **SafeZone*** pipelines early warning SPI predictions. **Figure 13** shows the iron throw and the SPI on the same chart. The initiation of the SPI correlates very well with the initiation of corrosion for events on the left-hand side of the chart. The magnitude of the SPI and the magnitude of the iron throw are not expected to correlate, instead on only the initiation. Other indicators then predict the severity of the expected corrosion. Towards the left side of the chart, chlorides were systematically higher and could not be effectively reduced below the needed 4 ppm threshold as described above, thus the salt deposition process remained active. On the right-hand side of the chart, however, more aggressive desalting led to improved ability to lower chlorides with caustic. This allowed the nearly complete elimination of detected corrosion, well before the run was completed.

When active corrosion potential was first detected by the **SafeZone*** engine, refinery operations gave approval to increase caustic to the upper limit to drive chlorides low enough to stop induced salting and subsequent corrosion. Unfortunately, this action alone was not enough to mitigate the event. This is because as chlorides were decreased and approached 8 ppm, the sodium spec in the residua was reached and no further caustic could be used. At this point, several other options were evaluated, leading the team to decide upon using a 2-degree temperature increase as the next best move. Several options were theoretically computed and ranked by the **SafeZone*** platform, as the minimum move needed to push the system from the salting zone to the non-salting zone. Because of the small change in tower temperature needed after that other actions were taken, the SPI's upwards pressure was largely reduced. By simultaneously lowering and balancing neutralizer feed as the event passed and the corrosive species left the system, good corrosion control was maintained throughout the event.

The weighted list of factors and information quantitatively displayed by the **SafeZone*** platform allowed the best balance of options to be chosen each day by the team to move the unit as minimally needed to mitigate corrosion to the greatest degree possible. Later, as mentioned above, further manipulations at the desalter allowed chlorides to be driven even lower and corrosion was nearly eliminated, making the remainder of the opportunity run a large and well controlled success. **Figure 14** shows the margin report during the event. It can be seen how the various indicators interplay during the case. It is important to note that keeping traditional KPIs inside traditional ranges were not sufficient to effectively

*Trademark of SUEZ; may be registered in one or more countries

mitigate this transient event. This was especially true because in this case, the local teams had no a-priory knowledge or experience running this challenging crude.

Conclusion

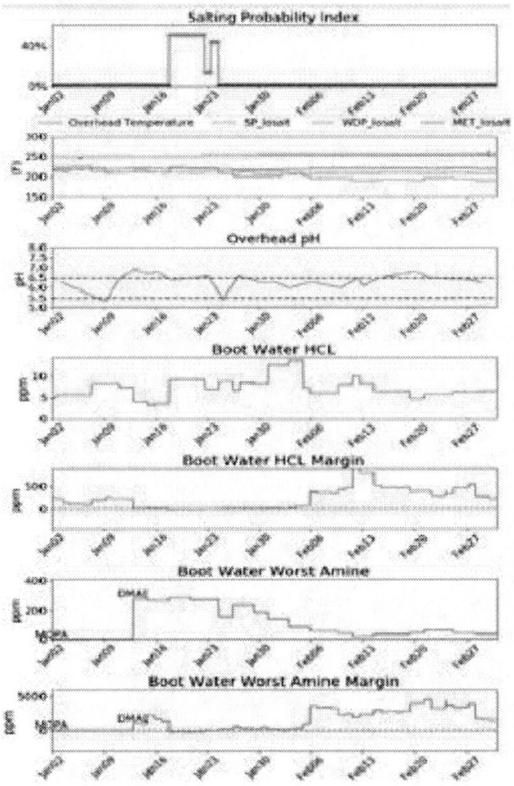

Figure 14: SafeZone Margin report. Stacked plots showing how the chlorides and pH were well controlled during the run. When the neutralizer (DMAE) was introduced to stabilize pH, volatile acid attack was transferred to amine hydrochloride attack. The technique allowed all the most key factors to be identified, evaluated, and balanced to achieve a smooth and successful run.*

The new techniques and infrastructure described here have been shown to proactively detect a wide range of previously un-observed corrosion events and enable their effective and practical mitigation as they are occurring. Furthermore, the signal processing techniques used to derive the SPI enable the specific drivers and response curves of key variables affecting corrosion and production to be quantitatively evaluated and balanced in a proactive and team oriented way. Advanced analysis of the information output by the platform is proving to enable a wide range of novel and key actions to be quickly and effectively deployed towards a range of mitigation solutions well beyond the traditional. Using traditional low frequency manual techniques to derive salt points or corrosion rates would not have been able to detect the event in question, detail its drivers, or affect any actions in the required time scales. In such cases previously, events have often remained hidden until a surprise shutdown and inspection are utilized due to unplanned shutdowns. In many cases, by the time such a unit is opened, the damage cannot be recovered and most of the evidence showing the stream of dynamic corrosion episodes giving rise to the overall problem are no longer understandable.

MANCHESTER
1824
The University of Manchester

Fouling Modelling for Crude Oil Heat Exchanger Networks

José Loyola-Fuentes
Prof. Robin Smith
Dr. Megan Jobson
Centre for Process Integration
The University of Manchester

17 AIChE Annual Meeting

MANCHESTER
1824
The University of Manchester

Contents

- Introduction
- HEN and Fouling Modelling
- Case Study
- Conclusions and Future work

17 AIChE Annual Meeting

MANCHESTER 1824
The University of Manchester

Introduction

17 AIChE Annual Meeting

MANCHESTER 1824
The University of Manchester

Fouling in Crude Oil Refineries

- Undesired accumulation of solid material
- In crude oil refineries, mostly located within preheat trains
- Associated costs include capital, energy, maintenance, among others
- Almost 50% of fouling related costs attributed to distillation unit (US$ 13 per year for 100,000 bbl/day production)

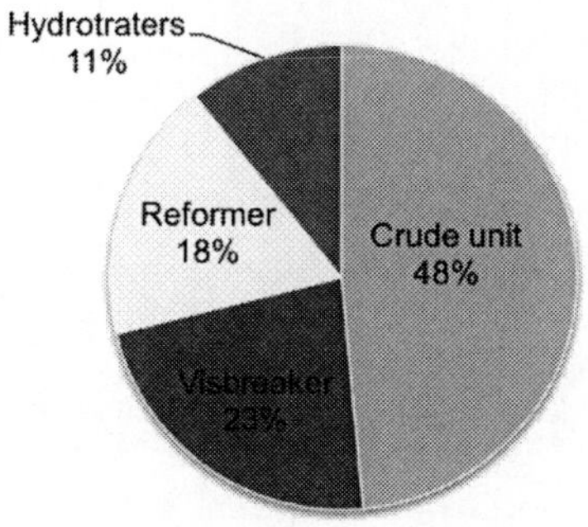

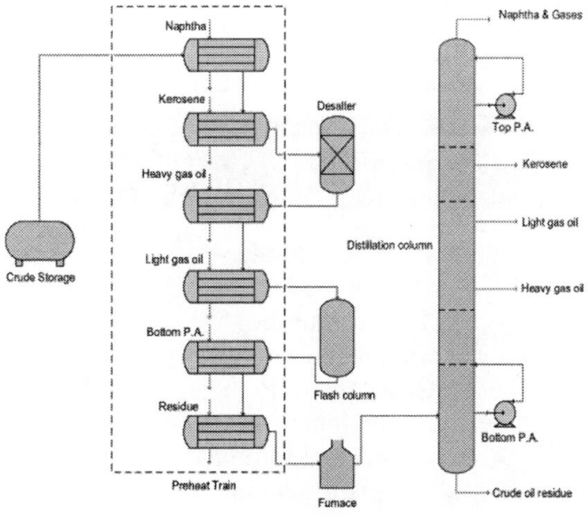

Coletti, F., Joshi, H. M., Macchietto, S. and Hewitt, G. F. (2015). Chapter One: Introduction, in *Crude Oil Fouling.* Boston: Gulf Professional Publishing, 1-22

17 AIChE Annual Meeting

Fouling Rate – Definition

- ❑ Fouling rate: Dynamic change in fouling resistance

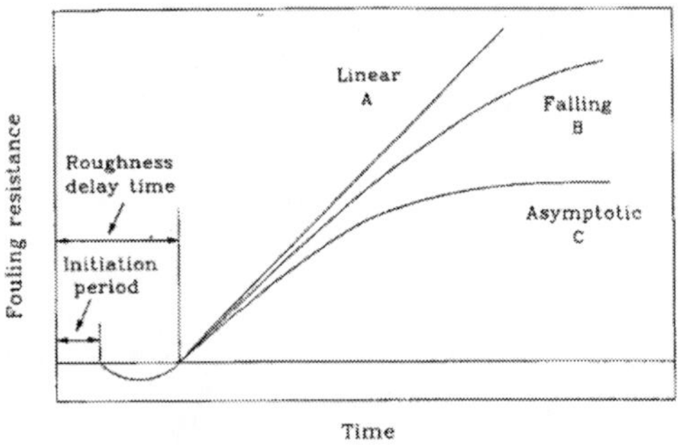

Fouling Rate = Formation - Removal

- ❑ Operational conditions set formation and removal rates
- ❑ Certain combinations can mitigate fouling occurrence
- ❑ Different effects for different fouling mechanisms
- ❑ Occurrence limited to existence of fouling threshold

Ebert, W. A. and Panchal, C. B. (1995). Analysis of Exxon crude-oil-slip stream coking data, in Panchal, C. B., Bott, T. R., Somerscales, E. F. C. and Toyama, S., (eds.) *Fouling Mitigation of Industrial Heat-Exchange Equipment*, 18th-23th June 1995. San Luis Obispo, California. USA: Begell House

17 AIChE Annual Meeting

Fouling Rate – Modelling

- ❑ Linear rate

$$\frac{dR_f}{dt} = \alpha_2'$$

Model parameters need to be determined

- ➢ dR_f/dt: Fouling rate
- ➢ Re: Reynolds number
- ➢ Pr: Prandtl number
- ➢ E_A: Activation energy
- ➢ R: Gas constant
- ➢ T_{film}: Film temperature
- ➢ T_{wall}: Wall temperature
- ➢ τ: Wall shear stress
- ➢ α, β, γ: Fouling model parameters

- ❑ Chemical reaction rate
- ➢ Ebert and Panchal

$$\frac{dR_f}{dt} = \alpha_1' Re^{\beta'} exp\left(\frac{-E_A'}{RT_{film}}\right) - \gamma' \tau$$

- ➢ Panchal et. al

$$\frac{dR_f}{dt} = \alpha_1' Re^{\beta} Pr^{-0.33} exp\left(\frac{-E_A'}{RT_{film}}\right) - \gamma' \tau$$

- ➢ Polley et. al

$$\frac{dR_f}{dt} = \alpha_1' Re^{-0.8} Pr^{-0.33} exp\left(\frac{-E_A'}{RT_{wall}}\right) - \gamma' Re^{0.8}$$

Ebert, W. A. and Panchal, C. B. (1995). Analysis of Exxon crude-oil-slip stream coking data, in Panchal, C. B., Bott, T. R., Somerscales, E. F. C. and Toyama, S., (eds.) *Fouling Mitigation of Industrial Heat-Exchange Equipment*, 18th-23th June 1995. San Luis Obispo, California. USA: Begell House.
Panchal, C.B., Kuru, W.C., Liao, C.F., Ebert, W.A. and Palen, J.W., 1999. Threshold conditions for crude oil fouling. Understanding heat exchanger fouling and its mitigation, 273, p.282.
Polley, G.T., Wilson, D.I., Yeap, B.L. and Pugh, S.J., 2002. Evaluation of laboratory crude oil threshold fouling data for application to refinery pre-heat trains. Applied Thermal Engineering, 22(7), pp.777-788.

17 AIChE Annual Meeting

MANCHESTER 1824
The University of Manchester

Fouling Rate – Modelling

- ❑ Fouling considerations in heat exchanger networks
  - ➢ Assessment for operational conditions for fouling mitigation
  - ➢ Optimisation of cleaning schedules
  - ➢ Design and retrofit for fouling mitigation
- ❑ However
  - ➢ How to determine each fouling model parameter?
  - ➢ How to account for different fouling mechanisms and different crudes?
  - ➢ How to identify the dominant fouling mechanism on each side of heat exchangers?
  - ➢ How to get a representative and free of error sample for reliability of output results?

↓ Use of operational data

↓ Accounting for shell and tube side contributions

17 AIChE Annual Meeting

MANCHESTER 1824
The University of Manchester

Fouling Rate – Split Contributions

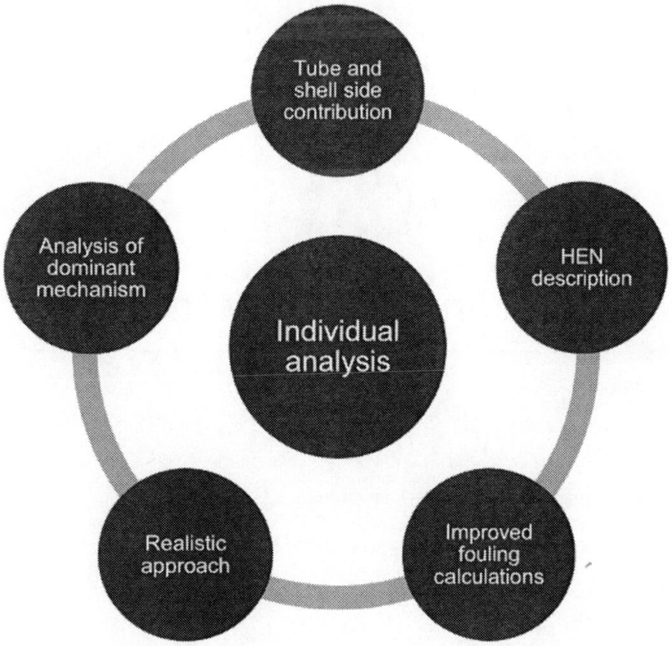

17 AIChE Annual Meeting

MANCHESTER 1824
The University of Manchester

Research Objectives

To determine fouling rate models in crude oil preheating systems from online reconciled data using parametric regression

1) Parametric fitting of fouling models for shell and tube sides on heat exchanger networks
2) Application of fitted models for predicting fouling behaviour
3) Identification of faulty measurements by means of data reconciliation and gross error detection
4) Optimisation of cleaning schedules and operational conditions for fouling mitigation

MANCHESTER 1824
The University of Manchester

HEN and Fouling Modelling

General Modelling Scheme

HEN modelling

- ➢ Matrix representation based on graph theory
- ➢ Accounts for fouling dynamics and contributions from shell and tube side

Data reconciliation

- ➢ Mitigation of random and gross errors
- ➢ Identification of faulty instruments
- ➢ Optimisation based on non linear programming (NLP)

Parametric regression

- ➢ Calculation of fouling model parameters for HEN on each side
- ➢ Optimisation approach based on Genetic Algorithm and deterministic solvers

HEN Simulation and Fouling Modelling

❑ HEN simulation: Matrix representation (de Oliveira Filho et.al, 2007)

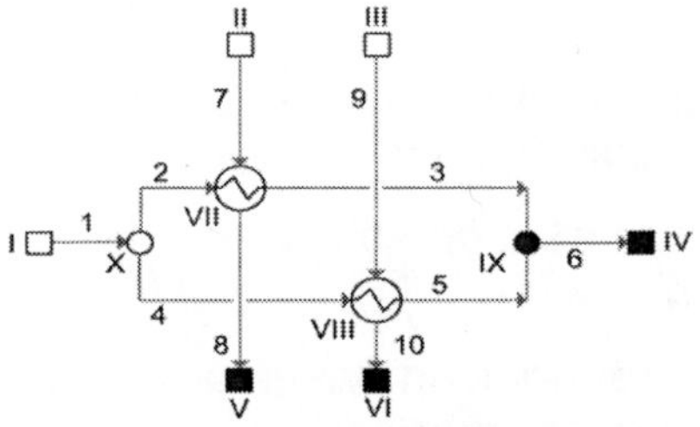

❑ 3 supply and demand units

❑ 2 heat exchanger

❑ 1 mixer, 1 splitter

$$M = \begin{bmatrix} M^{PS} \\ M^{PD} \\ M^{HE} \\ M^{MX} \\ M^{SP} \\ M^{PR} \end{bmatrix} = \begin{bmatrix} M_c^{PS} & M_h^{PS} \\ M_c^{PD} & M_h^{PD} \\ M_c^{HE} & M_h^{HE} \\ M_c^{MX} & M_h^{MX} \\ M_c^{SP} & M_h^{SP} \\ M_c^{PR} & M_h^{PR} \end{bmatrix}$$

	1	2	3	4	5	6	7	8	9	10
I	-1									
II							-1			
III									-1	
IV						1				
V								1		
VI										1
VII		1	-1				1	-1		
VIII				1	-1				1	-1
IX			1		1	-1				
X	1	-1		-1						

Mass and energy balance are solved using matrix *M*

MANCHESTER 1824
The University of Manchester

HEN Simulation and Fouling Modelling

- Modelling for fouling contributions

$$R_f = \frac{1}{U_D} - \frac{1}{U_C} = R_{fi} \cdot \left(\frac{d_o}{d_i} \right) + R_{fo}$$

- Different fouling models can be used based on heat exchanger location and operating conditions

$$R_f \Big|_{n+1} = R_f \Big|_n + \Delta t \cdot \frac{dR_f}{dt} \Big|_n$$

- U_C: Clean overall heat transfer coefficient
- U_D: Fouled overall heat transfer coefficient
- R_{fi}: Tube side fouling resistance
- R_{fo}: Shell side fouling resistance
- h_i: Tube side heat transfer coefficient
- h_o: Shell side heat transfer coefficient

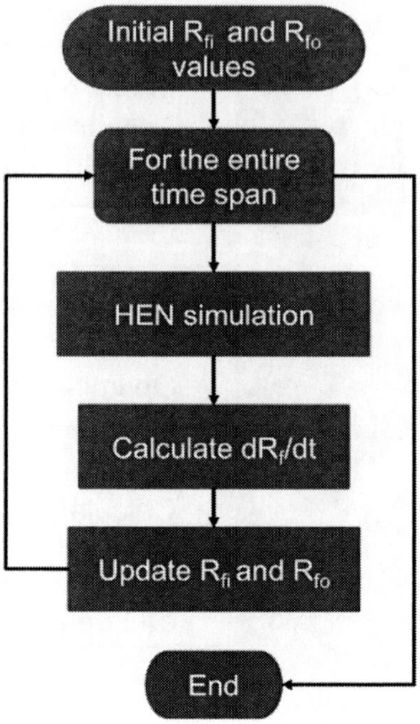

MANCHESTER 1824
The University of Manchester

HEN Simulation and Fouling Modelling

- Data reconciliation

$$\begin{aligned} \underset{y_R}{\text{minimise}} \quad & (y_M - y_R)^T \psi^{-1} (y_M - y_R) \\ \text{subject to} \quad & f(y_R) = 0 \\ & g(y_R) \geq 0 \\ & y_R^L \leq y_R \leq y_R^U \end{aligned}$$

- Gross error detection

$$\begin{aligned} \underset{y_R}{\text{minimise}} \quad & (y_M - y_R)^T \psi^{-1} (y_M - y_R) \\ \text{subject to} \quad & f(y_R - B_\xi \omega_\xi) = 0 \\ & g(y_R - B_\xi \omega_\xi) \geq 0 \\ & y_R^L \leq y_R \leq y_R^U \end{aligned}$$

- Where:
  - y_M: Measured value
  - y_R: Real value
  - B_ξ: Position matrix for single and multiple gross errors
  - ω_ξ: Gross error magnitude
  - ψ: Covariance matrix

- Mass and energy balance are used as process constraints
- Data reconciliation solved as a NLP problem
- Gross error solution based on combinatorial approach
- Maximum number of gross errors limited by number of independent constraints

MANCHESTER 1824 The University of Manchester

HEN Simulation and Fouling Modelling

- Parametric fitting
  - Needs to account for parameters of different magnitudes
  - Needs to account for sensitivity on initial estimation
  - Needs to avoid local optima
- Minimisation of error between simulated and predicted values

$$\text{minimise} \quad \sum_{i=1}^{n} \frac{\left(y_i^{fit} - y_i^{msr}\right)^2}{y_i^{msr}}$$

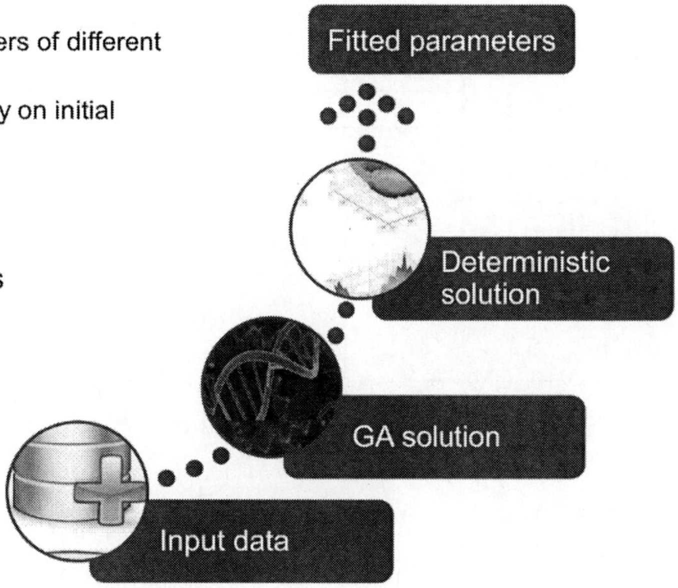

MANCHESTER 1824 The University of Manchester

Case Study

MANCHESTER 1824
The University of Manchester

Case study – Crude Oil Preheat Train

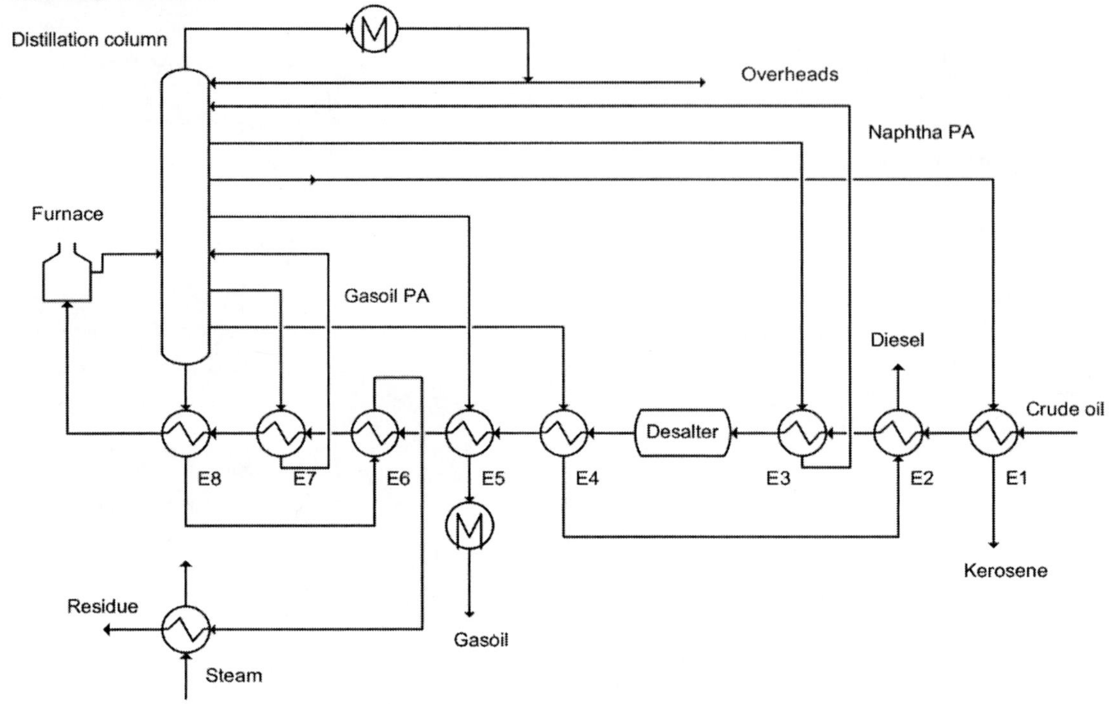

MANCHESTER 1824
The University of Manchester

Case study – Crude Oil Preheat Train

Problem formulation

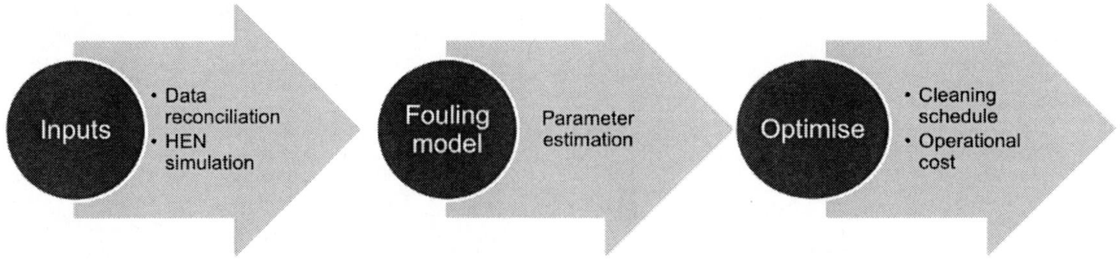

- ➢ Measured flow rates and temperatures
- ➢ Linear fouling rate at cold end of preheat train
- ➢ Polley's model at hot end for tube side
- ➢ Heat exchangers specified by area
- ➢ Hot and cold utilities specified by heat duty

MANCHESTER 1824
The University of Manchester

Case study – Crude Oil Preheat Train

Base case: No cleaning actions on Preheat Train

Total operating cost (MM$/y)	Time period (y)	Temperature drop (°C)
17.34	1	55.91

- ❑ Furnace inlet temperature decreases
- ❑ Furnace input duty increases
- ❑ Decreasing in energy recovery
- ❑ Fouling deposits accumulate over time

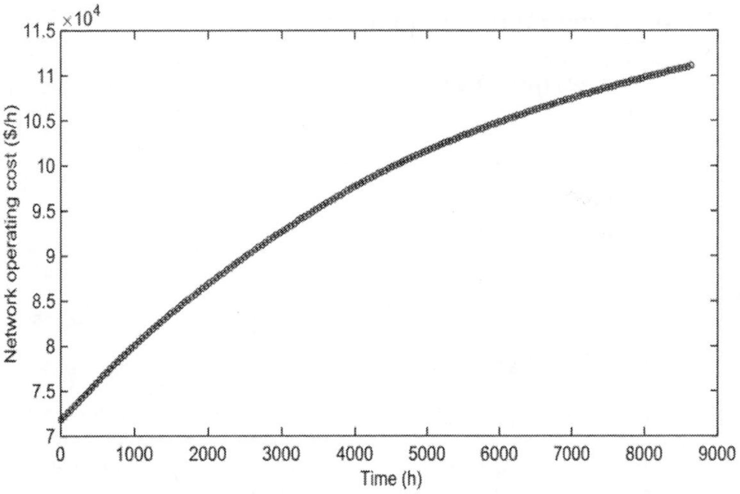

17 AIChE Annual Meeting

MANCHESTER 1824
The University of Manchester

Case study – Crude Oil Preheat Train

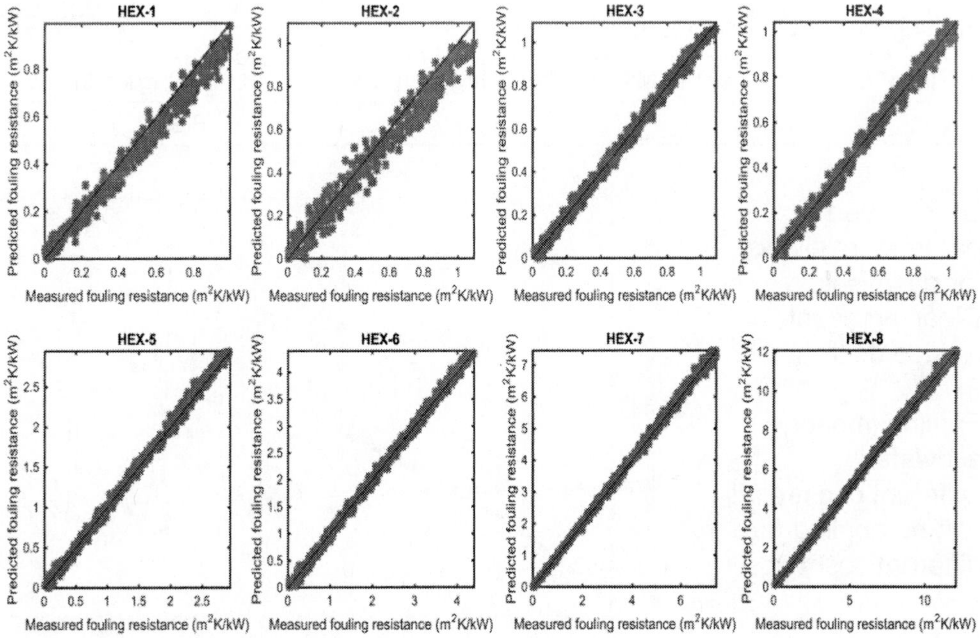

17 AIChE Annual Meeting

Case study – Crude Oil Preheat Train

- Parameters for simulations (base):

α'_1 ($m^2K\ kW^{-1}h^{-1}$)	E'_A ($kJ\ mol^{-1}$)	γ' ($m^2K\ kW^{-1}h^{-1}$)	α'_2 ($m^2K\ kW^{-1}h^{-1}$)
1.00e+06	48.00	1.50e-09	5.42e-05

- Fitted parameters for optimisation

	α'_1 ($m^2K\ kW^{-1}h^{-1}$)	E'_A ($kJ\ mol^{-1}$)	γ' ($m^2K\ kW^{-1}h^{-1}$)	α'_2 ($m^2K\ kW^{-1}h^{-1}$)
E1	-	-	-	5.04e-05
E2	-	-	-	5.09e-05
E3	-	-	-	5.41e-05
E4	-	-	-	5.40e-05
E5	5.74e+03	16.38	1.50e-07	4.09e-07
E6	9.98e+07	67.09	1.62e-08	2.58e-04
E7	3.83e+04	32.80	1.20e-07	1.97e-04
E8	3.49e+04	31.25	1.42e-07	6.04e-05

17 AIChE Annual Meeting

Case study – Crude Oil Preheat Train

Optimised case: Optimal cleaning schedule

Total operating cost (MM$/y)	Cleaning actions	Cost savings (%)
16.13	11	6.98

- Use of fitted parameters allows for optimisation
- Cleaning events reduce decrease in FIT
- Fouling impact is alleviated
- Different constraints can be applied for different scenarios

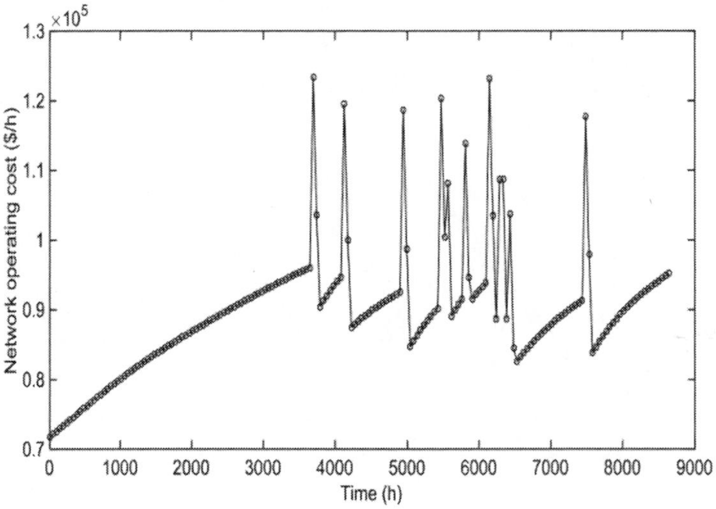

17 AIChE Annual Meeting

MANCHESTER
1824
The University of Manchester

Conclusions and Future work

MANCHESTER
1824
The University of Manchester

Conclusions and future work

- ❑ Fouling can lead to significant impact in thermal and hydraulic performance
- ❑ Fouling can be modelled by using fouling rate models and determining their specific parameters
- ❑ Including fouling into design and optimisation of heat exchanger networks allows for mitigation strategies
- ❑ A methodology has been developed for stablishing fouling models using reconciled data
- ❑ Fouling from shell and tube side are determined by means of different fouling model parameters
- ❑ Fitted parameters are used for optimisation of cleaning schedules, leading to operational savings of 7%

- ➢ Use of industrial data for testing rigor of methodology
- ➢ Identification of minimum bias magnitude for isolation of gross errors
- ➢ Application of methodology for operational optimisation and retrofit of heat exchanger networks

MANCHESTER
1824
The University of Manchester

Acknowledgements

To CONICYT (Chile) for the financial support provided

CONICYT
Ministerio de Educación
Gobierno de Chile

17 AIChE Annual Meeting

MANCHESTER
1824
The University of Manchester

Thank you

17 AIChE Annual Meeting

509054 Solve this! Fundamental Approach to Problem Solving in Industrial Processes I (Invited Talks)

Abstract:

Solving complex plant problems is an everyday exercise that affects all projects and budgets, and it is very rarely introduced to students at any level. In fact, by the time young engineers are few years into working in industry, most have lost the engineering skills that they developed in school. We hope to encourage future engineers to retain those skills and learn the importance of plant problem-solving.

502715 Mitigating Risks for Refiners Processing Opportunity Crudes

Abstract:

The refining industry has changed noticeably over the past several years with an ample supply of opportunity crude oils available, which include light tight oils and heavy sour crudes, and sometimes high TAN crudes. Opportunity crudes have been around for years, but only until recently has the abundance of these discounted crudes shifted the behavior of refiners.

With the use of opportunity crude oils comes new processing challenges; Automation technology is advancing to meet these challenges to ensure refiners are successfully alleviating these new issues. Advances in real time crude analysis and crude unit monitoring and control allow refiners to maximize opportunity crude processing while mitigating issues like accelerated fouling and corrosion.

Refiners on the coast will typically have access to more opportunity crudes and can process 50 or more different crude oils in a year. Because crude oils can be very different in properties, refiners try to match the required crude oil composition for their refinery configuration; usually by blending two or more crude oils. To make matters more challenging, sometimes crude oils purchased to process are crudes that the refiner has no experience processing.

Crude blending operation has become more important with the ability to establish consistent crude feed properties. Steady operation enables better optimization of the crude unit by pushing against the appropriate constraints which often includes maximizing the quantity of discounted opportunity crude oils blended.

This presentation will highlight several refiners that have made modifications to effectively process opportunity crudes while mitigating risks related to crude incompatibilities, high TAN crudes, and varying crude oil feed properties.

Headworks International Inc.

2017 AIChE Annual Meeting
Session 242
Shale Oil, Oil Sands and other Heavy Fuels:

Impact of Opportunity Crudes on Desalter Operation and Wastewater Treatment in a Refinery

Somnath Basu, PhD, PE
Headworks International
Houston, TX

Contents

- Introduction
- Desalter Operating Principle
- Refinery Wastewater Treatment
- Opportunity Crudes and their Impacts
- Methodology for Process Monitoring and Control
- Opportunity Crudes on Wastewater Treatment Case Studies
- Remedial Measures
- Conclusion
- Acknowledgement

Introduction

- Opportunity crudes are newer types of crudes that have not been traditionally processed in the past
- These are available in the market in abundance and at discounted rates compared to benchmark prices
- But these are difficult to process because of their special characteristics
- Desalters and wastewater treatment systems play a critical role in overcoming many of challenges
- Refiners have tried blending these crudes with traditional ones to balance their properties, but:
  - Sometimes the crudes are not compatible
  - Leads to formation of solid or semi-solid asphalt materials that precipitate in desalters

Desalter Operating Principle

- Crude is a mixture of hydrocarbons
- Crude is valuable source of fuels and petrochemical feedstocks after refining
- Crudes contain many impurities that can impede the refining process
- Desalters are the first equipment to handle and process crudes
- It makes intimate contact between crude and water by thorough mixing to:
  - Transfer inorganic salts and solids from crude to water as brine stream as much as possible
  - Separate crude and water in two layers

Desalter in Petroleum Refinery

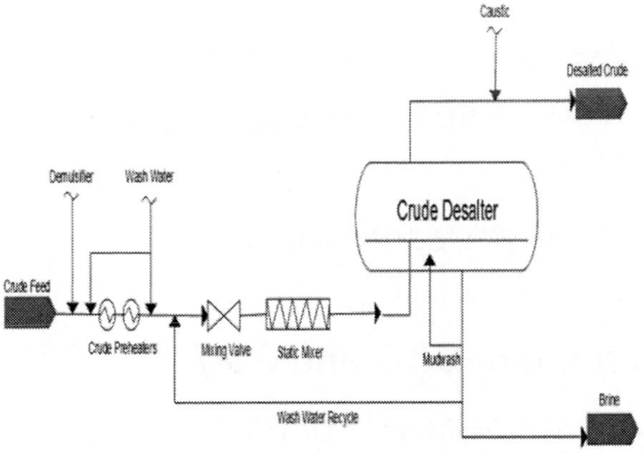

- Crude is preheated (105 – 150 °C) and mixed with wash water (5 – 8% v/v)
- Passed through a mixing valve and static mixer
- In desalter, oil and water separate in top and bottom layers with an emulsion layer (rag layer) in between
- Thick emulsion layer indicates poor oil-water separation
  - Large concentration of water in oil phase and oil in water phase
  - Excessive water in desalted crude causes high energy consumption in heating and pumping, and potential corrosion by dissolved salts
  - Excessive oil under-carry in brine leads to large oil and grease (O&G) and large COD loads to wastewater treatment

Typical Refinery Wastewater Treatment

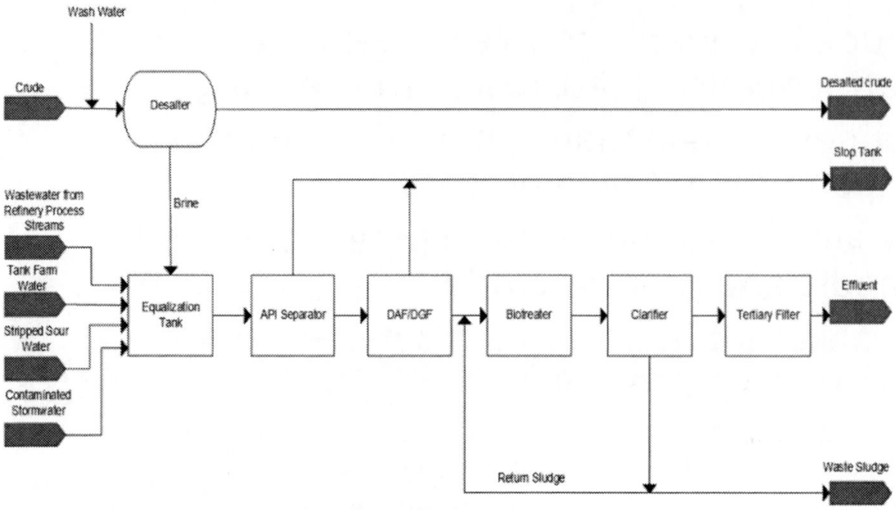

Wastewater Treatment

- Desalter brine mixed with other wastewater sources in an equalization tank
- Wastewater fed to API separators for primary oil-water separation and suspended solids removal
- Dissolved (or induced) air or gas separator for secondary separation
- Biological treatment to remove O&G and COD
- Clarifiers to remove more suspended solids
- Tertiary filtration, if needed, to meet discharge requirements

Opportunity Crudes and their Impacts

- Opportunity crudes disturb desalter operations due to constituents that refineries are not typically designed to process
- Some examples of opportunity crudes:
  - Heavy crudes from Canadian Rockies
  - Doba crude from West Africa
  - Light tight oil (LTO) from North American shale plays
- Characteristics of concern:
  - API gravity
  - Viscosity
  - Naphthenic acids content
  - Metals concentrations, especially Ca and Fe
  - Filterable solids (FS)
  - Amines content

Western Canada Crudes

- Typically heavy crudes with API Gravity of 19-22
- Have high viscosity, which causes challenges for refiners:
  - Demand high energy in piping transportation
  - Cause poor mixing leading to reduced salt removal from crude
  - Emulsion formation in desalter leading to poor oil-water separation
- Have high filterable solids (FS)
  - Difficult to separate in desalter because of th
  - Grow the "rag layer" inside desalter leading to poor oil-water separation

- High concentration of sulfur compounds
  - Lead to large COD and nitrification (TKN) loads
- High Total Acid No. (TAN)
  - High concentration of naphthenic acids cause corrosion of metals
  - Not sufficiently broken down by biological treatment

Headworks® INTERNATIONAL

Doba and other West African Crudes

- Heavy Crude with API Gravity of 21
- Also have high TAN
- Have high metals content
  - Causes high potential for scaling and deposition in wastewater treatment equipment when carried over with brine
  - Large excess of calcium that can poison FCC catalysts if carried over with desalted crude
  - Contribute to high conductivity of crudes leading to poor oil-water separation in desalter
  - Stabilizes the desalter emulsion layer preventing oil and water in the rag layer to break loose

Headworks® INTERNATIONAL

Light Tight Oil (LTO)

- LTOs are crudes derived from hydrofracturing of shale formations
- Hydrofracturing and horizontal drilling have changed the global supply and demand over the last few years
- US is the largest producer of LTO grade crudes

(Source: US Energy Information Administration, 2015)

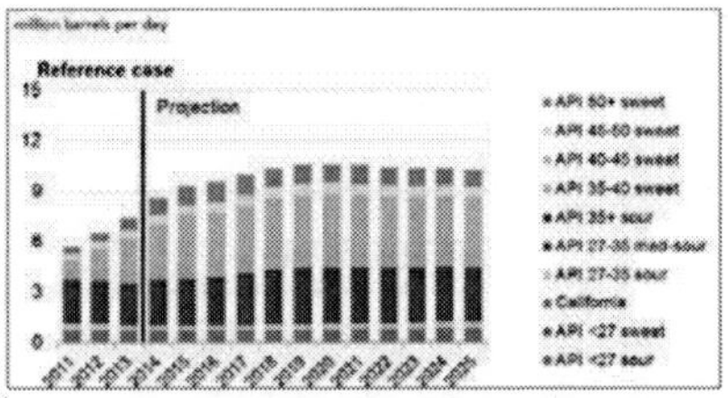

(Source: US Energy Information Administration, 2013)

Rank	Country	Shale Oil Proven Reserves, billion bbl
1	Russia	75
2	United States	58
3	China	32
4	Argentina	27
5	Libya	26
6	Australia	18
7	Venezuela	13
8	Mexico	13
9	Pakistan	9
10	Canada	9
Total Global Recoverable Shale Oil Reserves = 345 billion bbl		

Headworks INTERNATIONAL

Light Tight Oil (LTO)

- LTOs are light crudes with high API Gravity of >42
- Refiners blend LTOs with heavier crudes to balance properties, but they are not always compatible
- Challenges of LTO crudes:
  - Incompatible with certain heavy crudes, leading to precipitation of asphaltenes
  - High filterable solids
  - Entrained H_2S that requires addition of amines for handling and transportation
  - High concentration of paraffin wax leading to fouling of equipment at low temperatures

Headworks INTERNATIONAL

Effects of Opportunity Crudes

- Low API Gravity: difficult handling, processing, and transportation
- High Acidity: high COD and affects microbiological population
- High Metals Concentration: Calcium and iron are present both in particulate and dissolved forms
- High Sulfur: Corrosion potential, health and safety risks
- High Filterable Solids (FS): clogging equipment, reduction of active volume of desalters, and promoting stabilization of rag layers
- High Amines Concentration: part amines partition with water phase and leave with brine to wastewater, create high COD and nitrification loads

Target	Light Crudes	Heavy Crudes
Salt Removal Efficiency, %	> 90 in a single stage, > 95 in a double stage desalter	> 90 in a single stage, > 95 in a double stage desalter
Salt in Desalted Crude, ptb	< 2	< 2
Water carryover in Desalted Crude, % (v/v)	< 0.3	< 0.7
Oil Under Carry in Brine, ppm	< 200	< 1000

Methodology for Process Monitoring and Control

- API Gravity of crudes measured by hydrometer (ASTM Method D1298)
- Filterable solids measured by filtration through filter of 0.45μm (ASTM Method D4807)
- Water in oil determined by colorimetric titration (ASTM Method D4928)
- Chloride in crude measured by Potentiometric titration (ASTM Method D6470)
- Chloride in water measured by Ion Specific Electrode
- Oil in water analyzed by ASTM Method D3921
- TAN determined by titration to neutralize sample with standard KOH solution, and expressed as mg of KOH/g of crude
- Portable Electric Desalters (PED) are benchtop units that simulate operation of desalter to demonstrate salt removal potential, oil-water separation, and emulsion formation of a given crude
- For wastewater treatment, parameters monitored for API Separator and DAF are O&G and TSS
- For biotreatment, parameters monitored are O&G, TSS, BOD, COD, and reactor basin MLSS and MLVSS

Opportunity Crudes on Wastewater Treatment Case Studies

- The effects of the contaminants resulting from processing opportunity crudes are shown in three Case Studies
- The studies cover a wide range of refinery locations, and crude sources and types

Case 1

- US gulf coast refinery that processes a heavy sour crude of 21.5 API Gravity and 3.4% sulfur content
- Effluent requirements are routinely met, with effective tertiary filters
- However, foaming was severe and clarifiers became non-functional on occasions
- Foam sample indicated sludge bulking due to proliferation of Nocardia
  - Filamentous bacterial population that floats creating a thick layer of foam on surfaces of aeration tanks and secondary clarifiers (photomicrograph of S. Pineformis Nocardia after 1000 x illumination of Gram Stained sample)
- Possible reasons for growth:
  - High concentration of oil
  - Excessively high solids retention time (SRT) in aeration basins

Case 2

- North American refinery processing LTO from North Dakota
- Free oil escaped wastewater treatment on an occasion and appeared as a sheen of oil on the surface of receiving water
- Review of operations data indicated excessive oil carried over with brine
  - Due to the desalter failing to separate oil from water
- There was a high concentration of FS, which settled down in desalter and treatment plant equipment
- This caused reduction of active volumes and HRT of all process equipment, leading to escape of oil with treated effluent
- A new desalter system was installed to handle situation

Case 3

- Refinery in the Far East using Arab Light Sweet crude
- They blended with medium heavy sour crude with 28 API Gravity and 2.4% sulfur content
- Salt removal by desalter dropped from 80% to 50%
- Removed the heavy crude from the blend and restored original salt removal efficiency

Remedial Measures for Desalters

- Control formation of emulsion and growth of resulting rag layer by:
  - Characterization of crude to evaluate potential for emulsion formation due to presence of compounds
  - Select correct type and dosage of demulsifying chemicals to break emulsion
  - Follow standard operation procedures for desalter by monitoring its operation
    - Closely monitor oil-water interface and presence of rag layers
    - Determine and maintain mudwashing frequency for high FS crudes
- Acids are effective in treating amines, but create a corrosive environment in desalter
- Demulsifiers will not give best performance of desalter if proper operation procedures are not followed

Remedial Measures for Wastewater Treatment: MBBR

- Some opportunity crudes contain large quantities of naphthenic acids which leave as refractory COD
- Discharge of COD may need moving bed bioreactor (MBBR) process
- MBBR has superior BOD and COD removal rates to activated sludge
- Advantages of using MBBR are:
  - Resilient to peak flows and shock loads
  - Attached growth biomass is resistant to toxic and shock loads
  - Effective in degrading more refractory COD than can be obtained by traditional biological treatment processes
  - Biomass are less vulnerable to bulking and foaming

Moving Bed Bioreactor (MBBR): Typical Media

MEDIA SPECIFICATIONS

Media	Protected Area	Total Area	Weight	Diameter	Depth	Specific Gravity	Material
 AC 920	680 m^2/m^3 207 f^2/f^3	920 m^2/m^3 280 f^2/f^3	-146 kg/m^3 9.1 lb/ft^3	13 mm X 13 mm 0.51 in x 0.51 in	13 mm 0.51 in	0.94-0.98	HDPE
 AC 700	550 m^2/m^3 168 f^2/f^3	700 m^2/m^3 213 f^2/f^3	-105 kg/m^3 6.6 lb/ft^3	22 mm 0.87 in	8 mm 0.32 in		
 AC 515	485 m^2/m^3 148 f^2/f^3	515 m^2/m^3 158 f^2/f^3	-142 kg/m^3 8.9 lb/ft^3	22 mm 0.87 in	16 mm 0.63 in		
 AC 450	402 m^2/m^3 122 f^2/f^3	450 m^2/m^3 137 f^2/f^3	-120 kg/m^3 7.5 lb/ft^3	21 mm 0.83 in	16 mm 0.63 in		

Conclusions

- Opportunity crudes provide refiners opportunities for margin enhancement
- However, these crudes present difficult challenges for processing due to their special characteristics
- Desalters are important in ensuring that contaminants are effectively removed
- Poor desalter performance can lead to:
  - Corrosion of refining process equipment
  - Catalyst poisoning
  - Poor performance of wastewater treatment, potentially leading to discharge permit violations
- Challenges of opportunity crudes can be overcome by:
  - Close monitoring and control of desalters
  - Enhancement of biological treatment performance by MBBR
  - Routine sampling and analytics
  - Proper chemical dosing program

Headworks® INTERNATIONAL

11000 Brittmoore Park Dr.
Houston, TX 77041

Phone: 713.647.6667
Fax: 713.647.0999
E-mail: hw@headworksintl.com

www.headworksinternational.com

498479 CFD Applied to SHALE OIL Pyrolisys

Abstract:

Abstract

The use of oil is intensive in the architecture of the world energy matrix. Consumption has only increased in the last decades, which has been causing the search for alternative routes of generation. Discovery of shale oil/gas and its recent commercial exploitation at industrial scale may, in a few years, change the current axis of unilateral dependence on it. However, much remains to be done about the technology behind the obtaining process, since such substance is confined in rocks. The shale pyrolysis has been investigated numerically under non-isothermal conditions. This reactive flow process, dominated by diffusion mechanism and chemical kinetics, in a porous moving bed uses a vertical cylindrical geometry with gas internal distributor and fluid injections. The main idea was to reproduce the regions, respectively from the top to the base, formed along the reactor. The shale drying takes place in the inlet zone, due to the thermal exchange that occurs because of the local fluid temperature, once the material enters the reactor at room temperature. Chemical reactions happen shortly after drying zone; kerogen decomposition and consumption generating bitumen, which is broken in products and residues (pyrite and coke). A three dimensional, turbulent and multiphase mathematical modeling and numerical simulation (CFD) of the shale thermal degradation were presented. Mass, heat and momentum balances were formulated following the Eulerian approach. All chemical species were subject to consumption and formation reactions of the process on space and time domain, under first-order kinetic. Empirical correlations guarantee the closure of the model and its numerical solution. The obtained results are evaluated and compared with the literature. An understanding of all these chemical reactions and their rates, as well as process regions can help in design and security minimizing costs, including those associated with meeting environmental regulations.

Future oil sands production costs and GHG emissions based on emerging technologies

Experience Nduagu[1]*, Alpha Sow[1], Evar Umeozor[1,2], Dinara Millington[1]

[1]Canadian Energy Research Institute (CERI), Calgary, AB T2L 2A6, Canada.

[2]Department of Chemical and Petroleum Engineering, University of Calgary, Calgary, AB, Canada

Abstract

Competitiveness challenges in a low oil price environment, evolving climate change policies and pressures to reduce environmental footprints indicate that high costs and greenhouse gas (GHG) emissions associated with bitumen production from oil sands are no longer permissible in the long run. It has been acknowledged by stakeholders that the oil sands industry must innovate to be competitive. On the other hand, there has been increasing effort towards development of new technologies to address these challenges. Some of these efforts have progressed from ideas or concepts to field and commercial demonstration stages of research and development (R&D). To our knowledge, the cost and emissions reduction potentials of these technologies have not been assessed or quantified in a comprehensive and systematic manner. This study fills that gap by estimating future oil sands bitumen production costs and GHG emissions by determining the potential impacts of new and emerging technologies and processes to reduce costs and GHG emissions. We identify emerging technology options that can be deployed in the oil sands industrial sector in the short-term, focusing on in-situ steam assisted gravity drainage (SAGD) process-based projects, spanning surface and sub-surface operations. Then the resulting impacts of adopting these technologies in oil sands in-situ operations and processes are assessed. Results show that near-commercially ready technologies, which can be deployed for bitumen production in 5-7 years, can achieve dramatic reductions in costs and emissions simultaneously. By combining surface and sub-surface technologies, bitumen production costs and emissions can be reduced by 34-40 percent and 64-80 percent, respectively.

Keywords: Canadian oil sands, bitumen, production costs, greenhouse gas emissions, CO_2.

* Corresponding author; Email: enduagu@ceri.ca

1.0 Introduction

167.9 billion barrels of recoverable deposits of bitumen in Western Canada makes Alberta's oil sands the third-largest oil reserves in the world after those of Venezuela and Saudi Arabia. Oil sands, which is highly viscous, sticky mixtures composed of clay, water and bitumen, are hosted in subsurface reservoirs. The constituent bitumen has a viscosity of 200,000 – 2 million centipoise (cp), akin to the viscosity of a peanut butter (Canadian Heavy Oil Association, 2013). Because oil sands deposits are buried underground, they are either mined or extracted in-situ. Thereafter, bitumen is separated from the sand and water, and afterwards upgraded for downstream processing in refineries. Thus, bitumen extraction from oil sands, its processing and upgrading involve various units and processes, which consequently pose considerable investment and operational challenges. Though enormous oil sands resource in Alberta presents significant prospects for economic growth and energy security, its extraction and processing bring significant costs and environmental hurdles.

It must also be acknowledged that research and development of new technologies targeted at addressing the costs and GHG emissions challenges facing bitumen extraction and processing have intensified with promising outcomes apparent. Many of these efforts are still at low technology readiness levels whereas others have progressed from concepts to field and commercial demonstration stages. A few studies (Council of Canadian Academies, 2015, Findlay, 2016, Sleep et al., 2017) cataloged the technologies being developed in this area, and in Sleep *et al.* (2017), expert elicitation is used to project future performances of oil sands processes. In cases where some quantitative assessments are done, only a part of the process chain was considered. To our knowledge, no study has assessed the potentials of technologies to cut costs and reduce emissions across the bitumen extraction and processing operations which include both surface and sub-surface operations. Though it seems obvious that progress is being made through research and development of technologies in the process segments, detailed cost and GHG assessments are needed to show if expected outcomes will be marginal or step-change improvements.

Thus, this study fills that gap by estimating oil sands bitumen production costs and GHG emissions by first determining the potentials of new and emerging technologies and processes to reduce costs and emissions. We identify emerging technology options that can be deployed in the oil sands industrial sector in the short-term, focusing on in-situ process-based projects, covering surface and sub-surface operations. Then the resulting impacts of adopting these technologies in oil sands in-situ operations and processes are quantified.

2.0 Emerging technologies

Existing in-situ technologies such as Cyclic Steam Stimulation (CSS), Steam-Assisted Gravity Drainage (SAGD) and solvent-assisted SAGD are in commercial application. In-situ process segments can be grouped into two broad categories: surface and sub-surface operations as shown in Table 1.

Table 1. In-situ oil sands bitumen recovery (surface and sub-surface operations) and upgrading challenges and emerging technologies.

	Sub-surface	Surface				Upgrading
	Reservoir (RES)	**Business Mgt. & Analytics (BM)**	**Steam Generation (SG)**	**Water & Wastes (WWT)**	**Well & Wellpads (WWP)**	**Upgrading (UPG)**
Challenges	Enhancing oil recovery through bitumen viscosity reduction	Measuring and analyzing processes, and prescribing best solutions	Producing better steam quality and quantity with lower costs and emissions	Separating oil/water and others materials with reduced costs and wastes	Reducing time and cost for drilling and wells	Reducing diluent requirement and producing higher value end product
Technologies Assessed	Pure solvent	Digital FEED	Rifled pipes –based OSTG	Skim tank	Lean drilling	Full upgrading
	Steam Solvent	Sensors	Direct contact steam Generation	Induce gas floatation	Standardization	Cavitation
	Electro magnetic heating	Steam flood management	Solid Oxide fuel cell CoGen	Reverse osmosis filter	Modularization	Catalyst-based upgrading
	Surfactant	Digital regulation	Forced-circulation steam generation	Evaporator	Horizontal Drilling	Desulfuration
	In-situ Thermal	Predictive maintenance	Gas turbine OTSG	Ion exchanger		Deep conversion
		Production optimization	Blowdown boiler	Electrocoagulation		Thermal cracking
				Lime softening		

In this section, a brief description of unit operations and processes for surface and/or sub-surface applications and emerging technologies for reducing costs and emissions in these areas are presented.

2.1 Surface operations

The surface operations are primarily made up of WWT, SG and WWP (see Table 1). BM application in surface operations is related to digitalization of processes such as engineering procurement and construction (EPC). Figure 1 illustrates the connectivity between the different parts of the in-situ production process. Steam produced from boilers (e.g. drum boiler, OTSG or HRSG) is injected into the sub-surface (an oil reservoir) through a well bore (in the WWP segment). The produced mixture of bitumen, water, clay and other materials is sent to a separator where bitumen is recovered. Oil-water separation is achieved through free water knock-out, skim tank and gas flotation steps, and oil removal filter.

As indicated in Table 1, the goal of water/waste treatment technologies is to realize the difficult task of separating the produced oil completely from the water and to do so at lower costs and emissions. It is also important that the recovered water meet the specifications to be reused for steam generation. Major emerging technologies in this area are presented in Table 1.

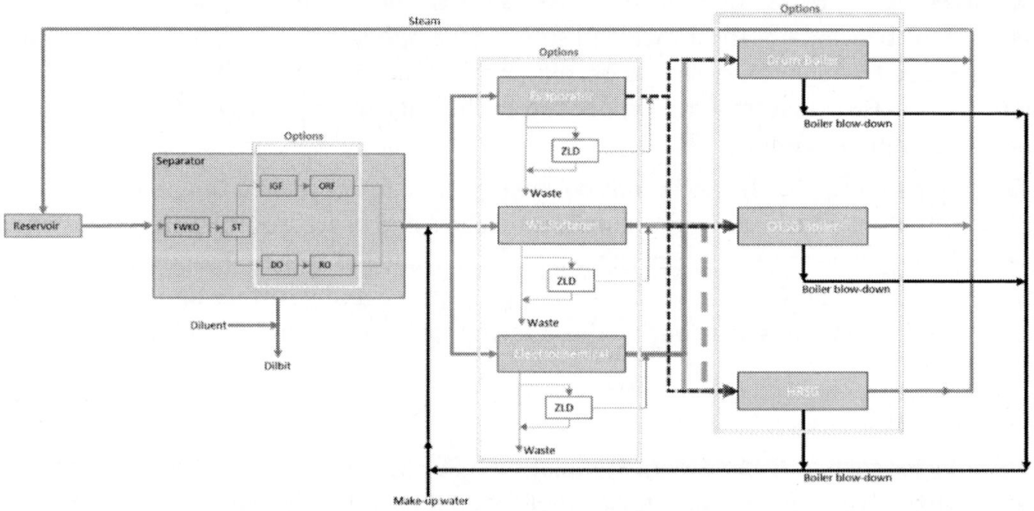

Figure 1. Surface facilities superstructure for in-situ bitumen production. Technologies and processes that are alternatives to each other are represented as options. Notations: FWKO – free water knock out; ST-skim tank; IGF-induced gas floatation; ORF-oil removal filter; DO – De-oiling; RO-reverse osmosis; ZLD-zero liquid discharge, WL-warm lime; OTSG-once-through steam generator; HRSG-heat recovery steam generator.

Steam generation segment uses natural gas combustion and boiler technologies to convert water to superheated steam. Large amounts of steam are required for in-situ thermal bitumen extraction processes. This implies that the steam generation segment is critical to cost and emissions reductions which can be realized through reductions in combustion and steam generation inefficiencies. Currently, Once-Through Steam Generators (OTSGs) are the dominant type of boilers used in SAGD plants as they are more robust and can handle feed water with higher TDS content (<8000 ppm). The major challenges in steam generation and the technologies being developed to address them are listed in Table 1.

Wells and wellpads (WWP) are the linkage between surface and sub-surface. Through wells, steam and/or solvents are injected into the bitumen reservoir and also, mobilized bitumen is pumped to the surface. Wells and pads are, therefore, an important part of the in-situ bitumen extraction process and they are expensive as well as take time to complete. However, as shown in Table 1, different time saving and cost reducing methods are being introduced.

BM segment is made up of lean business methods, intelligent business management and data analytics. The benefits of digital oilfield technology have been underscored by recent studies (2015 BP Technology Outlook[1] and IHS CERA[2]). They project a 13% cost reduction, a 2% to 8% increase in productivity with an operating expense reduction of 5% to 25% and a capital expenditure reduction from 1% to 10% (depending on site localization). Some digital technologies are currently applied or tested in the oil sands industry – some of these are listed in Table 1 and assessed in this study.

2.2 Sub-surface operations

RES and BM make up sub-surface operations. Unlike in the conventional crude oil production where the reservoir pressures are sufficient to push the mobile, less viscous oil through a production well to the surface, in bitumen extraction from oil sands, a recovery mechanism is required by which the viscosity of the material is lowered to the point where it can flow to the production well (Butler and Mokrys, 1989).

---

[1] British Petroleum, 2015. BP Technology Outlook: Technology choices for a secure, affordable and sustainable energy future. Available at http://www.bp.com/content/dam/bp/pdf/technology/bp-technology-outlook.pdf

[2] Edward, T. Adding Value from Digital Oilfield. Available at http://s428994907.websitehome.co.uk/stepchange/wp-content/uploads/2010/05/adding-value-from-digital-oilfield-HP-20102.pdf

Bitumen mobilization can be done through one or a combination of chemical, thermal or physical stimulation (see Table 1). These in-situ methods improve bitumen mobility to an extent that it can be moved easily to the production wellbore and produced at the surface. The stimulation and mobilization mechanisms result in high capital and operation costs and are inherently energy and GHG emissions intensive. Technologies that are applied in the sub-surface for oil exploration and to mobilize bitumen from the oil sands are listed in Table 1.

Under the BM segment, steam flood data and analytic are used to optimize steam injection and maximize bitumen production.

2.3 Upgrading

A conventional in-situ the base plant produces bitumen which must be upgraded or diluted to make it pipeline ready by meeting pipeline specifications: 20-22 API gravity, 3-3.6 wt.% sulfur (CEPA, 2011) and <350 centistoke viscosity. Diluents are usually added to the bitumen with the final product becoming dilbit (dilute bitumen) or when bitumen is upgraded, the resulting product becomes synthetic crude oil (SCO). Partial upgrading technologies are providing alternative options to diluent addition and producing higher value product (see Table 1). Partial upgrading assessment is not covered in this paper, but readers can refer to detailed studies on the subject (Castañeda et al. 2014, Nduagu et al., 2017, Svrcek et al. 2016).

3.0 Methodology

We consider a conventional SAGD facility with a steam-oil ratio (SOR) of 3 m^3/m^3 and a production capacity of 30,000 barrels per day (bbl/day) as a base case. SOR is measured as the volume of steam (cubic meters cold water equivalents) injected into the reservoir to produce a unit volume (in cubic meter) of bitumen. For processes where pure solvents are used as a replacement for steam, solvent-to-oil ratios (volume of solvent, in cold volume equivalents) injected into the reservoir to produce a unit volume of bitumen. The SOR indicates how efficient injected steam or solvent is utilized in the reservoir for mobilizing bitumen and weeping the mobilized bitumen to the production well. It is also a rough indication of the efficiency of an in-situ bitumen extraction process, therefore, the higher the SOR, the less efficient the process is and vice versa.

Supply cost and GHG emissions calculations

Supply cost is calculated as the minimum constant dollar price required to recover all capital expenditures, operating costs, royalties, and taxes, as well as to earn a specified return on investment of 10% (nominal return is 12% with a 2% inflation rate). The supply cost is used as a measure of economic viability in this study whereas process emissions at the plant are calculated to show the GHG emissions intensity of bitumen extraction and processing on a well to plant gate basis. On this basis, the base plant emits 61.4 $kgCO_2e$/bbl (kilogram CO_2 per barrel of bitumen).

Capital costs for the base plant is calculated based on the most recent projects (2013-2015) and the closest upcoming ones (2016-2019). This gives an average capital cost of C$39,760 per flowing barrel (C$/bbl/day) or a total installed cost of C$1,192 Million for a 30,000 bbl capacity over a plant line of 30 years. The supply cost of the SAGD base plant is calculated to be $43.31/bbl and the capital cost component is $19.65. The capital cost component is 16% lower than that reported in a previous CERI study (Millington and Murillo, 2017). This is consistent with what was reported by CanOil (2016). The base year for economic calculation is 2015 and a construction period of 3 years is assumed. The Canadian and Alberta provincial tax and royalty regimes are applied.

The technologies assessed are those that are near-commercial which can be deployed within the next 5-7 years. This stage of development is above a technology readiness level (TRL≥6), where the technology is either piloted or field tested in a relevant environment. The potential of surface and sub-surface technologies to reduce the supply costs ($43.31/bbl) and GHG emissions (61.4 $kgCO_2e/bbl$) of the base plant is assessed. Also, the cumulative effect of adopting viable technology options are shown.

4.0 Results and discussions

The costs and emissions results obtained from the economic and GHG emissions modeling of the surface and sub-surface technologies are presented in this section. Results of technologies that fall under the same segment are combined and presented as ranges and benchmarked against the supply cost and GHG emissions of the base plant.

4.1 Surface operations

Supply cost results obtained from technologies for both surface and sub-surface operations are presented in Figures 2. These results come from applying emerging technologies in the segment assessed but keeping other process segments consistent with the SAGD Base technology design and make-up.

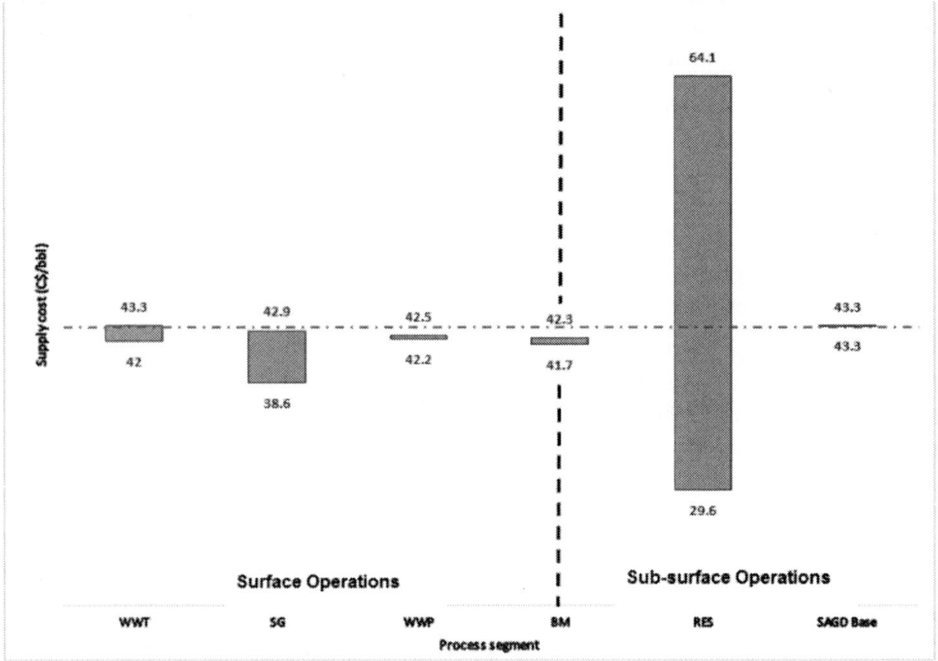

Figure 2. Range plot showing potential to reduce supply costs by emerging technologies applicable to surface and sub-surface operations

Where necessary, changes are made where other segments are altered due to the application of the assessed technology.

Results show that technologies deployable for surface operations reduce the supply cost of base plant by 3-12%. The highest contributor to this reduction is the steam generation segment which is driven by steam solvent process applied to a green field facility. Of the technologies assessed, the technology with the highest costs reduction under this segment is steam generation via the use of solid oxide fuel cells (SOFCs) for cogeneration of steam and electricity. SOFCs generates energy directly by chemically reacting a fuel and oxygen, rather than by combustion; thus, SOFCs have a high efficiency. Typical SOFC combined heat and power efficiency can reach 80% when operating temperature up to 1000°C. However, SOFCs integration with oil sands facilities has not been demonstrated previously and could face significant deployment challenges.

Modest performances are realized from other surface technologies, and overall, sub-surface operations deliver significantly higher cost reductions.

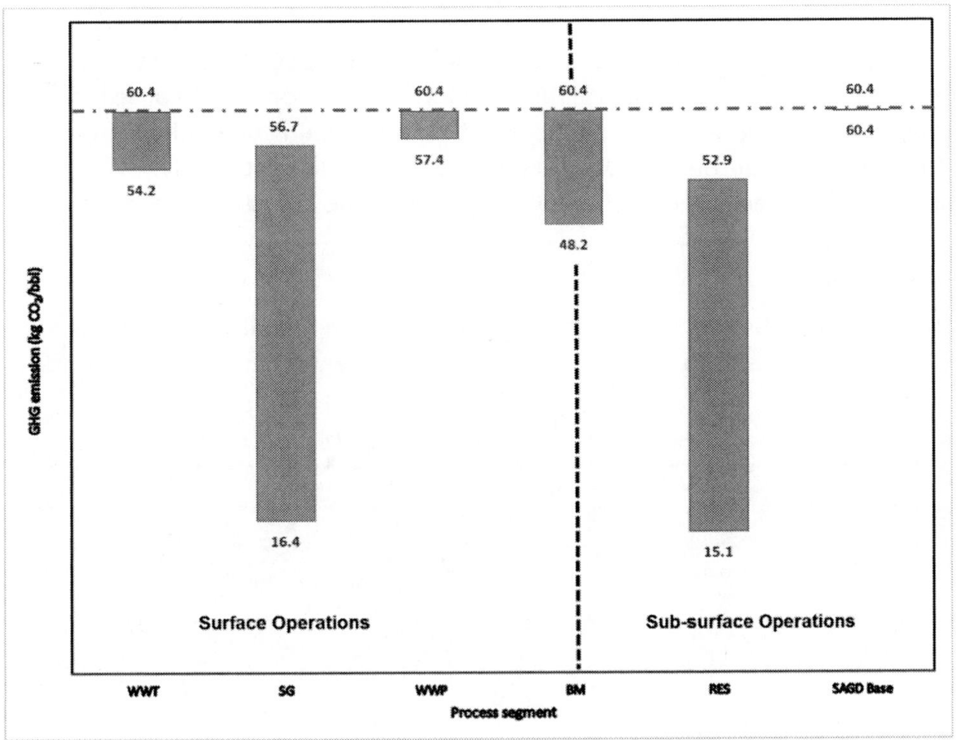

Figure 3. Range plot showing potential to reduce GHG emissions by emerging technologies applicable to surface and sub-surface operations

Technologies for surface applications reduce direct plant emissions considerably. Figure 3 is a range plot showing that the GHG emissions of the base plant (60.4 kgCO_2e/bbl bitumen) are reduced by up to 73% through the steam generation process. This improvement is attributed to direct contact steam generation (DCSG). The DCSG co-injects flue gas and steam into the reservoir and based on modeling results, can sequester 30-60% of the CO_2 emissions in the reservoir. In addition, the DCSG design enables thermal efficiencies of up to 98%. However, some technical and economic issues related to fouling, cost of oxygen for oxy-combustion, fate of CO_2 "stored" in the reservoirs still need to be addressed.

Other surface technologies show potential to reduce GHG emissions, albeit marginally. For example, the WWP and WWT technologies reduce emissions 5% and 10%, respectively. No significant emissions reduction is achieved through the surface BM technologies.

4.2 Sub-surface operations

Results show that the technologies applied for sub-surface operations outperforms those of surface technologies by as much as 3 times in supply costs reduction. On the other hand, not all sub-surface technologies reduce supply costs – results show that some of them can increase the costs significantly (Figure 2). The highest contributor to cost reduction achieved in sub-surface is the RES segment, which is driven by steam solvent processes applied to a green field facility. Steam solvent processes inject a mixture of steam and light hydrocarbons e.g. propane and butane into the reservoir. Steam solvent mixtures are more effective than steam in bitumen mobilization and also results in significant production uplift, up to 38%. However, a production uplift of 9% is required for a solvent process to perform better than the base plant. Other reservoir technologies such as pure solvent methods can reduce the supply cost by about 8% in addition to benefits of partially upgraded product, which a higher value bitumen.

Similar maximum emissions reductions obtained for surface technologies are also realized by technologies for sub-surface applications. As shown in Figure 3, the GHG emissions of the base plant are reduced by up to 75% through the RES segment with pure solvent as the enabling technology. Besides the RES, the BM reduces GHG emissions of the base plant by up to 17% through data analytics for steam flood and production optimization. However, this technology also need to be broadly deployed for performance validation.

4.3 Optimal technology combinations

In Table 2 we present optimal configurations scenarios that can be applied in the oil sands process infrastructure for optimal results. One scenario for a brown field development whereas the rest is for green field facilities.

Table 2. Optimal technology integration scenarios that result in maximum costs and emissions reductions

<table>
<tr><th></th><th colspan="4">SURFACE FACILITIES</th><th colspan="2">SUB-SURFACE FACILITIES</th></tr>
<tr><th>BROWN FIELD</th><th>WWP</th><th>WWT</th><th>SG</th><th>BM</th><th>BM</th><th>RES</th></tr>
<tr><td>Steam Solvent based Scenario</td><td></td><td>Waste and Water treatment</td><td></td><td></td><td>Steam flood management</td><td>Steam Solvent</td></tr>
<tr><th>GREEN FIELD</th><th></th><th></th><th></th><th></th><th></th><th></th></tr>
<tr><td>Steam Based Scenario with DCSG</td><td rowspan="5">Well pad standardization</td><td rowspan="2">Evaporator</td><td>No-CoGen</td><td rowspan="5">EPC Digitalization</td><td rowspan="5">Steam flood Management</td><td>Steam and CO_2 mixture</td></tr>
<tr><td>Steam Based Scenario with SOFC</td><td>CoGen</td><td>Steam</td></tr>
<tr><td>Steam-solvent Scenario (FTB w/o DCSG)</td><td rowspan="2">Chemical Water treatment process</td><td>No-CoGen</td><td rowspan="2">Steam Solvent</td></tr>
<tr><td>Steam-solvent Cogen Scenario (SOFC)</td><td>CoGen</td></tr>
<tr><td>Solvent Based Scenario</td><td></td><td>No-CoGen</td><td>Pure Solvent</td></tr>
</table>

The potential costs and emissions reductions from the different compatible technologies within and across the process segments are shown in Figure 4. The brown field scenario reduces supply cost and GHG emissions by 16% and 60%, respectively with sub-surface technologies contributing the most. This means that for brown fields to realize meaningful costs and emissions results simultaneously, the major focus should be on sub-surface technologies.

Green field scenarios tell a different story: surface and sub-surface technologies should be considered important for delivering simultaneous reductions in both supply costs and emissions.

Two technology configurations stand out as best performing scenarios for deep reductions in both supply costs and GHG emissions. These are the Steam-solvent and Steam-solvent Cogen scenarios – the performance of both configurations is driven mostly by outstanding performance of surface and sub-surface technologies.

Steam-solvent and Steam-solvent Cogen scenarios reduce supply costs by 34-40% and GHG emissions by 63-65%. If deeper GHG reduction is the goal, the Solvent-Based scenario, which achieves an 80% GHG emissions reduction becomes the preferred option.

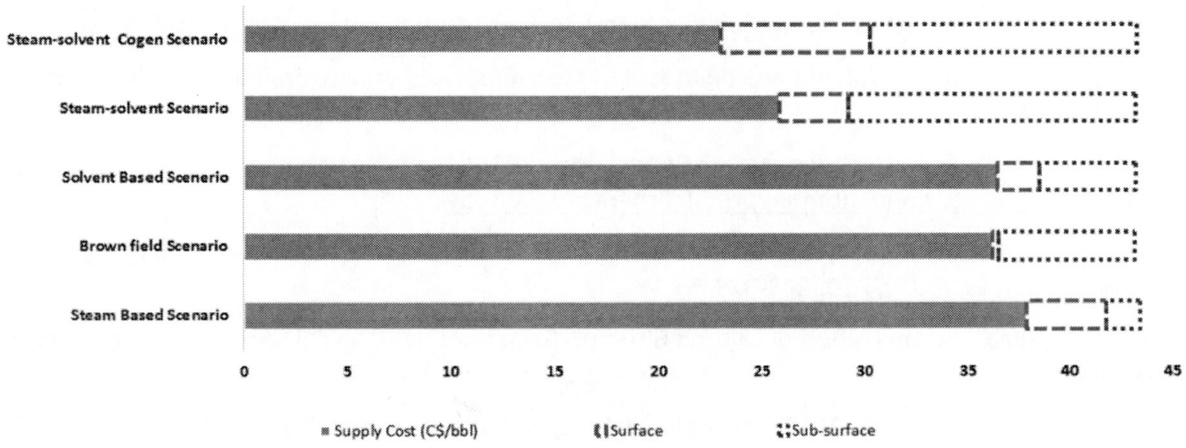

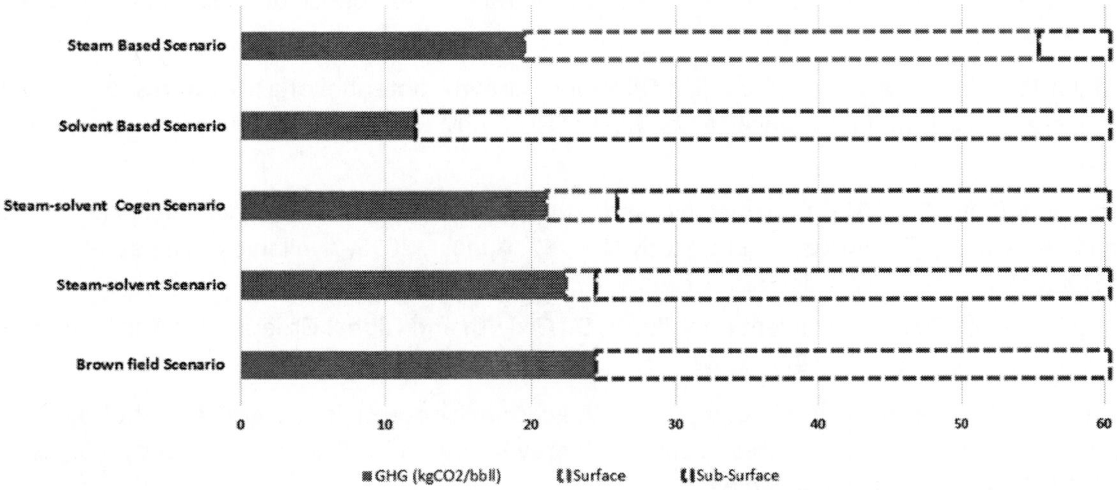

Figure 4. Potential supply costs (Up) and GHG emissions (Down) reductions through technology combinations in surface and sub-surface operations.

Otherwise, the optimal Steam-solvent and Steam-solvent Cogen scenarios should be the choice for deep cuts in both costs and emissions. Therefore, the scenarios presented offers clear options that meet different cost and emissions reduction objectives.

Conclusions

This paper provides a systematic and comprehensive techno-economic and GHG assessment study which shows the extent emerging technologies can address costs and emissions challenges of bitumen production. It can be concluded from the results and analysis that significant opportunities exist to reduce costs and emissions using emerging surface or sub-surface technologies. When technologies for surface and sub-surface operations are combined, even greater reductions in supply costs (up to 40-46%-points) and GHG emissions (up to 63-80%-points) are realizable. It is however, all of the technologies need further de-risking through pilot and field demonstration studies.

References

AER, 2012. ST98: Alberta's energy reserves & supply/demand outlook. Available online at https://www.aer.ca/data-and-publications/statistical-reports/st98

Alberta Energy, Facts and Statistics. Available at http://www.energy.alberta.ca/OilSands/791.asp

Butler, RM, Mokrys, IJ, 1989. Solvent analog model of steam-assisted gravity drainage. AOSTRA J. of Research. 1989, 5, 17-32.

CanOIl, 2016. CanOIl's Producers Benchmark Report, May 2016. Available at http://www2.jwnenergy.com/summaryproducerbenchmarkstudy

Castañeda, LC, Munoz, JA, Ancheyta, J, 2014. Current situation of emerging technologies for upgrading of heavy oils. Catalysis Today, 220, 248-273.

CEPA, 2011. Pipeline Transportation of Diluted Bitumen from the Canadian Oil Sands. Canadian Energy Pipeline Association.

Council of Canadian Academies, 2015. Technological prospects for reducing the environmental footprints of Canadian oil sands. The expert panel on the potential for new and emerging technologies to reduce the environmental impacts of oil sands development. The Council of Canadian Academies, Ottawa, ON, Canada.

Findlay, PJ, 2016. The Future of the Canadian Oil Sands: Growth potential of a unique resource amidst regulation, egress, cost, and price uncertainty. Report No. WPM 64. Oxford Institute for Energy Studies, Oxford, UK.

Millington, D, Murillo, CA. 2017. Canadian oil sands supply costs and development projects (2015-2035). Canadian Energy Research Institute (CERI) Study No. 152. August, 2015. Available online at http://resources.ceri.ca/PDF/Pubs/Studies/Study_152_Full_Report.pdf

Nduagu, EI, Gates ID, 2015. Unconventional Heavy Oil Growth and Global Greenhouse Gas Emissions. Environ. Sci. Technol. 2015, 49, 8824–8832

Nduagu, E, Sow, A, Umeozor, E, Millington, D., 2017. Economic potentials and efficiencies of oil sands operations: Processes and technologies. Canadian Energy Research Institute Report, Study No. 164. March, 2017. Calgary, Canada. Available at http://resources.ceri.ca/PDF/Pubs/Studies/Study_164_Full_Report.pdf

PTAC, 2012. Assessment of Innovative Applications of Electricity for Oil Sands Development. JACOBS Consultancy, Available at www.ptac.org/attachments/1425/download

Sleep, S, McKellary, MM, Bergerson, JA, MacLean, HL. 2017. Expert assessments of emerging oil sands technologies. J. Cleaner Prod. 2017, 144, 90-99.

Svrcek, B, Flint, L, Remesat, D, Penner, R, Guo, J, 2016. Partial Upgrading Background Review "White Paper" In Support of The National Partial Upgrading Program (NPUP). A Report prepared by Revamping & Optimising (ROI) Inc. for AI-EES. AI-EES Contract #2280.

498055 Asphaltene Precipitation in Bitumen-Diluent Blends

Abstract:

In this talk we will discuss experiments that explore the effect of diluent composition on asphaltene precipitation in bitumen blends. Naphtha is commonly used as a diluent in the bitumen extraction process, and a laboratory analog consisting of known solvent blends (e.g., heptol) will aid in the design of process modification and testing procedures. Furthermore, characterizing the influence of deviations in naphtha composition will aid in the control of varying feedstock and open opportunities for specialty or designer bitumen diluents. In this study, heptane-toluene blends (heptol) and naphtha-solvent blends were used as diluents to change the solvency power in diluted bitumen. Our goal was to measure the influence of the solvent blend on the solubility of asphaltenes to better understand their phase behavior and precipitation rate. Testing was predominantly performed at ambient conditions, but the influence of elevated temperature on the diluent solvency power was also explored. Because solvent blends and naphtha have different critical points and chemical properties, a heptol blend that represents naphtha at ambient condition may not be an appropriate analogy at elevated temperature. The results of this investigation provide an improved understanding of asphaltene phase behavior in diluent bitumen under the process conditions and will aid in design of novel bitumen extraction procedures.

493139 A Tool for Understanding Foaming Tendencies, Coke Morphology, and Quench Effectiveness - Real Time Visualization into Pilot Delayed Coker Drum

Abstract:

Bitumen and extra heavy oils are a major source of unconventional oils. The challenge with processing these types of oils is the high proportion of components boiling at greater than 1000°F i.e. the un-distillable vacuum residue. This un-distillable fraction is typically sent to the delayed coker. During the delayed coking process operators face problems such as foaming, changing coke morphology, or hot spots during decoking that can adversely affect the profitability and safety of the process. In commercial operations, the sources of these problems are difficult to identify. Industry is thus in need of quality data related to these problems to optimize their coke drum performance which may be otherwise costly and time consuming to obtain.

This paper will discuss how a real-time tool, a moving gamma-ray densitometer, attached to a pilot scale delayed coker is used in conjunction with other information obtained from the same unit for investigating and understanding foaming tendencies in the coke drum, coke morphology and quench effectiveness which is related to hot spot formation. This direct measurement device; a gamma-ray densitometer coupled with an automatic lifting system provides the density of the coke drum contents along its length every 1.5 minutes. It allows for real time visualization of the changes occurring inside the drum as the thermal reactions progress. From these time-dependent density scans, foaming events, foam collapse, and re-foam are easily identified and modeled as a function of operating conditions and feed properties. The effectiveness of antifoam agents are also evaluated using this tool. Mesophase to coke transitions, as well as coke densification (which is predominant in mixed morphology coke) are also modeled using data obtained from this densitometer along with pressure drop data and capillary pressure measurements on coke samples. Real-time quench effectiveness is evaluated using a combination of the temperature profile of the drum and the density scans which gives an indication for dense regions in the coke bed.

This continuously moving and scanning vertical gamma densitometer attached to a delayed coker pilot, providing real time visualization has proved to be a versatile and useful research tool that aids in understanding and modeling events that occur while changing operating parameters in a delayed coke drum. The data may be able to predict previously un-predictable events which could increase safety of operation and/or lower Capex and Opex.

494423 Shale Oil As Steam Cracking Feedstock: GC × GC Characterization and COILSIM1D Modeling

Abstract:

Even though, shale oil is commonly considered as a potential alternative energy resource, the complex chemical composition of the feedstock imposes a challenge for refining and in particular for petrochemical processes. To assess the potential of shale oil utilization as a steam cracking feed and co-feed, a detailed compositional analysis of a pyrolysis shale oil sample is performed and steam cracking product yields are estimated using the COILSIM1D software [1]. The complex chemical composition was unraveled by GC × GC analysis separation coupled to various detectors, i.e. time-of-flight mass spectrometer, flame ionization detector, sulfur and nitrogen chemiluminescence detector, allowing to quantify also hetero-atomic molecular families. High amounts of unsaturated hydrocarbons are characteristic for shale oil produced via kerogen pyrolysis. In this case the olefin concentration was shown to be as high as 18 wt.%. Cycloalkanes and monoaromatics are present in significant concentrations, i.e. 17 wt.% and 11 wt.%, respectively. On the other hand, paraffin content of 8 wt.%, makes shale oil far less advantageous for steam cracking compared to conventional feedstocks. Furthermore, considerable concentration of oxygenates was detected, mainly belonging to chemical families of benzendiols (13 wt.%), hydroxybenzofurans (10 wt.%), phenols (4 wt.%) and ketones (9 wt.%). Nevertheless, shale oil quality is considered satisfactory, as sulfur (0.7 wt.%) and nitrogen (300 ppmw) content is low and the carbon number range is favorable, i.e. C_3 to C_{26}. The obtained chemical composition, in which the compounds were grouped according to carbon number and molecular family, was used to define the feedstock in COILSIM1D. In order to apply the latter the reaction network was extended to oxygen containing species before yields of steam cracking products were simulated by applying the single event microkinetic modelling methodology [2]. Product yields for steam cracking of pure shale oil, shale oil blended with naphtha (20:80 wt.%) and kerosene (20:80 wt.%) at typically industrially relevant coil outlet temperatures (800, 820 and 840°C) and steam dilution ratios (0.5, 0.75 and 1 kg/kg) were obtained.

References

1. S. P. Pyl, C. M. Schietekat, M. F. Reyniers, R. Abhari, G. B. Marin, K. M. Van Geem, "Biomass to olefins: Cracking of renewable naphtha" Chemical Engineering Journal, 176– 177, pp. 178– 187, 2011

2. K. M. Van Geem, M. F. Reyniers and G. B. Marin, "Challenges of Modeling Steam Cracking of Heavy Feedstocks" Oil & Gas Science and Technology – Rev. IFP, 63, pp. 79-94, 2008

487345 Characteristics of Unconventional and Heavy Oils: Their Economic and Environmental Impacts in Production and Processing

Abstract:

Production of conventional oils (generally with API>20) is less costly and less pollutant to unconventional oils which include shale oil (or kerogen oil), oil sands (or bituminous sands) and heavy oils. In addition, liquid fuels produced from coal or gas to liquid fuel may also be considered under unconventional oil category. As production of conventional oil is in decline, production of unconventional oil becomes increasingly important in meeting future world energy demand.

The world conventional oil resources is estimate around 1,300 billion barrels of oil in place while total world shale oil resources is estimated around 5 trillion barrels. In addition the amount of heavy oil and tar sands is estimated at 3 to 4 trillion barrels. Even with 20% recovery from these unconventional resources, proven unconventional oil resources exceed from those of conventional sources.

The nature and constituents of shale oil, oil sands and heavy oils are significantly different from those of light and conventional oils [1]. Production, processing and consumption of these energy sources cause environmental issues such as carbon emission, atmospheric pollutants (acid rain, smog, ozone, etc.) as well as hazardous wastes and aqueous effluents. Processing cost of unconventional oil is also higher than conventional oils [2,3,4]. In this presentation at first the differences between conventional and unconventional oils especially from compositional and elemental points of views are discussed. Some important characteristics of unconventional oils such as metals, sulfur, nitrogen, asphaltene contents, estimation methods and their impacts on the selection of processing method and related costs will be discussed. Finally world refining investment patterns in relation with processing of heavy oil and residues will be analyzed and discussed.

References:

[1] Riazi, M. R., "Characterization and Properties of Petroleum Fractions," ASTM International, Conshohocken, PA, USA, 2007. 435p.

[2] Riazi, M. R., Eser, S., Agrawal, S. S., Pena Diez, J. L., "Petroleum Refining and Natural Gas Processing" ASTM International, Conshohocken, PA, USA, 2013. 811p.

[3] Riazi, M. R., (Ed.) "Exploration and Production of Petroleum and Natural Gas" ASTM International, Conshohocken, PA, USA, 2016. 739p.

[4] Riazi, M. R., Gupta, R., (Ed.) "Coal Production and Processing Technology" CRC Press, Taylor & Francis, FA, USA, 2015. 535p.

502557 Numerical Evaluation of a Fluid Catalytic Cracking Unit with Internal Baffles

Abstract:

Fluid Catalytic Cracking (FCC) units play a very important role in an oil refinery, because they convert heavy fractions (vacuum distillates or some vacuum residues) to gasoline, C3-C4 cuts and petrochemicals. These units and the catalysts used are in continuous evolution. They must adapt to market changes: gasoline yield and/or quality maximization, petrochemicals production, conversion of residues, environmental requirements. New riser configurations require that the phases have an increasingly short contact time, being separated after 1 s or 2 s of flow. Regenerators operating at a higher combustion temperature allow a higher recovery of the catalyst. The catalyst entering the reactor warmer requires that the mixture between the phases be extremely efficient, since a poor thermal exchange between the phases results in high temperature points and, consequently, the favoring of thermal cracking reactions. Products formed by thermal cracking have lower value than those formed by catalytic route, and therefore regions with temperatures too high must be avoided. Once the gas-solid flow in risers can develop a "core-anulus" pattern, in which catalyst particles are more concentrated near the walls of the reactor, the inclusion of internals can enhance the phases mixture. However, it can also increase its pressure drop, which is a limiting factor to the process. Therefore, the aim of this study is to evaluate the performance of an industrial riser type reactor, with the inclusion of airfoil-shaped ring-type baffles in order to improve the gas-solid flow. Computational fluid dynamics (CFD) techniques are applied to simulate in a three-dimensional geometry, in which the turbulent three-phase flow was calculated in a transient formulation. Gases (air and steam) and catalyst particles were the two Eulerian phases simulated, and a third Lagrangian phase was composed of liquid oil droplets which were evaporated during the process. Results indicate that it is possible to enhance the gasoline production without causing an expressive increase in the pressure drop of the unit.

17AIChE
Annual Meeting, Minneapolis, MN

A Group Contribution Method for the Prediction of the Mid Infrared (MIR) Absorption Spectra of Species Involved in Fluid Catalytic Cracking (FCC)

Sandra López-Zamora, Alejandro Molina, Hugo de Lasa

UNIVERSIDAD NACIONAL DE COLOMBIA SEDE MEDELLÍN
Western UNIVERSITY · CANADA
FRUN Flujos Reactivos de la Universidad Nacional de Colombia

Outline

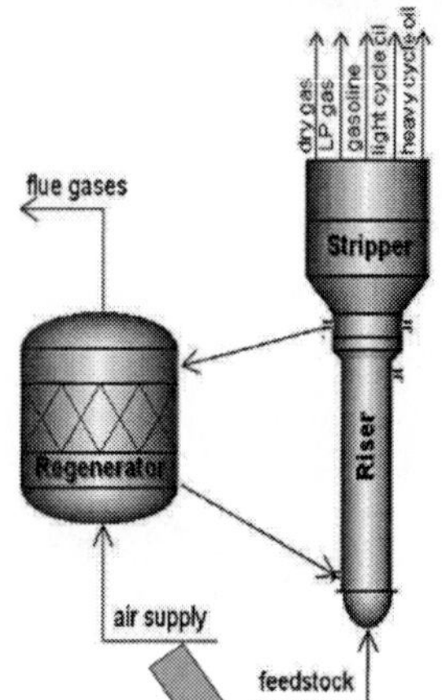

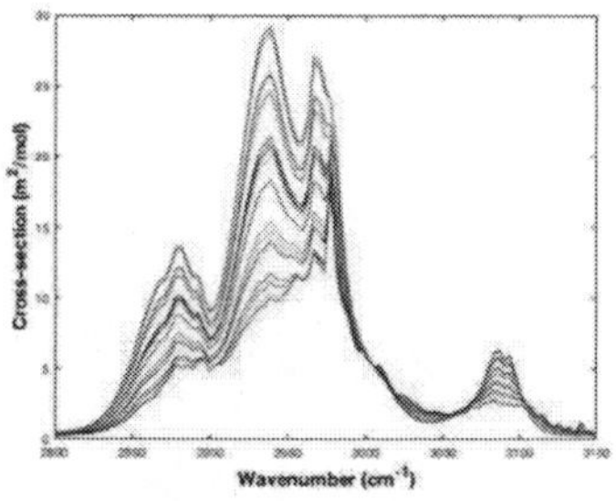

Validation of the Group contribution method

Introduction

Effect of C-H bonds number in absorbance

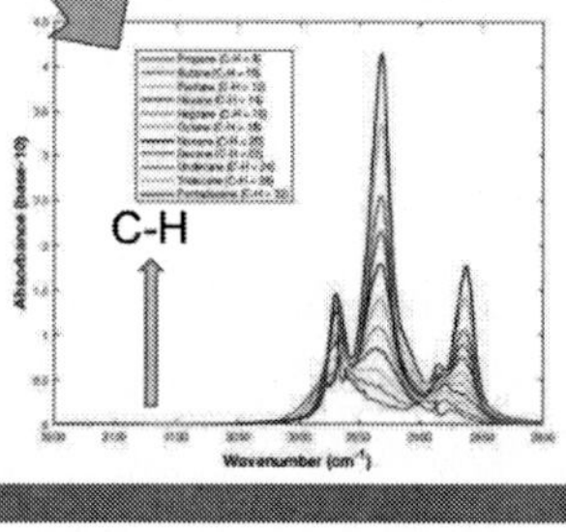

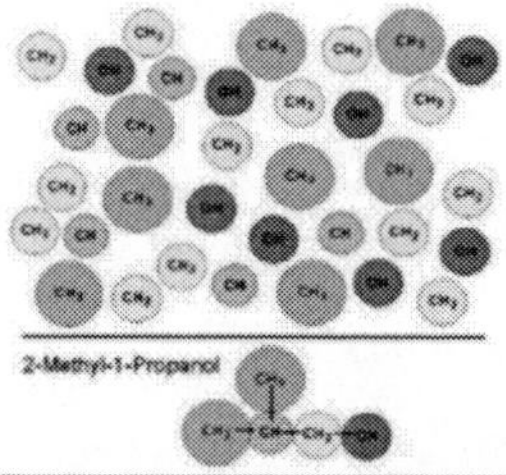

Group contribution method

UNIVERSIDAD NACIONAL DE COLOMBIA SEDE MEDELLÍN
Western UNIVERSITY · CANADA
FRUN Flujos Reactivos de la Universidad Nacional de Colombia

Fluid Catalytic Cracking (FCC)

- Fossil fuels are extensively employed in transportation, industry and electricity generation
- Over 95% of the energy in the transportation sector is derived from petroleum
- Fluid catalytic cracking is one of the most important and profitable process technologies in the refinery

Image from: elfinanciero.com.mx & 123rf.com

UNIVERSIDAD NACIONAL DE COLOMBIA SEDE MEDELLÍN — Western UNIVERSITY · CANADA — FRUN Flujos Reactivos de la Universidad Nacional de Colombia

Fluid Catalytic Cracking (FCC)

- The feed to the riser is basically composed of C30−C40 hydrocarbons that are cracked to more value molecules
- The reaction zone in FCC involves:
  - ✓ Catalyst stripper
  - ✓ Two coupled reactors: Riser (or downer) and regenerator
- Operating conditions
  - ✓ Riser: 3 atm, Pressure drop: 35-62 kPa, Temperature: 550 °C
  - ✓ Downer:2.5 atm, Pressure drop: 20 kPa, Temperature: 550 °C

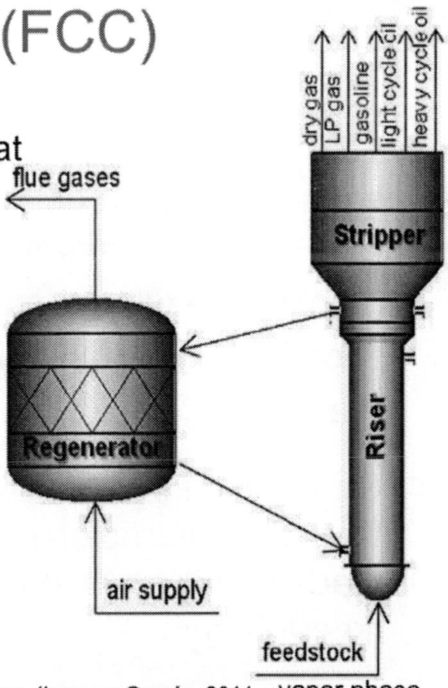

Image from: Jimenez-García, 2011

UNIVERSIDAD NACIONAL DE COLOMBIA SEDE MEDELLÍN — Western UNIVERSITY · CANADA — FRUN Flujos Reactivos de la Universidad Nacional de Colombia

FCC Reactions

- The feed to the riser is basically composed of C30–C40 hydrocarbons that are converted to smaller molecules.
- Cracking reactions are fast (<10 s).
- Reactants and products can be divided in: paraffins, olefins, naphthenes and aromatics
- Main reactions:
  - ✓ Cracking
  - ✓ Cyclization
  - ✓ Isomerization
  - ✓ H transfer
  - ✓ Dehydrogenation

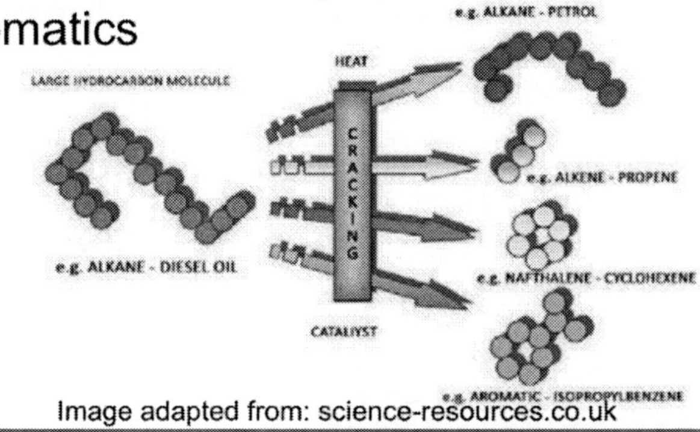

Image adapted from: science-resources.co.uk

FCC reactions

- FCC experiments miss the first seconds of the reaction.
- Some experiments have resolved reaction data down to seconds by quenching the reaction
- Short reaction time: 0.5 – 10 s
- Reaction characterization is made ex-situ with long analysis time >20 min

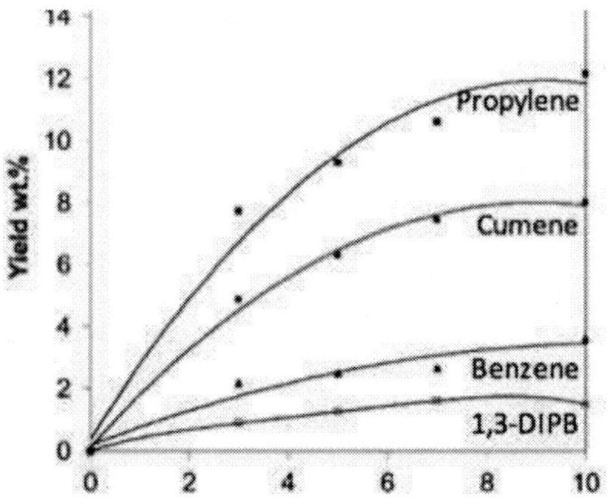

Catalytic cracking of 1,3,5-TIPB: change in the product yield with reaction time (Taken from Al-Khattaf et al. 2002)

UNIVERSIDAD NACIONAL DE COLOMBIA SEDE MEDELLÍN
Western UNIVERSITY · CANADA
FRUN Flujos Reactivos de la Universidad Nacional de Colombia

Reactions characterization

Conventional techniques

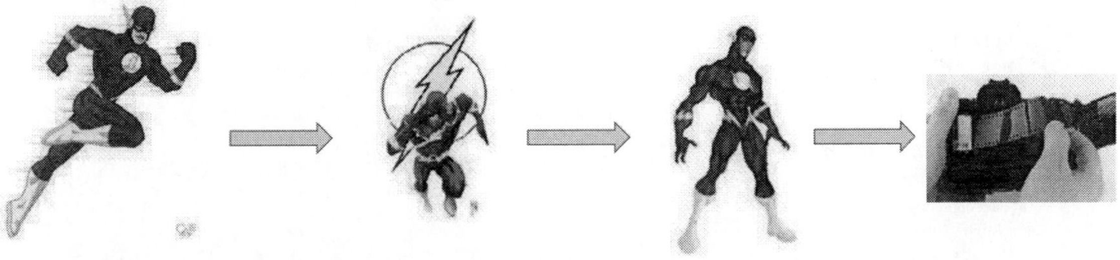

Laser techniques

UNIVERSIDAD NACIONAL DE COLOMBIA SEDE MEDELLIN
Western UNIVERSITY · CANADA
FRUN Flujos Reactivos de la Universidad Nacional de Colombia

Reactions characterization

Laser techniques

Useful for:

- Higher number of data during the first seconds of reaction
- Process control
- Computational fluid dynamics (CFD) validation → Time-resolution

UNIVERSIDAD NACIONAL DE COLOMBIA SEDE MEDELLIN
Western UNIVERSITY · CANADA
FRUN Flujos Reactivos de la Universidad Nacional de Colombia

Previous works

➢ Several studies present the evaluation of hydrocarbons by the use of laser techniques

PARAFFINS

- C-H stretch vibrational absorption can be exploited.
- n-Heptane mole fraction → IR HeNe laser absorption
- Iso-octane → 2948 cm^{-1} HeNe laser
- N-dodecane and n-decane → Tunable mid-infrared laser using difference frequency-generation (DFG) near 2941cm^{-1}

OLEFINS

- Ethylene and propylene → widely-tunable rapid-scanning Mid-IR Laser Spectrometer: 3283 - 2812 cm^{-1}
- Ethylene → CO_2 gas laser near 943 cm^{-1}
- Propene → External cavity quantum-cascade laser (ECQCL) system
- 1,3-butadiene and propylene → Tunable diode laser (TDL): 891 cm^{-1}

AROMATICS

- Toluene → Differential-absorption LIDAR, third harmonic of the Nd:YAG laser
- Aromatic compounds and particles in an atmospheric-pressure opposed-flow flame of ethylene → Laser Induced Emission
- polycyclic aromatic hydrocarbons (PAH) → Laser-induced fluorescence, Nd:YAG Laser in the fourth harmonic (266 nm) and a nitrogen laser at 337 nm

➢ In the referred literature no study applied laser diagnostics to a mixture such as the one for FCC process

UNIVERSIDAD NACIONAL DE COLOMBIA SEDE MEDELLÍN

FRUN Flujos Reactivos de la Universidad Nacional de Colombia

Molecules interaction with light

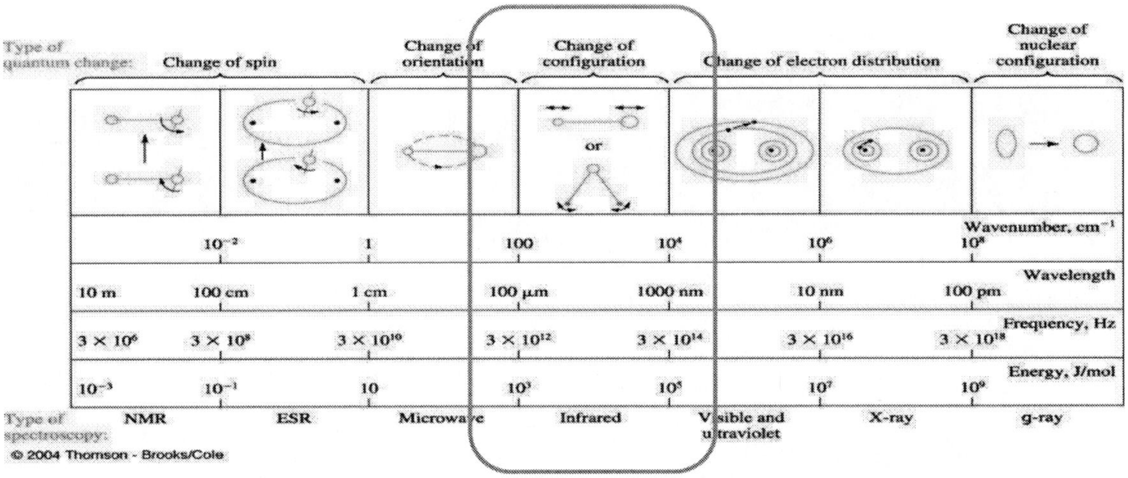

Image from: en.wikipedia.org/wiki/Infrared_spectroscopy

UNIVERSIDAD NACIONAL DE COLOMBIA SEDE MEDELLÍN

Western UNIVERSITY · CANADA

FRUN Flujos Reactivos de la Universidad Nacional de Colombia

IR spectra

- A change of configuration in the molecule bonds occurs
- Molecules absorb specific frequencies that are characteristic of their structure: resonant frequencies
- The absorption of IR radiation can be used to identify and study chemicals
- The behaviour of hydrocarbons in mid-infra red (2.5 - 40 μm) is of interest

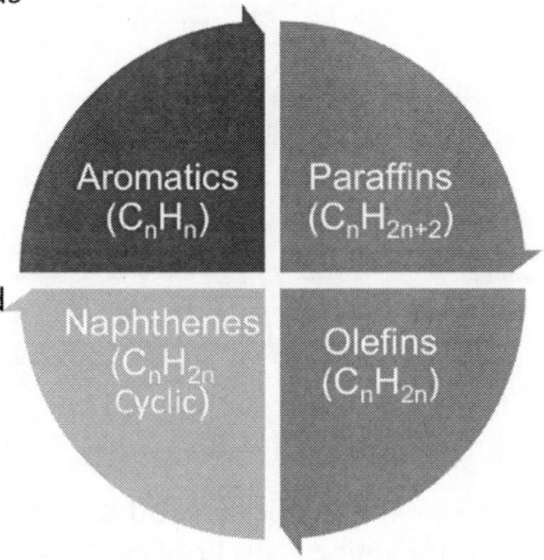

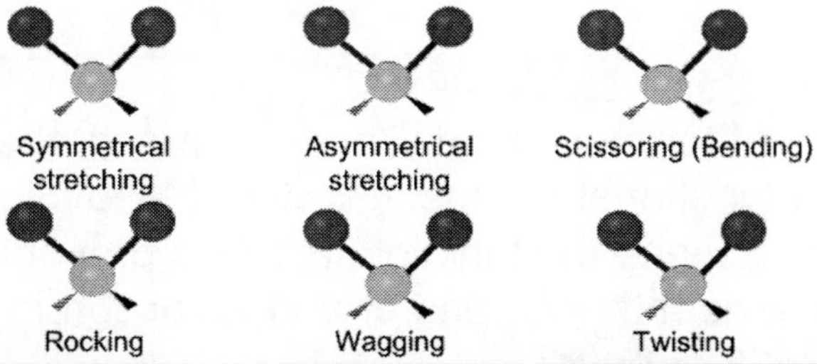

Image from: en.wikipedia.org/wiki/Infrared_spectroscopy

UNIVERSIDAD NACIONAL DE COLOMBIA SEDE MEDELLÍN · Western UNIVERSITY · CANADA · FRUN Flujos Reactivos de la Universidad Nacional de Colombia

Absorbance: Beer law

According with Beer law, cross section can be related to absorbance:

$$\alpha = -ln\left(\frac{I}{I_0}\right) = \sigma c L \qquad \sigma = \frac{\alpha}{cL}$$

where I_o is light intensity, I is light intensity after interaction with molecules, σ is cross section, c is concentration and L the pathlength

In a given region the integrated band intensity can be obtained:

$$\psi = \int_{\lambda_0}^{\lambda} \sigma d\lambda$$

UNIVERSIDAD NACIONAL DE COLOMBIA SEDE MEDELLÍN · Western UNIVERSITY · CANADA · FRUN Flujos Reactivos de la Universidad Nacional de Colombia

MIR characterization of hydrocarbons

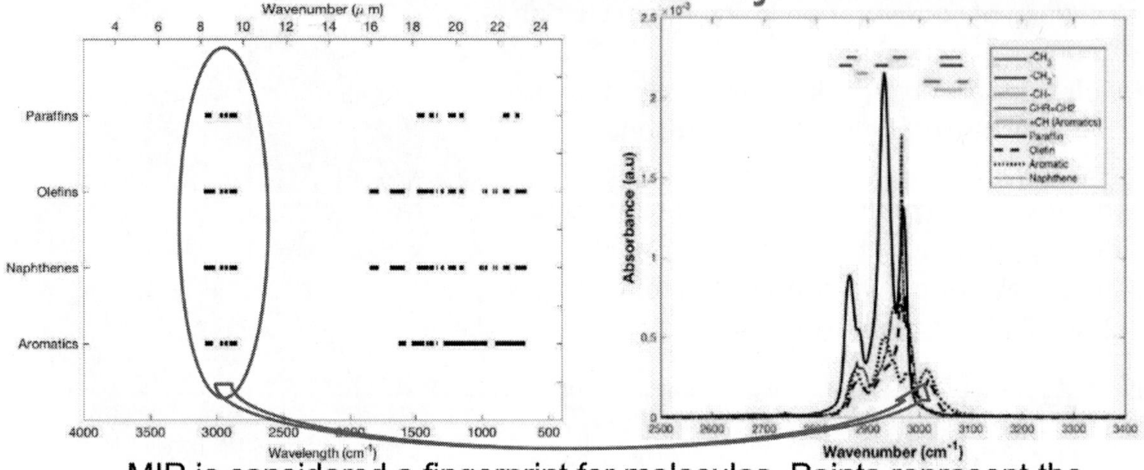

MIR is considered a fingerprint for molecules. Points represent the interaction with light of different functional groups. It is possible to use the region 3200-2800 cm^{-1} to characterize the functional groups present in hydrocarbons: paraffins, olefins, naphthenes and aromatics according to the C-H bonds

UNIVERSIDAD NACIONAL DE COLOMBIA SEDE MEDELLÍN

Western UNIVERSITY · CANADA

FRUN Flujos Reactivos de la Universidad Nacional de Colombia

MIR behaviour: 3200-2800 cm^{-1}

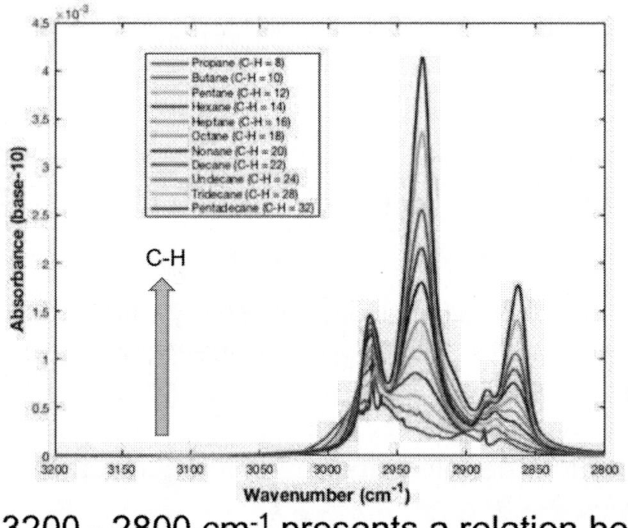

The region 3200 - 2800 cm^{-1} presents a relation between the number of C-H bonds and the absorption.
Spectra are taken from PNNL laboratory (Sharpe, 2004)

UNIVERSIDAD NACIONAL DE COLOMBIA SEDE MEDELLÍN

Western UNIVERSITY · CANADA

Sharpe (2004), Klingbeil (2007), Alfarae (2013), Campbell (2014)

FRUN Flujos Reactivos de la Universidad Nacional de Colombia

Group contribution method (GCM)

- Group contribution methods use the chemical structure of components to predict the properties of a given compound
- Since Lyndersen (1955) and Joback (1987), group contribution methods have been employed for the prediction of thermodynamic properties
- We proposed to use a group contribution method for the prediction of cross section in the region of interest.

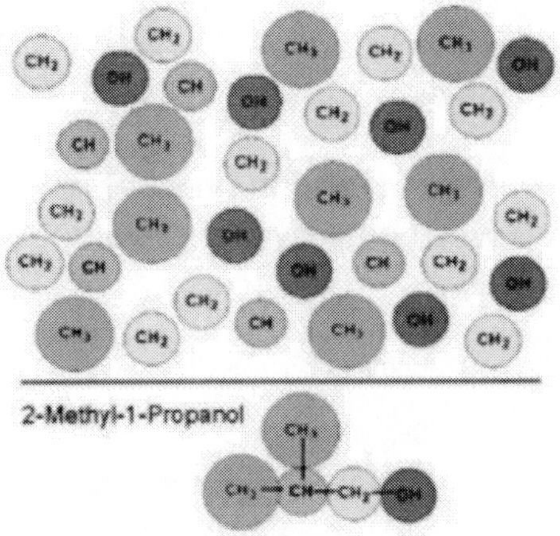

Graphical representation of a group contribution method

Image from:
https://en.wikipedia.org/wiki/Group_contribution_method

UNIVERSIDAD NACIONAL DE COLOMBIA SEDE MEDELLÍN

Western UNIVERSITY · CANADA

FRUN Flujos Reactivos de la Universidad Nacional de Colombia

Group contribution method (GCM)

- Mathematical model
- Linear behaviour of functionalities
- Statistical parameters should be consider
- Validation and applications
- Experimental setup

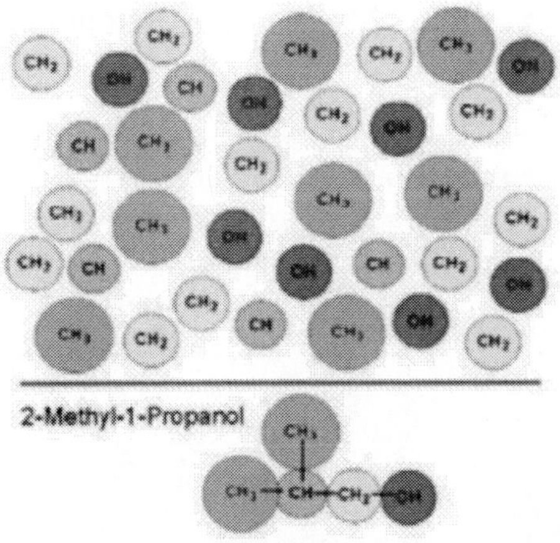

Graphical representation of a group contribution method

Image from:
https://en.wikipedia.org/wiki/Group_contribution_method

UNIVERSIDAD NACIONAL DE COLOMBIA SEDE MEDELLÍN

Western UNIVERSITY · CANADA

FRUN Flujos Reactivos de la Universidad Nacional de Colombia

Group contribution method (GCM)

- 15 parameters related to the kind of C-H bond are employed for the prediction of cross section
- Hydrocarbons present in PNNL database (Sharpe, 2004) were employed: 65 compound

9 parameters:

- ✓ $-CH_3$
- ✓ $-CH_2-$
- ✓ $>CH-$
- ✓ $>C<$

Simple bond ➡

- ✓ $-CH_3$ (α)
- ✓ $-CH_2-$ (α)

- ✓ $=CH_2$
- ✓ $=CH-$
- ✓ $=C<$

Double bond ➡

- ✓ $-CH_3$ (α)
- ✓ $-CH_2-$ (α)

- ✓ a=CH
- ✓ a=C<

Aromatic ring ➡

- ✓ $-CH_3$ (α)
- ✓ $-CH_2-$ (α)

6 parameters related to α type carbons:

$$\sigma = \sum_{i=1}^{15} n_i \gamma_i$$

UNIVERSIDAD NACIONAL DE COLOMBIA SEDE MEDELLÍN — Western UNIVERSITY · CANADA — FRUN Flujos Reactivos de la Universidad Nacional de Colombia

Relationship between C-H number and cross section

A linear relationship between cross section (σ) and C-H number at HeNe wavelength (3.39μm) for paraffins, olefins, naphthenes and aromatics can be observed.

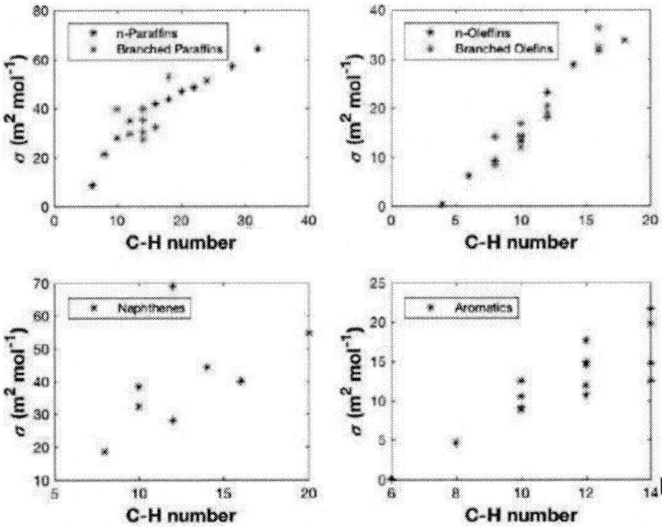

R^2=0.858 R^2=0.945

R^2=0.288 R^2=0.768

UNIVERSIDAD NACIONAL DE COLOMBIA SEDE MEDELLÍN — Western UNIVERSITY · CANADA — FRUN Flujos Reactivos de la Universidad Nacional de Colombia

Relationship between C-H number and integrated band intensity

Linear behaviour increases for integrated band intensity (Ψ) between 3200-2800 cm^{-1} and C-H number

R^2=0.959

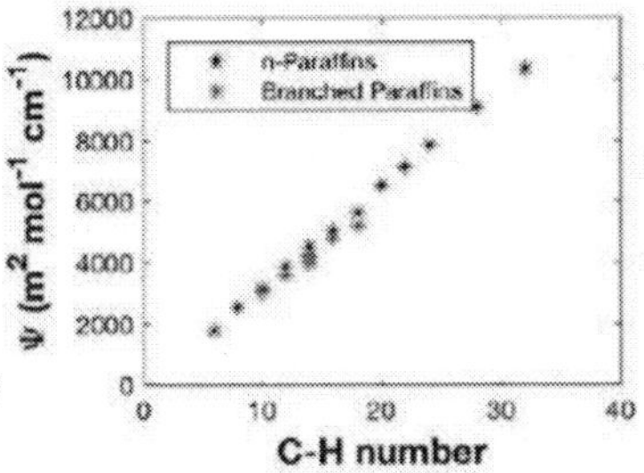

R^2=0.977

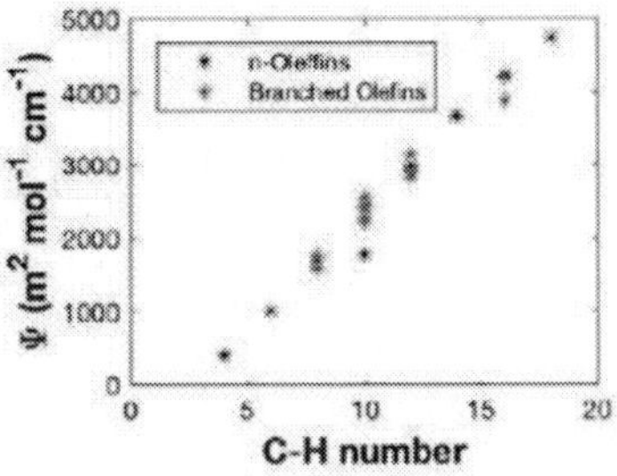

R^2=0.954

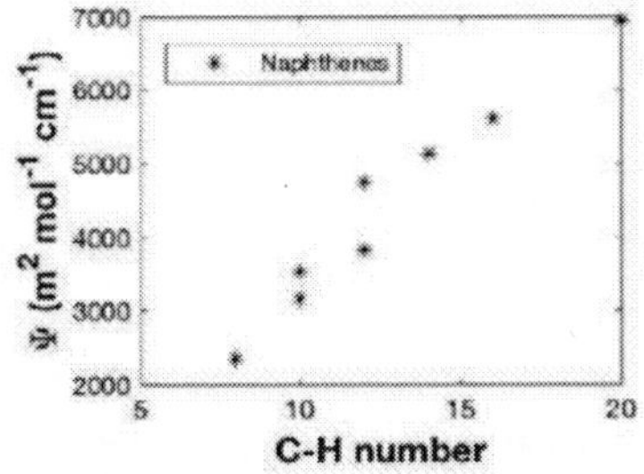

R^2=0.988

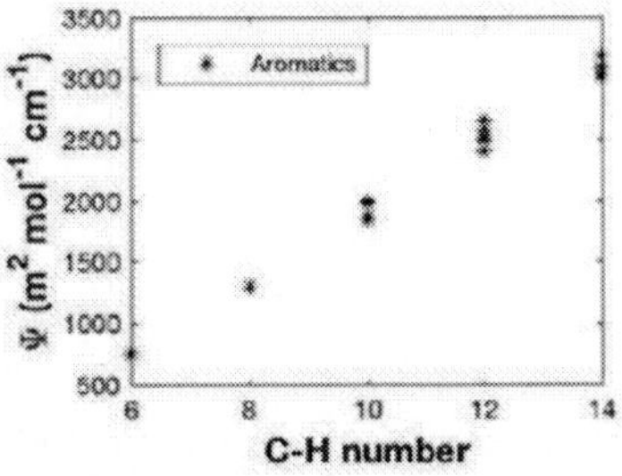

UNIVERSIDAD NACIONAL DE COLOMBIA SEDE MEDELLÍN   Western UNIVERSITY · CANADA   FRUN Flujos Reactivos de la Universidad Nacional de Colombia

GCM validation: pure components

The group contribution method (GCM) was employed for the prediction of paraffins, olefins, naphthenes and aromatics.

A good prediction for the spectra was obtained. Statistical moments of the different order (statistical moments of order one, two and three) had a R^2 of 0.97, 0.90, 0.78 respectively.

The prediction for integrated band intensities had a R^2=0.978

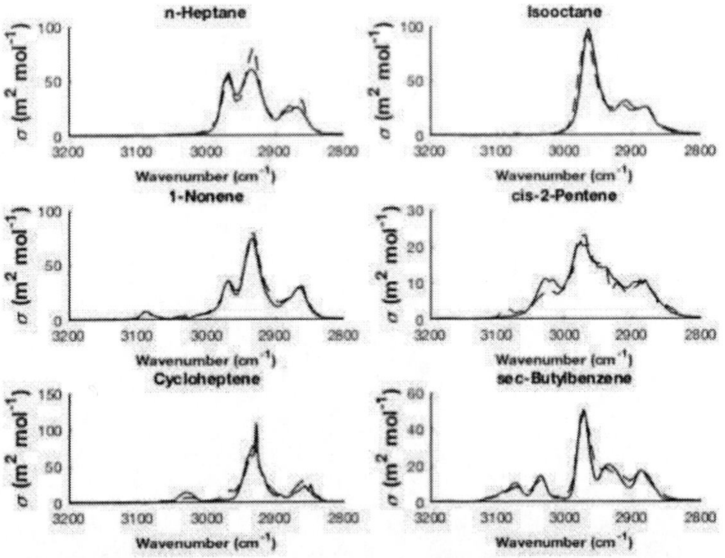

Cross section prediction.
-- Experimental (PNNL)
- Predicted

UNIVERSIDAD NACIONAL DE COLOMBIA SEDE MEDELLÍN   Western UNIVERSITY · CANADA   FRUN Flujos Reactivos de la Universidad Nacional de Colombia

GCM validation: Gasolines

Group Contribution Method was used for the prediction of absorption spectra of 21
The statistic summary for the prediction of absorption coefficients in the case of regular and premium gasolines present R^2>0.97 and RER>15

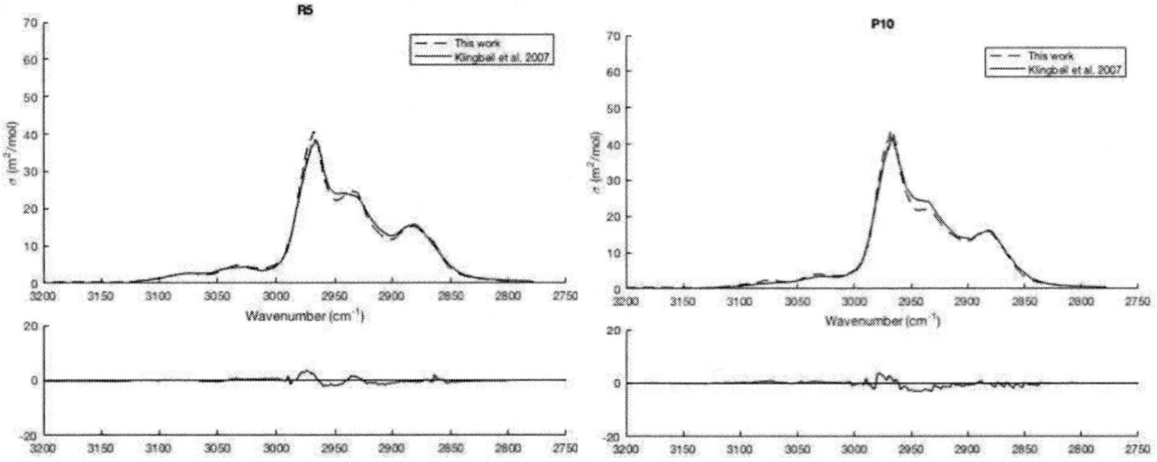

(a) Regular gasoline comprised of 20.6% n-paraffins, 32.8% isoparaffins, 18.6% olefins, 27.9% aromatics, 0% ethanol (b) Premium gasoline comprised of 8% n-paraffins, 49% isoparaffins, 4.7% olefins, 27.2% aromatics, 11.2% ethanol.

UNIVERSIDAD NACIONAL DE COLOMBIA SEDE MEDELLIN

Western UNIVERSITY · CANADA

FRUN Flujos Reactivos de la Universidad Nacional de Colombia

GCM validation: PIOA and RON

PIOA prediction for premium gasolines was possible with R^2=0.936, RER=13,56.
RON prediction for premium gasolines was possible with R^2=0.958, RER=14.11.

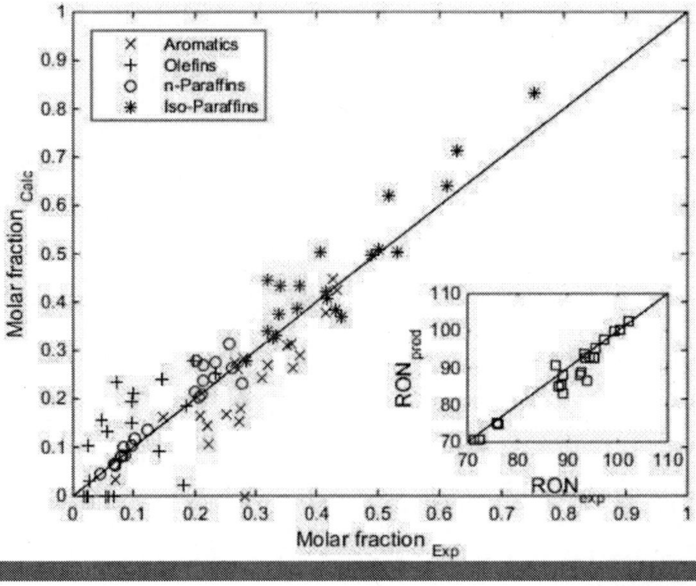

UNIVERSIDAD NACIONAL DE COLOMBIA SEDE MEDELLIN

FRUN Flujos Reactivos de la Universidad Nacional de Colombia

GCM application: Reaction progress

It is possible to employ the linear relationship between the C-H number and the integrated band intensity for calculating the reaction progress of catalytic cracking reaction

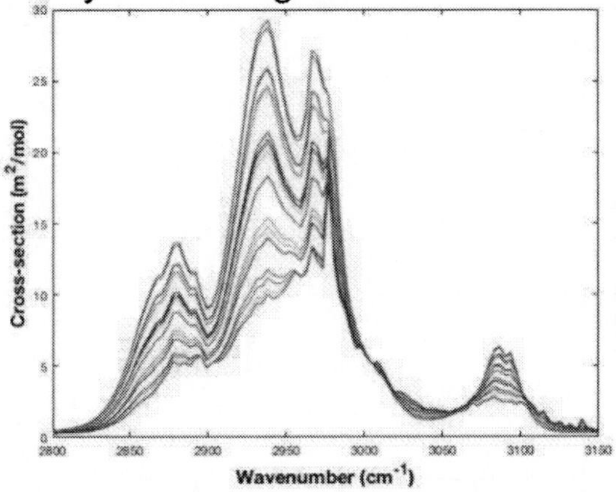

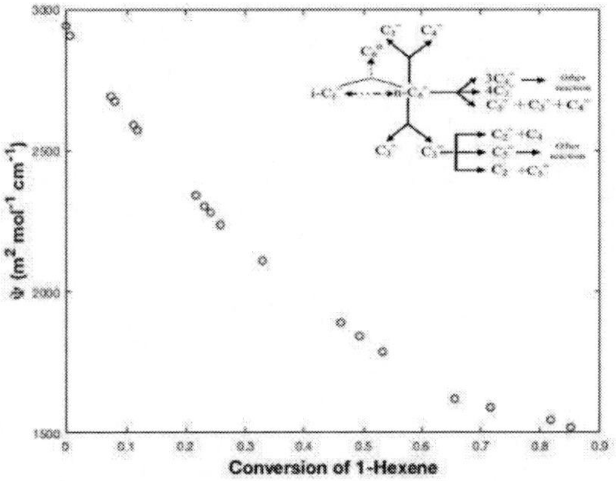

Model reaction: Catalytic cracking of 1-Hexeno. a) change in spectra with time b) Relationship between integrated band intensity (Ψ) and conversion

UNIVERSIDAD NACIONAL DE COLOMBIA SEDE MEDELLÍN

Western UNIVERSITY · CANADA

FRUN Flujos Reactivos de la Universidad Nacional de Colombia

GCM application: Reaction progress

It is possible to employ the linear relationship between the C-H number and the integrated band intensity for calculating the reaction progress of catalytic cracking reaction

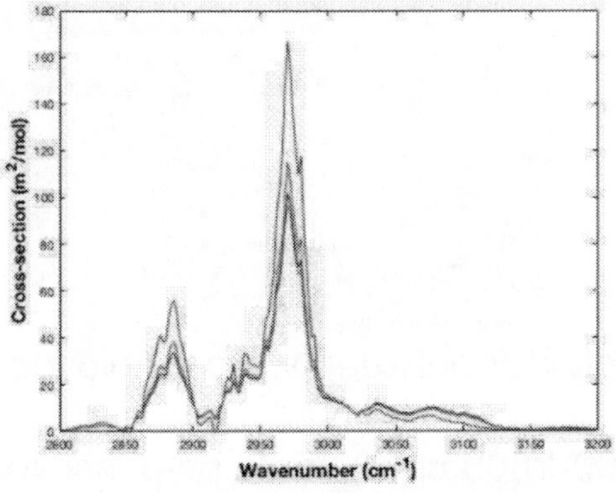

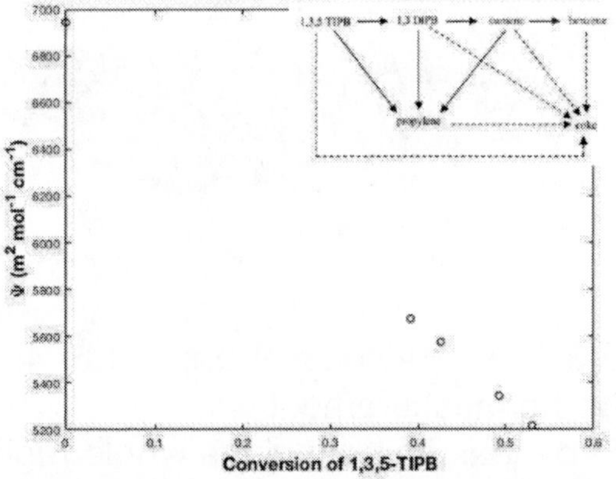

Model reaction: Catalytic cracking of 1,3,5-Triisopropylbenzene. a) change in spectra with time b) Relationship between integrated band intensity (Ψ) and conversion

UNIVERSIDAD NACIONAL DE COLOMBIA SEDE MEDELLÍN

Western UNIVERSITY · CANADA

FRUN Flujos Reactivos de la Universidad Nacional de Colombia

GCM application: Reaction progress

In laboratory, the use of 3.39 μm HeNe laser to estimate the reaction conversion is possible

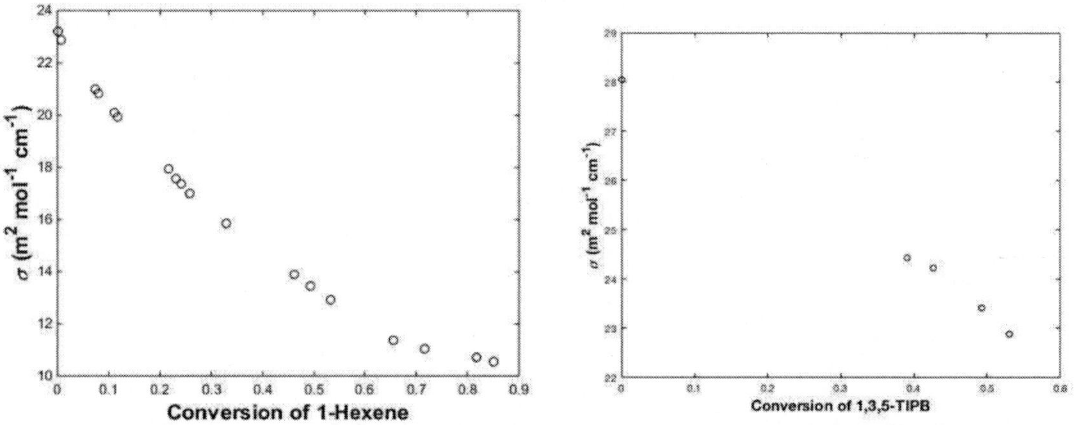

Model reactions: Catalytic cracking of a) 1-Hexeno and b) 1,3,5-TIPB in the HeNe wavenumber

Experimental setup

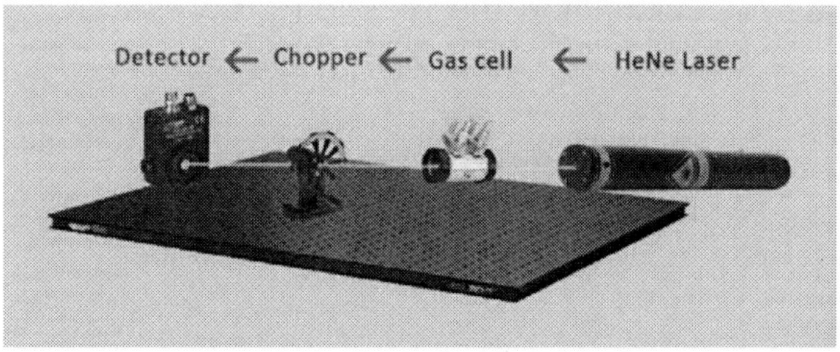

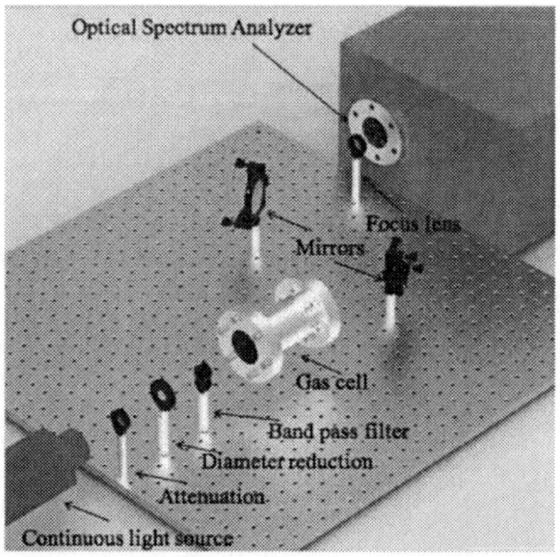

- ✓ A 3.39μm HeNe laser and a 1.5- 4.8 μm detector between were employed for the measurement
- ✓ The measurement in the whole region (3200-2800 cm^{-1}) could be possible with a high energy continuous light source and an optical spectrum analyzer

UNIVERSIDAD NACIONAL DE COLOMBIA SEDE MEDELLÍN
Western UNIVERSITY · CANADA
FRUN Flujos Reactivos de la Universidad Nacional de Colombia

Experimental results

- ➢ HeNe was employed for the measurement of absorption cross section of pure compounds
- ➢ It is possible to validate GCM predictions for 3.39 µm wavelength

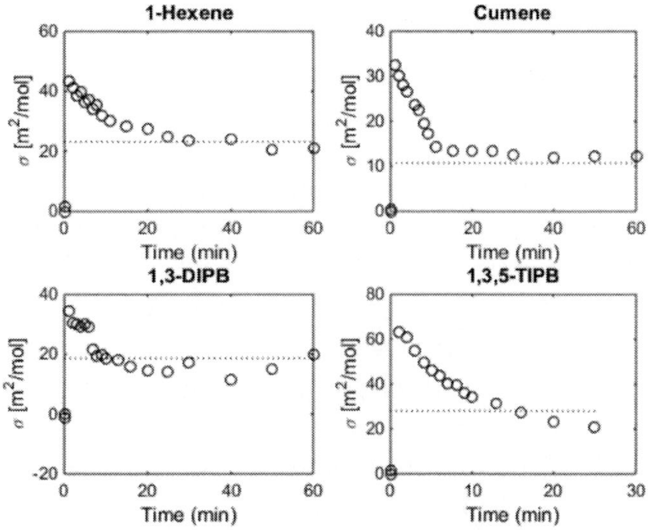

Experimental results for absorbanc at HeNe wavelenght (Dotted line corresponds to theoretical value)

Future work

- Measurement of synthetic mixtures representing the catalytic cracking reaction
- In situ measurement of the reaction progress

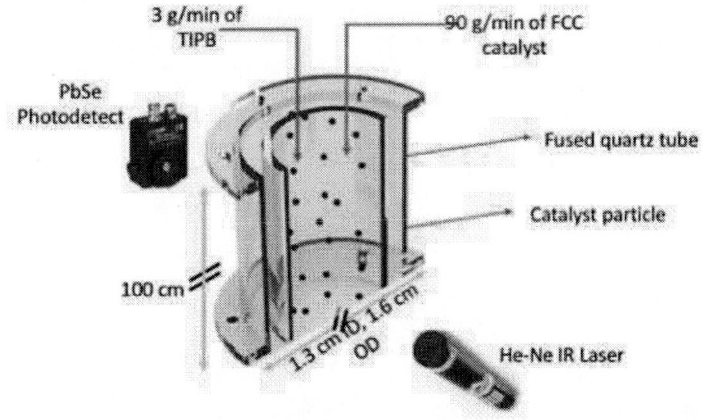

Conclusions

- A group contribution method for the prediction of absorption spectra of paraffins, olefins, naphthenes and aromatics in the region 3200-2800 cm^{-1} was proposed.
- A good prediction for the paraffins, olefins, naphthenes and aromatics by using the GCM was obtained.
- Validation of GCM by gasoline spectra prediction was presented obtaining $R^2 > 0.97$ and RER>15
- Application of GCM for PIOA and RON prediction was possible
- The linear behaviour of absorption coefficient and C-H number was used for the conversion prediction of catalytic cracking reaction
- An experimental setup for the use of this methodology was presented

Acknowledgements

The authors would like to acknowledge

- The financial support of the Colombian Science Foundation (COLCIENCIAS), the Colombian State Oil Company (ECOPETROL S.A.) under contract No. 0423 2013
- Emerging Leaders in the Americas Program (ELAP) Canadian Exchange Program.

UNIVERSIDAD NACIONAL DE COLOMBIA SEDE MEDELLÍN
Western UNIVERSITY · CANADA
FRUN Flujos Reactivos de la Universidad Nacional de Colombia

References

S. Al-Khattaf, The role of diusion in alkyl-benzenes catalytic cracking, Applied Catalysis A: General 226 (2002) 139-153.

M.A. Alrefae, Mid-IR Absorption Cross-Section Measurements of Hydrocarbons, King Abdullah University of Science and Technology, 2013.

M.F. Campbell, K.G. Freeman, D.F. Davidson, R.K. Hanson, FTIR measurements of mid-IR absorption spectra of gaseous fatty acid methyl esters at T=25-500 C, J. Quant. Spectrosc. Radiat. Transf. 145 (2014) 57–73.

G. Jiménez-García, R. Aguilar-López, R. Maya-Yescas, The fluidized-bed catalytic cracking unit building its future environment, Fuel 90 (2011) 3531–3541.

Klingbeil, A. E. (2007). MID-IR Laser absorption diagnostics for hydrocarbon vapor sensing in harsh environments. Standford University.

Sharpe, S. W., Johnson, T. J., Sams, R. L., Chu, P. M., Rhoderick, G. C., & Johnson, P. A. (2004). Gas-phase databases for quantitative infrared spectroscopy. Applied Spectroscopy, 58(12), 1452–61. http://doi.org/10.1366/0003702042641281

UNIVERSIDAD NACIONAL DE COLOMBIA SEDE MEDELLÍN
Western UNIVERSITY · CANADA
FRUN Flujos Reactivos de la Universidad Nacional de Colombia

Previous works

- Qualitative or quantitative determination of hydrocarbons
  - Methane (CH_4) (Sepulveda-Jauregui et al. 2012)
  - Ethane (C_2H_6) (Serdioutchenko et al. 2004)
  - Ethylene (C_2H_4) (Bamford et al, 2007)
  - Acetylene (C_2H_2) (Serdioutchenko et al. 2004)
  - Propylene (C_3H_6) (Harward et al. 2004)
  - 1,3-butadiene (C_4H_6) (Harward et al. 2004)
  - Toluene (Porter 2009)
  - n-dodecane (Porter 2009)
  - n-decane (Porter 2009)
- Only a small number of experiments reported the successful evaluation of hydrocarbon with optical techniques
- In the referred literature no study applied laser diagnostics to a mixture such as the one for FCC process

UNIVERSIDAD NACIONAL DE COLOMBIA SEDE MEDELLÍN Western UNIVERSITY · CANADA FRUN Flujos Reactivos de la Universidad Nacional de Colombia

References

A. Sepulveda-Jauregui, K. Martinez-Cruz, A. Strohm, K. M. Walter Anthony, F. Thalasso, A new method for field measurement of dissolved methane in water using infrared tunable diode laser absorption spectroscopy, Limnology and Oceanography: Methods 10 (2012) 560-567.

D. J. Bamford, D. J. Cook, S. J. Sharpe, A. D. Van Pelt, Widely tunable rapid scanning mid-infrared laser spectrometer for industrial gas process stream analysis, Applied Optics 46 (2007) 3958.

C. N. Harward, R. E. Baren, M. E. Parish, Determination of molecular parameters for 1,3-butadiene and propylene using infrared tunable diode laser absorption spectroscopy, Spectrochimica Ada Part A: Molecular and Biomolecular Spectroscopy 60 (2004) 3421-3429.

A. Serdioutchenko, I. Moller, H. Soltwisch, Infrared diode laser absorption Spectroscopy of C_2H_2 and C_2H_6 in capacitively coupled methane RF discharges, Spectrochimica Acta Part A: Molecular and Biomolecular Spectroscopy 60 (2004) 3311-3318.

J. M. Porter, Laser Based Diagnostics for Hydrocarbon Fuel in the Liquid and Vapor Phases, Ph.D. thesis, Stanford University, 2009.

UNIVERSIDAD NACIONAL DE COLOMBIA SEDE MEDELLÍN Western UNIVERSITY · CANADA FRUN Flujos Reactivos de la Universidad Nacional de Colombia

Statistical parameters

Average absolute error

$$\varepsilon = \frac{1}{N}\sum_{i}\left|\sigma_{Pred} - \psi_{Exp}\right|$$

Average signed error or BIAS

$$\varepsilon_{sgn} = \frac{1}{N}\sum_{i}\left(\psi_{Pred} - \psi_{Exp}\right)$$

Average relative absolute error

$$\xi = \frac{1}{N}\sum_{i}\left|\frac{\psi_{Pred} - \psi_{Exp}}{\psi_{Exp}}\right|$$

Average relative signed error

$$\xi_{sgn} = \frac{1}{N}\sum_{i}\left(\frac{\psi_{Pred} - \psi_{Exp}}{\psi_{Exp}}\right)$$

Root mean squared error of prediction

$$RMSEP = \sqrt{\frac{1}{N}\sum_{i}\left(\psi_{Pred} - \psi_{Exp}\right)^2}$$

Standard Error of Prediction

$$SEP = \sqrt{\frac{1}{(N-1)}\sum_{i}\left(\psi_{Pred} - \psi_{Exp} - BIAS\right)^2}$$

Range Error Ratio

$$RER = \frac{Range}{SEP}$$

$RER \geq 4$, the calibration is acceptable for sample screening; $RER \geq 10$, it is acceptable for quality control; and $RER \geq 15$, it is good for quantification (Hayes2015)

UNIVERSIDAD NACIONAL DE COLOMBIA SEDE MEDELLÍN

Western UNIVERSITY · CANADA

FRUN
Flujos Reactivos de la Universidad Nacional de Colombia

GCM validation: Gasolines

The statistic summary for the prediction of absorption coefficients in the case of regular and premium gasolines present $R^2>0.97$ and RER>15

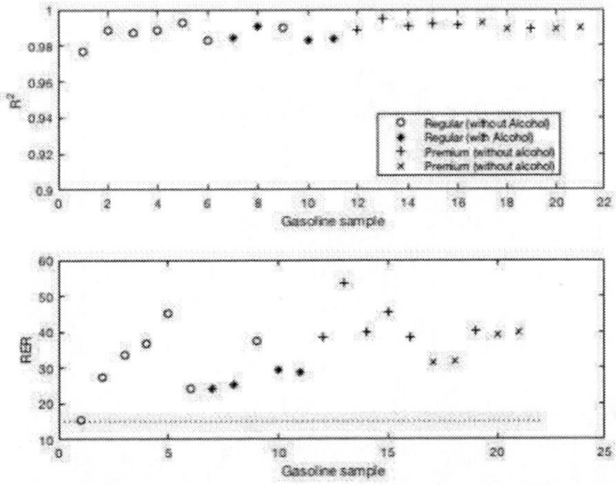

UNIVERSIDAD NACIONAL DE COLOMBIA SEDE MEDELLÍN

FRUN
Flujos Reactivos de la Universidad Nacional de Colombia

504589 Fundamental research on phase equilibrium of extraction desulfurization

Abstract:

Fundamental research on phase equilibrium of extraction desulfurizaiton of FCC gasoline

Yuhao Zhang, Liang Zhao, Jinsen Gao, Chunming Xu

State Key Laboratory of Heavy Oil Processing, China University of Petroleum (Beijing), Beijing, China, 102249

Due to the increasing awareness of protecting environment, the great demand for ultralow sulfur contents in motor gasoline has become increasingly significant. Nowadays, many technologies have the ability to produce clean gasoline to meet the requirement of sulfur content (<10 ppm). Approaches of desulfurization could be divided into hydrodesulfurizaiton and non-hydrodesulfurization. Wherein, the hydrodesulfurization processes were well developed and widely applied in the whole world, such as ISAL, Octgain, Prim-G^+, SCAN Fining, and CDTech [1-5] processes. Part of non-hydrodesulfurization processes were also gathered much attention, such as adsorptive desulfurization, extraction desulfurization, oxidative desulfurization, and biological desulfurization[6-12]. Wherein, extraction desulfurization process was well developed because advantages of variety of solvents, high selectivity, low operating price, low energy consumption, eco-friendly, et al. Commonly, it was divided into solvent extraction[7,13] and extractive distillation[14-17] depending on the mechanism, such as Shen reported a gasoline desulfurization using solvent extraction[7] and GTC-BTX Plus technology[14] by extractive distillation. In spite of the theory of these two methods were different, the interaction of phase equilibrium between solvents and key components of FCC gasoline still plays a significant role to impact the effect of separation which could be express through liquid-liquid equilibrium or vapor-liquid equilibrium experiment.

In this research, the composition of different distillation fractions of FCC gasoline were analyzed and the feedstock of solvent extraction and extractive distillation were selected carefully. And then, particular investigated the properties of different kinds of single solvents using liquid-liquid equilibrium and vapor-liquid equilibrium experiment to found the fundamental of extraction desulfurization of FCC gasoline. Furthermore, an optimal compound solvent were selected using phase equilibrium experiment.

In conclusion, the effect of separation performance of different solvents were studied in several ways. Sulfolane shown a great interaction of aromatics and sulfides. Furthermore, a best compound solvent was determined which the sulfur separation rate was 75 % and has the ability to separate C6~C8 olefins into the raffinate containing a small quantity of aromatics. The results were shown in Figure 1 to Figure 3.

Figure 1 Selectivity of olefins and aromatics using different single solvents

k_A shows the ability to dissolve aromatics; θ shows the ability of selectivity between aromatics and olefins

Figure 2 Sulfur separation rate and raffinate yield using compound solvents

Figure 3 Olefin and aromatics distribution in raffinate using optimal compound solvent

Acknowledgement

This research is supported by National Natural Science Foundation of China (Grants 21336011, 21476260 and 21236009). And also supported by the Science Foundation of China University of Petroleum-Beijing (Grants 2462015YQ0311).

References

[1] Martinez, N.P. Meeting gasoline pool sulfur and octane targets with ISAL process. *NPRA Annual Meeting*, **2000**, San Antonio.

[2] Shih, S. S., Owens, P. J., Palit, S. Mobil's Octgain Process: FCC gasoline desulfurization reaches a new performance level. *NPRA Annual meeting*. **1999**, San Antonio.

[3] Baco, F., Debuisschert, Q. Prime-G^+ process, desulfurization of FCC gasoline with minimized octane loss. *Fifth International Conference on Refinery Processing, AIChE 2002 Spring National Meeting*. **2002**, New Orleans.

[4] Kaufmann, T., Kaldor, A., Stuntz G. Catalysis science and technology for cleaner transportation fuels. *Catalysis Today*. **2000**, 62 (1), 77-90.

[5] Babich, I.V., Moulijn, J.A.. Science and technology of novel processes for deep desulfurization of oil refinery streams: a review. *Fuel*. **2003**, 82 (6), 607-631.

[6] Khare et al. Transport desulfurization process utilizing a sulfur sorbent that is both fluidizable and circulatable and a method of making such sulfur sorbent. **2002**, US 5914292.

[7] Haitao Shen, Benxian Shen, Hao Ling. Desulfurization of fluid catalytic cracking gasoline by extractive distillation coupled with hydrodesulfurization of heavy fraction. *Energy & Fuels*. **2013**, 27, 5153-5160.

[8] Ke Ming, Zhou Aiguo, Zhao Zhensheng, Jiang Qingzhe, Song Zhaozheng. Alkylation of thiophenic sulfur technology in the production of clean fuel. *Chemical Industry and Engineering Progress*. **2006**, 25 (4), 357-361.

[9] Qiu Shi, Wu Liguang, Shi Guozhong, et al. Preparation and pervaporation property of chitosan membrane with functionalized multiwalled carbon nanotubes. *Ind Eng Chem Res*. **2010**, 49 (22), 11667-11675.

[10] Waserscheid P, Keim W. Ionic liquids new solutions for transition metal catalysis. *Angew Chem*. **2000**, 39 (21), 3772-3789.

[11] Xu Hang, Zhang Dandan, WU Fengmin, Cao Renqiang. Deep oxidative desulfurization of fuels based on [C(4)mimCl] CoCl2 ionic liquid oxone solutions at room temperature. *Fuel*. **2017**, 208, 508-513.

[12] Zhang Liang, Lin Xiaojuan, Wang Jinting, Jiang Feng, Wei Li, Chen Guanghao, Hao Xiaodi. Effects of lead and mercury on sulfate-reducing bacterial activity in a biological process for fuel gas desulfurization wastewater treatment. *Scientific Reports*. **2016**, 6, 1-10.

[C_8mim][BF_4] ionic liquid. *Separations*. **2007**, 53, 3108-3115.

[14] GTC Technology Corporation. Process of removing sulfur compounds from gasoline. US 6551502, **2003**.

[15] Shen H, Mei Z, Shen B, Ling H. The desulfurization of fluid catalytic cracking gasoline by extractive distillation. *Energy Sources Part A Recovery Utilization & Environmental Effects*. **2012**, 34 (2), 187-196.

[16] Tang Xiaodong, Wen Xiaohong, Wang Rui, Nie Tongyuan. Study on desulfurization of FCC light gasoline by extractive distillation. *Petroleum Processing and Petrochemicals*. **2013**, 44 (12), 77-80.

[17] Xiaochuan Jiang; Yi Nie; Chunxi Li; Zihao Wang Imidazolium-based alkylphosphate ionic liquids-A potential solvent for extractive desulfurization of fuel. *Fuel*. **2008**, 87, 79-84.

2017 AICHE Annual Meeting

The Effect of New Nanocomposite on Fluidity of Waxy Crude Oil in Low Temperature

China University of Petroleum – Beijing (CUPB)

***Huirong Huang**, Wei Wang*, Yanfen Ding, Zeheng Peng, Jing Gong**

w.wang@cup.edu.cn

ydgj@cup.edu.cn

CHINA UNIVERSITY of PETROLEUM, BEIJING
厚积薄发 开物成务

Background

- ☐ **Special problems:**

  Waxy crude oil, liquidity deterioration, pipeline blockage

- ☐ **Multi-phase mixed transport:**

  Oil, Gas, Water three-phase flow

- ☐ **Urgent need:**

  Improve the safety of marine pipelines

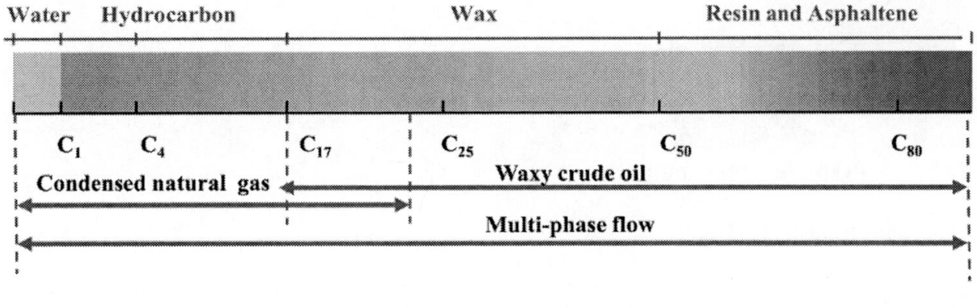

How to solve this problem with efficient ways ?

Critical Technology

- physical solution

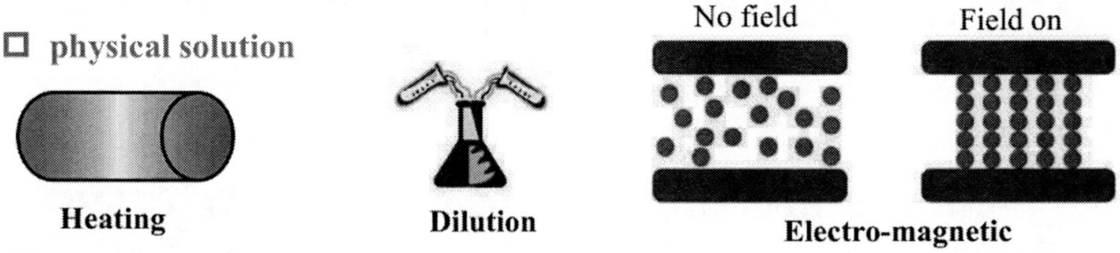

- chemical additives

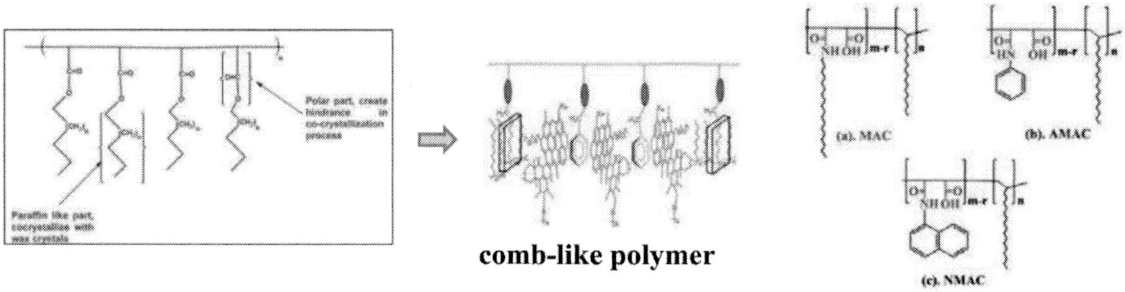

CHINA UNIVERSITY of PETROLEUM, BEIJING
厚积薄发 开物成务

How to solve this problem with efficient ways ?

Critical Technology

- chemical additives

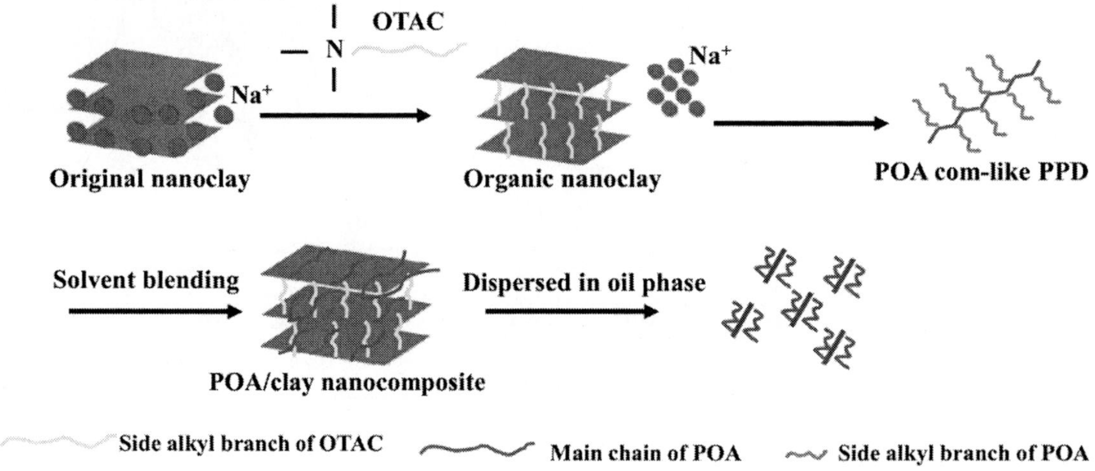

CHINA UNIVERSITY of PETROLEUM, BEIJING

厚积薄发 开物成务

Critical Technology

□ Interaction Mechanism

Crystal nucleation, Co-crystallization, Adsorption, Solubilization

Microscopic

Wax crystals

Mesoscopic

Macroscopic

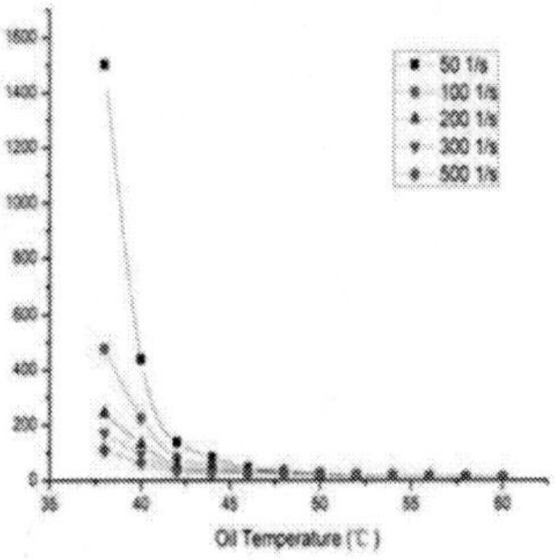

CHINA UNIVERSITY of PETROLEUM, BEIJING

厚积薄发 开物成务

Materials and Experimental Methods

□ Materials

a. **Mixed wax, D80, ethylene/vinyl acetate copolymers (EVA), Montmorillonite**

b. **The density of D80 is 0.795 kg/m³ (15℃)**

c. **The viscosity is 2.09mm²/s (25℃)**

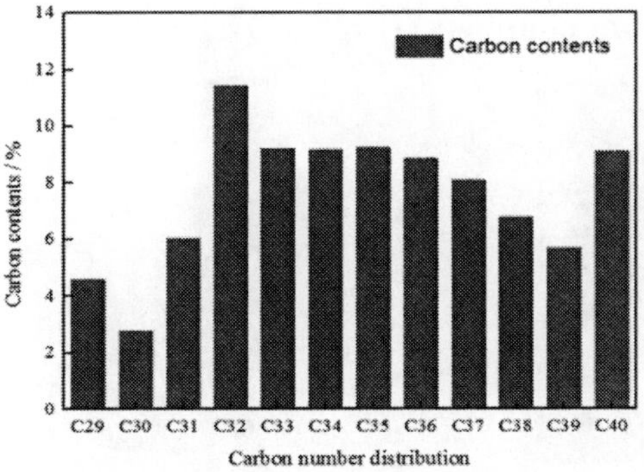

Carbon contents and distribution of mix wax

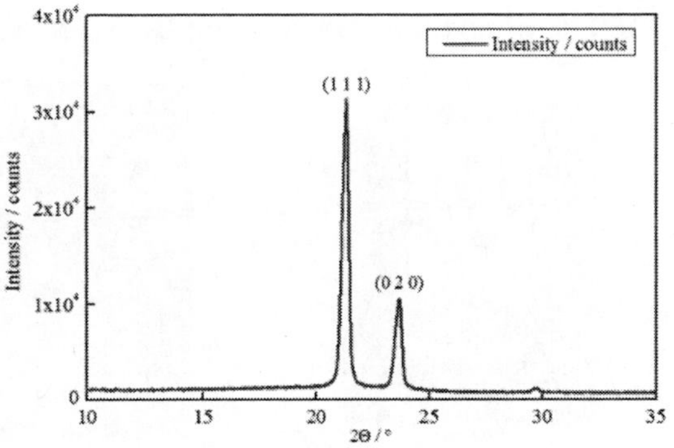

The XRD curve of mixed wax at room temperature

Materials and Experimental Methods

□ Materials

a. **Characterization of EVA and NPPD-2(composed of Montmorillonite and EVA)**

b. **The mass flow rate (MFR) of EVA and NPPD-2 is 20.5g/10min and 2.0g /10min**

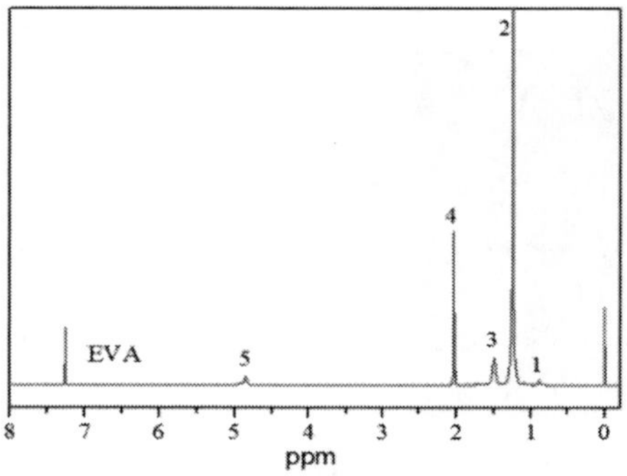

^{1}H NMR spectra of and EVA

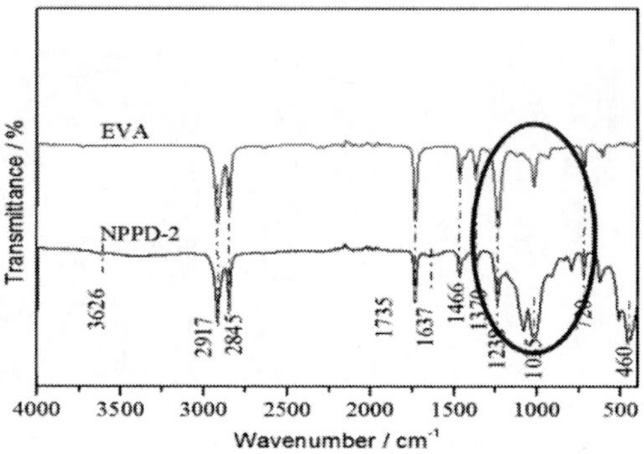

FTIR spectra of EVA and NPPD-2

CHINA UNIVERSITY of PETROLEUM, BEIJING
厚积薄发 开物成务

Materials and Experimental Methods

□ Methods

a. **The rheometer (Gel point, Yield stress, Viscosity) (Figure 1)**

b. **The X-ray diffraction (Lattice parameters) (Figure 2)**

c. **The polarizing microscope BX51(crystal morphology) (Figure 3)**

Figure 1

Figure 2

Figure 3

CHINA UNIVERSITY of PETROLEUM, BEIJING

厚积薄发 开物成务

The Effect of PPDs on Viscosity

PPD	shear rate/s^{-1}	Viscosity of model oil/ Pa.s	Doped oil 50ppm / Pa.s	Doped oil 100ppm / Pa.s	Doped oil 150ppm / Pa.s	Doped oil 200ppm / Pa.s
EVA	10.200	37.750	0.961	0.952	0.423	0.346
	50.000	1.888	0.082	0.073	0.060	0.056
	100.000	0.955	0.047	0.038	0.035	0.032
	150.000	0.735	0.038	0.029	0.029	0.026
	200.000	0.609	0.033	0.024	0.025	0.022
NPPD-2	10.200	37.750	0.349	0.418	0.299	0.290
	50.000	1.888	0.063	0.055	0.051	0.042
	100.000	0.955	0.044	0.039	0.034	0.025
	150.000	0.735	0.038	0.035	0.029	0.020
	200.000	0.609	0.034	0.032	0.026	0.017

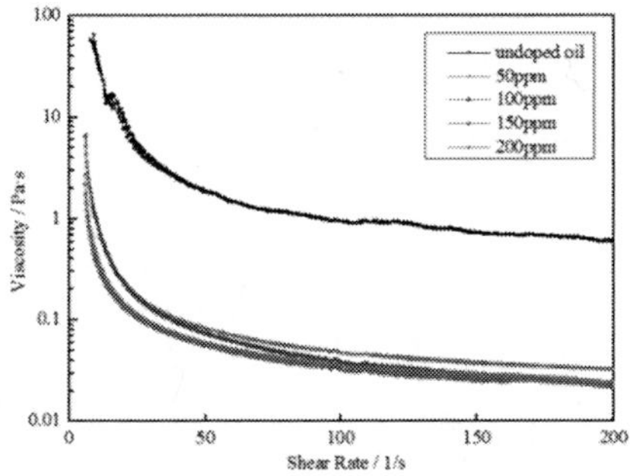

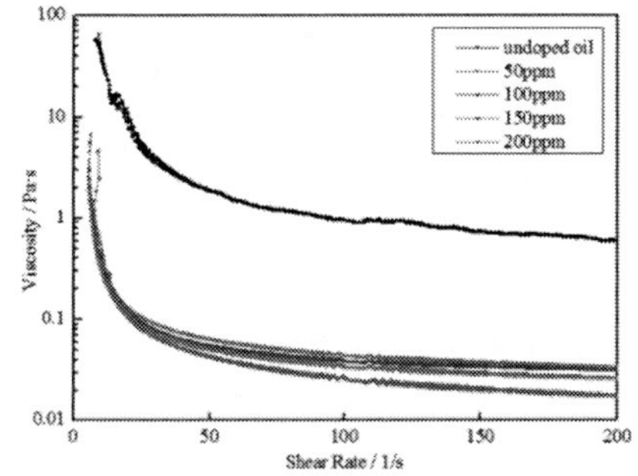

CHINA UNIVERSITY of PETROLEUM, BEIJING

厚积薄发 开物成务

The Effect of PPDs on Gelation

60℃ for 30 min

60℃ to 10 ℃(0.5℃/min), 0.01Pa

The gelation point represents the strength of the waxy network structure

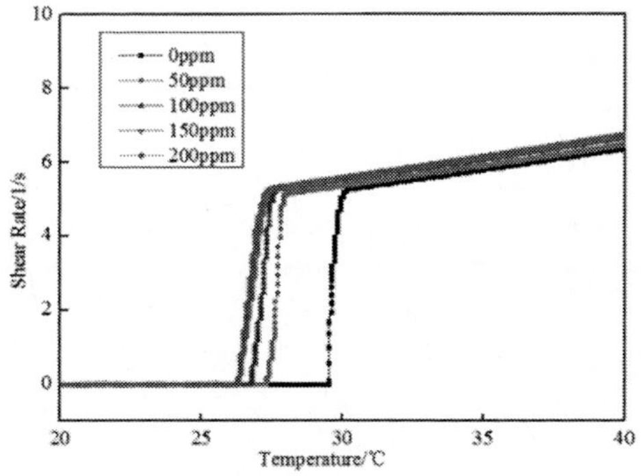

The gelation curves of undoped/doped oil with EVA

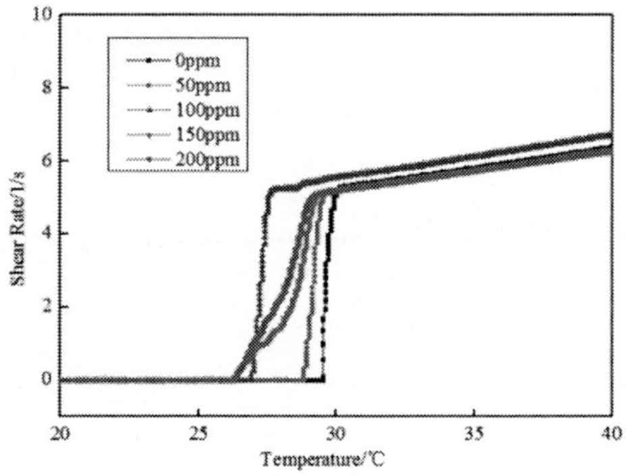

The gelation curves of undoped/doped oil with NPPD-2

CHINA UNIVERSITY of PETROLEUM, BEIJING

厚积薄发 开物成务

The Effect of PPDs on Yield Stress

60℃ for 30 min → 60℃ to 10 ℃(0.5℃/min) → 0 to 600Pa at the rate of 5Pa / min

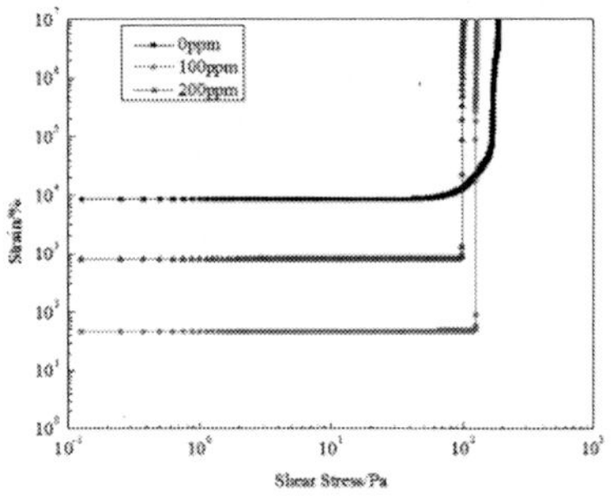

Yield curves（EVA）

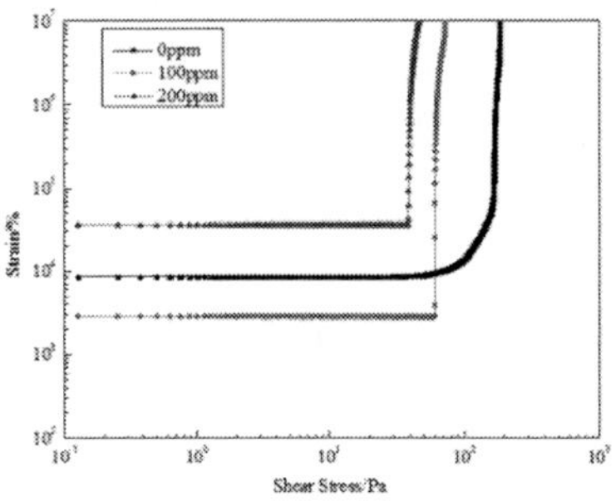

Yield curves（NPPD-2）

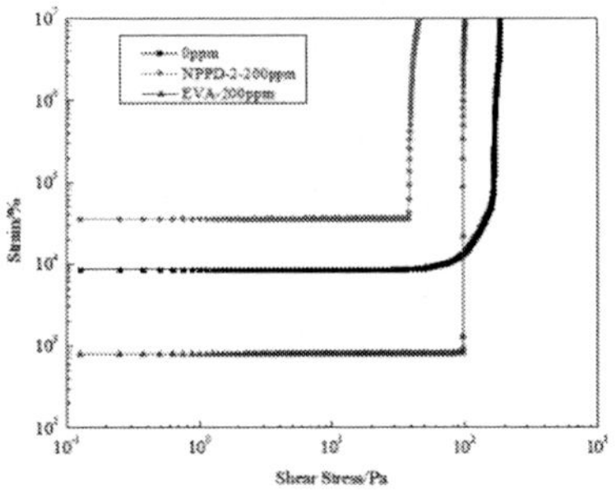

Yield curves（200ppm）

CHINA UNIVERSITY of PETROLEUM, BEIJING

厚积薄发 开物成务

The Effect of PPDs on WAT

$$r_c = \frac{2\gamma T^*}{\Delta H_f \Delta T}$$

γ is the interfacial tension, T^* is the solid-liquid equilibrium temperature expressed in kelvins, $\Delta T = T^* - T$ is the super-cooling, ΔH_f is the latent heat of fusion, and r_c is the radius of a critical nucleus.

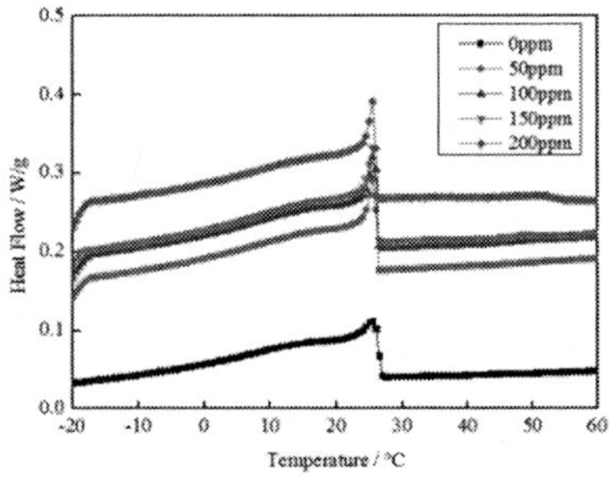

The WAT curves with EVA

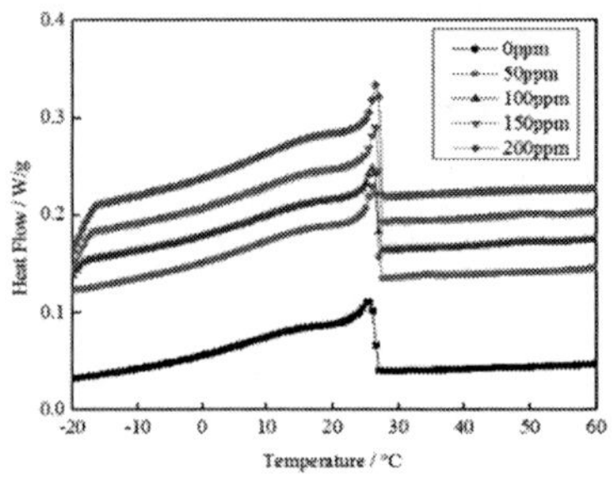

The WAT curves with NPPD-2

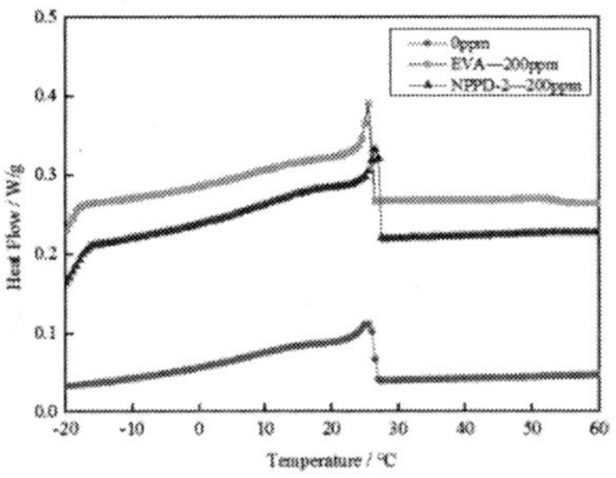

The WAT curves with at 200ppm

CHINA UNIVERSITY of PETROLEUM, BEIJING

厚积薄发 开物成务

The X-ray Diffraction

$$2d_{hkl}\sin\theta_{hkl}=\lambda \qquad c=l*d_{00l} \qquad a=h*d_{h00} \qquad \frac{1}{d_{hkl}^2}=\frac{h^2}{a^2}+\frac{k^2}{b^2}+\frac{l^2}{c^2}$$

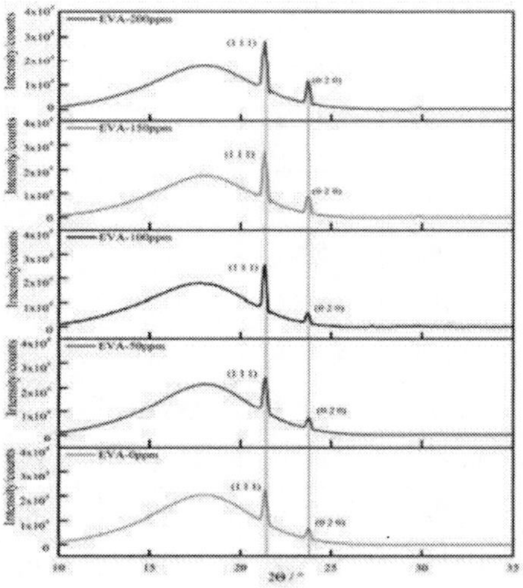

Doped with EVA

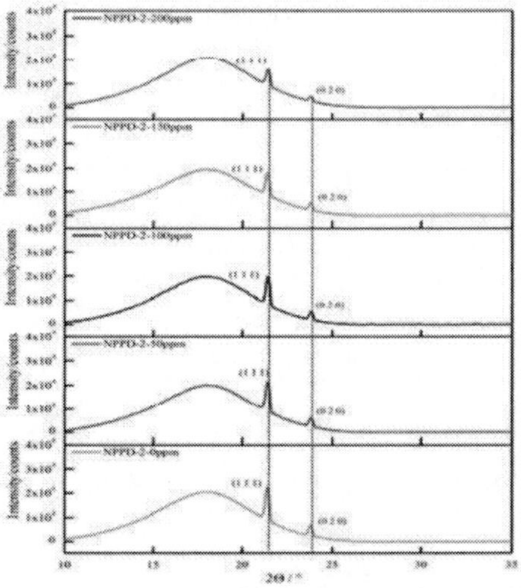

Doped with NPPD-2

CHINA UNIVERSITY of PETROLEUM, BEIJING

厚积薄发 开物成务

(2)

The X-ray Diffraction

	PPD concentration (ppm)	Peak (0 0 6)		Peak (110)		Peak (200)	
		2θ(°)	Intensity (counts)	2θ(°)	Intensity (counts)	2θ(°)	Intensity (counts)
undoped model oil	**0**	**6.166**	**485**	**21.371**	**16919**	**23.748**	**5633**
EVA	**50**	**6.408**	**774**	**21.370**	**18325**	**23.748**	**5916**
	100	**6.446**	**857**	**21.370**	**22045**	**23.705**	**5287**
	150	**6.477**	**901**	**21.332**	**22004**	**23.708**	**8154**
	200	**6.486**	**879**	**21.330**	**23089**	**23.708**	**10322**
NPPD	**50**	**6.489**	**1312**	**21.373**	**16114**	**23.784**	**5011**
	100	**6.489**	**1367**	**21.373**	**14916**	**23.785**	**4698**
	150	**6.487**	**749**	**21.376**	**12125**	**23.784**	**4083**
	200	**6.485**	**814**	**21.373**	**5885**	**23.784**	**3216**

(2)

The X-ray Diffraction

	PPD concentration(ppm)	d (006) (A)	d (110) (A)	d (200) (A)	a (A)	b (A)	c (A)
undoped model oil	0	14.3218	4.1543	3.7436	7.4872	4.9935	85.9308
EVA	50	13.7817	4.1545	3.7435	7.4870	4.9939	82.6902
	100	13.7016	4.1546	3.7503	7.5006	4.9900	82.2096
	150	13.6352	4.1618	3.7498	7.4996	5.0028	81.8112
	200	13.6153	4.1621	3.7498	7.4996	5.0033	81.6918
NPPD	50	13.6093	4.1539	3.7379	7.4758	4.9962	81.6558
	100	13.6100	4.1539	3.7379	7.4758	4.9962	81.6600
	150	13.6132	4.1534	3.7380	7.4760	4.9952	81.6792
	200	13.6193	4.1538	3.7381	7.4762	4.9959	81.7158

CHINA UNIVERSITY of PETROLEUM, BEIJING
厚积薄发 开物成务

The Observation of Polarized Microscope

- Cooling the sample from 60℃ to 25℃ with natural and recording the morphology of wax crystals at room temperature.

The picture of doped oil with NPPD-2

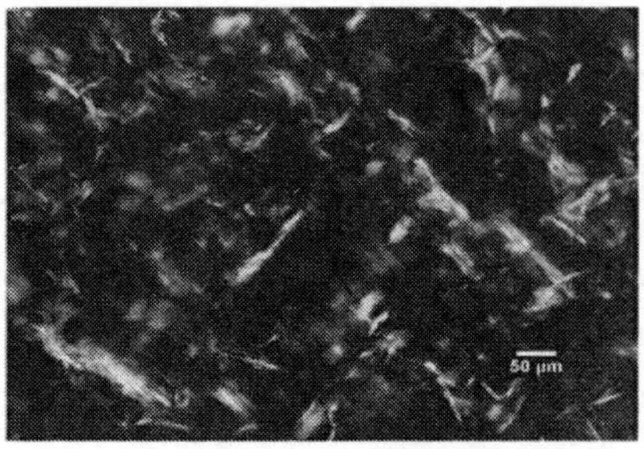

The picture of undoped oil

The picture of doped oil with EVA

CHINA UNIVERSITY of PETROLEUM, BEIJING

厚积薄发 开物成务

Interaction Mechanism

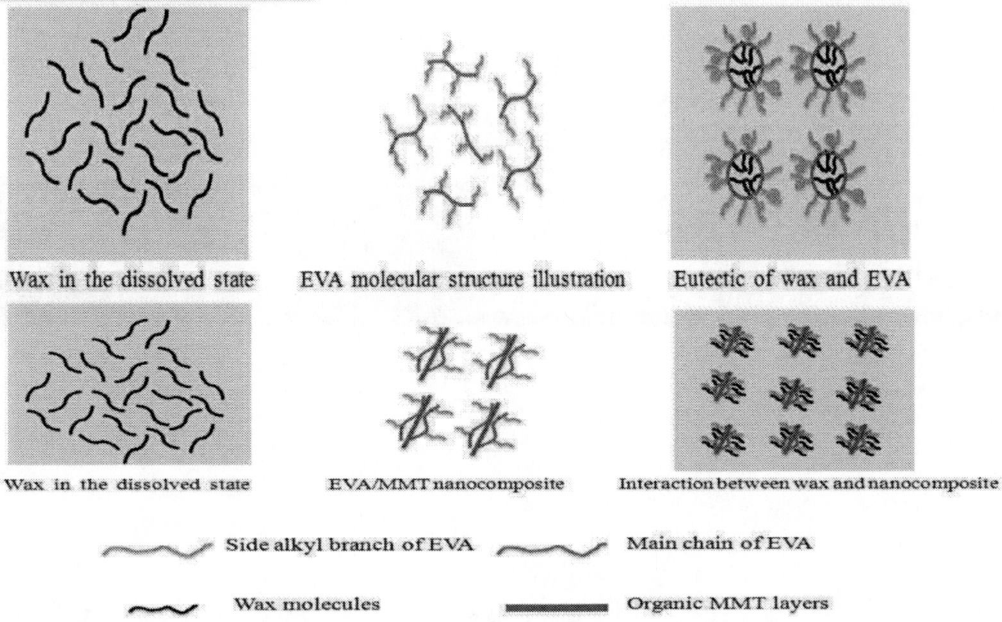

CHINA UNIVERSITY of PETROLEUM, BEIJING

厚积薄发 开物成务

Conclusion

- **The rheological parameters also demonstrate that NPPD-2 can reduce the gel point, and depress yield stress and viscosity of waxy crude oil more significantly than EVA.**
- **The WAT slightly decreases with the increase of EVA concentration, while the trend for NPPD is opposite.**
- **The X-ary diffraction displays that the intensity of the diffraction peak decreasing with the additive of NPPD-2 increasing， which indicates fewer wax crystals exit in doped oil.**
- **The interplanar spacing between the wax crystals decreases, which remarkably improves the low temperature fluidity of waxy crude oil.**
- **The morphology of the wax crystals observed by polarizing microscope shows fewer wax crystals and dispersed finely doped with NPPD-2.**

CHINA UNIVERSITY
of
PETROLEUM, BEIJING

Thanks for your attention!

488981 Co-Processing of Liquid Phase Pyrolysis Oil and Refinery Intermediates in a Continuous Hydrodeoxygenation Reactor

Abstract:

Introduction

Biomass pyrolysis is a suitable pathway for the production of 2nd generation biofuels.[1] During pyrolysis, one of the major products is pyrolysis oil. Due to its high water content, high corrosivity and other negative properties, pyrolysis oil needs intensive upgrading prior usage as fuel for combustion engines.[2] To achieve those standards, an upgrading step is necessary. One way to do so is hydrodeoxygenation (HDO).

BiomassPyrolysisRefinery[3]

An overall concept for the production of 2nd generation biofuels based on liquid phase pyrolysis (LPP) is the BiomassPyrolysisRefinery concept. In the first step, lignocellulosic biomass is pyrolysed in vacuum gas oil to produce an unpolar oil, pyrolysis oil and pyrolysis char through the bioCRACK process[4]. In the second step, the intermediate products are upgraded[5,6]. The bioCRACK process was successfully operated in pilot scale (100 kg/h lignocellulosic feed) over two years. HDO of LPP oil has already been performed continuously in the temperature range of 350°C to 400°C and at liquid hourly space velocities of 0.5 h^{-1} to 3 h^{-1}. To fulfill the requirements of a refinery integration, co-processing of LPP oil and refinery intermediates has to be investigated.

HDO of LPP oil and refinery intermediates

To perform co-processing in lab scale, a paraffin mix and LPP oil were used. HDO was carried out in a plug flow reactor of Parr Instrument Company on an inline-sulfided metal-oxide catalyst. Operating parameters were 121.1 bar hydrogen pressure and 400°C. Overall LHSV was 1 h^{-1}, with 75 wt.% paraffine mix and 25 wt.% LPP oil in the feed.

In Figure 1 the boiling ranges of the HDO products are compared to the boiling range of the paraffin feed. As the paraffin feed is already highly hydrogenated, only cracking reactions occurred. It can be seen, that 25 wt.% LPP oil didn't have a heavy impact on the product quality.

Figure 1: Boiling ranges of HDO products compared to the paraffin feed

1 A. Demirbas, *Appl. Energy*, 2011, **88**, 17–28.

2 Q. Zhang, J. Chang, T. Wang and Y. Xu, *Energy Convers. Manag.*, 2007, **48**, 87–92.

3 N. Schwaiger, R. Feiner, H. Pucher and L. Ellmaier, 2015, 1–8.

4 J. Ritzberger, P. Pucher and N. Schwaiger, 2014, **39**, 1189–1194.

5 H. Pucher, N. Schwaiger, R. Feiner, P. Pucher, L. Ellmaier and M. Siebenhofer, *Energy Res.*, 2014, **31**, 3205.

6 R. Feiner, N. Schwaiger, H. Pucher and L. Ellmaier, 2013, **43**, 1–6.

494343 Static and Dynamic Catalytic Adsorptive Desulfurization (CADS) of Real Diesel Using Low-Cost TiO_2/Fumed Silica

Abstract:

The catalytic adsorptive desulfurization (CADS) has been developed as a promising approach for ultra-deep desulfurization of diesel with low energy comsumption (Ren, X.; Miao, G.; Xiao, Z.; Ye, F.; Li, Z.; Wang, H.H.; Xiao, J. Catalytic Adsorptive Desulfurization Using TiO_2/SBA-15 under Mild Conditions. Fuel, *2016*, 174, 118-125.). In this work, low-cost fumed silica were screened as the supporting substrate for TiO_2 for effective CADS of real diesel (~35 ppm-S). Effect of TiO_2 loading and CADS temperature were optimized in batch reactor. The dynamic breakthrough performance was evaluated in fixed-bed reactor. The sorbent was regenerated by solvent wash followed by air oxidative treatment, and its regenerability was assessed in multiple adsorption-regeneration cycles. Results suggested that TiO_2/fumed silica (300-400 mesh) with higher surface area and larger amount of Si-OH groups showed higher desulfurization capacity. The TiO_2 loading was optimized to 13% with the maximized utilization of Ti catalytic sites and Si-OH adsorption sites on TiO_2/fumed silica for CADS. While the CADS temperature was optimized to 35°C with the balanced catalytic activity for the oxidation of dibenzothiophenes (and their derivatives, 4,6-DMDBT mainly) and minimized desorption of yielded dibenzothiophene sulfones. With a suitable O/S ratio of 5, and overall aspect ratio of 10, the TiO_2/fumed silica showed the maximal desulfurization capacity of 1.44 mg-S/g-A of real diesel at the breakthrough concentration of 5 ppm-S. Considering the effect of the dimension of fixed bed, the dynamic desulfurization performance with varied aspect ratio followed the order of H/R=10 > H/R=22 ~ H/R=5. The S-breakthrough concentration decreased with increased O/S ratio. The superior regenerability of TiO_2/fumed silica was suggested by no loss on desulfurization capacity was observed after six adsorption-regeneration recycles. The CADS using the low-cost TiO_2/fumed silica adsorbent with superior desulfurization capacity and remarkable stability provide an effective and economic approach for ultra-deep desulfurization of real diesel with low energy consumption.

Keyword: catalytic adsorptive desulfurization, TiO_2/fumed silica, diesel, dynamic adsorption.

494352 [O]-Induced Reactive Adsorptive Desulfurization of Liquid Fuel over Ag_Xo@SBA-15 Under Ambient Conditions

Abstract:

Development of clean energy production processes with low energy consumption is a central strategy for a sustainable and environmentally friendly planet. Here we report a novel [O]-induced reactive adsorptive desulfurization (RADS) approach for ultra-clean fuel production using multifunctional Ag_XO@SBA-15 adsorbent with and without the addition of air under ambient conditions. The adsorption capacity of Ag_XO@SBA-15 by RADS reached 4.4 times of that by conventional ADS at low feed sulfur concentration of 10 ppm-S; its adsorption selectivity was enhanced 2.4 orders of magnitude. Over 90% of sulfur was removed within 10 min (corresponds to the space volocity of 6 h^{-1}). The RADS mechanism was illustrated as a simultaneous oxidation (to sulfones) and selective adsorption of sulfones, where the nano-sized silver species detected at different valence states (AgO, Ag_2O and Ag) and the feed air played the versatile enabling roles as the oxidation catalyst and oxidant, respectively, and the surface silanol groups were suggested as the preferential adsoption sites.

Keywords: desulfurization, adsorption, fuel, silver, air

490671 Energy Saving from Process Design - a Service Oriented Architecture (SOA) Methodology

Abstract:

Energy saving is important and urgently needed in process manufacturing, which is energy intensive and takes ~75% of the total industrial energy use in US. To do the work, chemical engineers should take the lead role and process development & design has the greatest potential to achieve the goal.

However, it is very challenge to develop an optimal process design due to the complexity of the process. In general, there are four major challenges. First, the engineers or designers should have systematic knowledge and skills in process design. Second, we need to have an in-depth understanding of the whole studied process. Third, a successful process design implementation needs collaboration among designers, engineers, manufacturers, and operators. Fourth, a typical process design is a long-term development with dedicated specialists, which many companies especially small and medium sized companies could not afford.

To address the challenges, the author proposes a so called service oriented architecture (SOA) methodology. It is a framework to define the design scope and facilitate the work. Four case studies are presented which are based on the authorâ€™s industrial energy saving consulting work. It shows how the SOA methodology is applied into the real world and achieve accountable energy saving.

496499 Mathematical Modeling and Optimization of Commercial-Scale Catalytic Two-Stage, Ebullated Bed, Direct Coal Liquefaction Reactors

Abstract:

Due to the insecurity in crude oil supply and uncertainty in its prices, finding alternative fuel source is critical. Direct coal liquefaction (DCL) is one of the technically feasible solutions to convert coal to transportation fuels. However, this technology has not been widely commercialized due to the high capital investment, uncertainty in economic feasibility and several operating challenges. In the DCL technology, catalytic two-stage liquefaction (CTSL) unit with a configuration similar to the hydroprocessing reactor systems uses two ebullated bed reactors (EBR), in which coal reacts with the recycled solvent and hydrogen to produce syncrude. Since the conversion, and yield obtained in the CTSL reactors are probably the most important factor in the profitability of the DCL plants, the CTSL process must be optimized. A rigorous mathematical model of the CSTL unit is required for optimization. However, most of the models of the DCL reactors have been developed by assuming ideal reactor performance, with very simple reaction kinetics or by using simplified yield correlations developed using experimental observations.

In this work, a first-principles model of the CTSL reactors have been developed and validated by comparing with the commercial operating data. The ebullated-bed DCL reactors are simulated as axially dispersed reactor (ADR) with recycle stream based on rigorous reaction kinetics, hydrodynamics and mass and heat balances. Other standard unit operations, including heaters, pumps and flash drums, are modeled using mass and energy conservations. The capital cost of the reactors is estimated based on the pilot plant data, while that of other equipment items is estimated using Aspen Process Economic Analyzer (APEA). The CTSL unit is optimally designed to minimize an annuitized operating cost that includes operating and capital costs along with the revenue earned by selling the products. To summarize, the presentation will include the following: (1) two configurations of the CTSL unit- with and without heat integration, (2) rigorous modeling of the EBRs, (3) systems level models of the CTSL unit, (4) capital and operating cost estimates, (5) sensitivity studies of the key operating and design parameters, and (6) process optimization.

497031 Optimal Design of Pipeline System for Flare Minimization in Multiple Chemical Plants

Abstract:

Flare minimization (FM) in large-scale industry takes increasingly engineersâ€™ concerns nowadays. It is a win-win practice for not only environmental sustainability but also industrial profitability. A lot of researches have already focused on FM problems with many FM strategies for a single plant. However, there are few FM works focusing on multiple plantsâ€™ FM issues. Material exchanges (ME) among plants provide a new degree of freedom for material savings and flare reductions. For example, a start-up plant can borrow materials from another normal-working plant to help the plant compressor start-up, or a normal-working plant can receive off-spec materials from another plant undergoing the shutdown operation to help recover potential products. Such ME operations could be designed and scheduled based on plantsâ€™ operational flexibilities. Certainly, different plants may have different operating statuses when interplant ME are considered. The design of a pipeline system that satisfies different plant operating statuses is a must.

This work developed a general methodology for the optimal pipeline system design for FMs via interplant ME, which can be used for FMs under planed turnaround and emergency turnaround operations in multiple chemical plants. The operational scheduling together with the pipeline system design fully take interplant ME advantages while all possible plant operating scenarios have been considered. The methodology will provide a conceptual pipeline system design with the minimum capital cost, containing the information of detailed pipeline distributions and connections; control valve placements, and valve switching strategies under different interplant ME strategies for FMs. The efficacy of the developed method will be demonstrated by two case studies.

500436 Development of a Natural Gas to Industrial Chemicals Bioprocess Platform

Abstract:

The low cost and abundant supply of natural gas have been motivators for the development of many technologies that convert this feedstock into compounds of greater value. Intrexon has developed the first natural gas-to-liquids bioconversion platform powered by methantrophic fermentation, which applies synthetic biology to program bacteria to produce higher value materials of interest. Our innovative genetic toolbox allows us to engineer the biology of the methanotroph to more efficiently convert natural gas into a wide range of valuable chemicals and fuels. These applications include isobutanol, farnesene, 1,4-butanediol, 2,3-butanediol, isoprene and isobutyraldehyde – the building blocks for industrial and consumer products like fuels, synthetic rubber, acrylics, resins and spandex.

Using a methanotrophic bioprocess to produce fuels and chemicals has several advantages beyond the low cost and minimal processing of feedstock. Natural gas bioconversion offers scaling opportunities beyond traditional gas-to-liquid (GTL) facilities, as well as biomass feedstock facilities. Theoretical yields using the methanotroph and methane exceed that of yeast/sugar approaches by a wide margin. Biocatalytic processes that achieve high yields under milder processing conditions allow for reduced capital and operation expenditures, offering greater economic viability. The natural gas bioconversion also does not compete with food resources, unlike traditional biofuels that use sugar and corn as feedstock.

One molecule of interest in the methanotroph platform is isobutanol, a fuel for gasoline blending which has several positive attributes. Isobutanol is a clean burning and non-corrosive "drop-in" fuel which can be blended with gasoline and is compatible with the existing petroleum infrastructure. Currently, provisions allow for bio-butanol blending up to 12.5% with gasoline and ASTM specifications exist for sale as fuel. Every 1% of isobutanol blended with gasoline presents an $8 billion market opportunity. Combustion of isobutanol has cleaner emissions compared to gasoline by reducing hydrocarbon, carbon monoxide, nitrous oxide and sulfur dioxide production. Isobutanol also has advantages over first generation biofuels, such as ethanol, because of its higher energy density.

Intrexon Energy Partners was formed as a joint venture to commercialize the production of isobutanol from natural gas using a methanotroph. Gas fermentation processes were developed to evaluate growth of the bacteria and production of isobutanol from the microfermenation scale through the lab and pilot scales. The ability of our modified methanotroph to produce isobutanol has increased exponentially with time as the genetic toolbox was developed and early learnings in strain engineering were optimized into a design-build-test-learn cycle. A state of the art laboratory was custom built to bring research and development from a low throughput and manual process to a high throughput automated process. Increases in automation and throughput have positioned Intrexon as a leader in the genetic engineering of methanotrophs and accelerated the commercialization timelines of other chemical products.

To successfully bring this technology to commercial scale, consideration must be given not only to optimizing the metabolic pathway engineered into the methanotroph, but also to key issues in the fermentor design and downstream processing. Laboratory scale learnings were applied to select technology that was commercially feasible at demonstration scale. Since the substrate is a gas, the design of the fermentor must be optimized for effective mass transfer of the natural gas to the liquid phase for increased bioavailability. Downstream processing must take into consideration purification of the isobutanol and removal of fermentation byproducts and contaminants. Process selection must weigh the efficiency and selectivity of technology options, while at the same time, utilize as much mature technology as possible to minimize risk. The project team has performed numerous design reviews known as Front End Loading (FEL) to develop process criteria and scenarios

Beyond isobutanol, Intrexon's versatile platform strategy has accelerated the development of 2,3-butanediol, 1,4,butanediol and isobutyraldehyde platforms. Production of these intermediate chemicals has expanded the value of the natural gas conversion technology from the fuels market to consumer markets like tires and textiles. The commercial fermentation design optimization occurring in the pilot plant may also be applicable to other chemicals, shifting the focus of future products to downstream development.

495331 Comparison of Conventional and Newly Proposed Solvent Deasphalting Processes: Numerical Simulation Based on Energy and Cost Analysis

Abstract:

Oil consumption in non-OECD countries has been increasing, while the existing reserves of conventional oil keep decreasing. To replace conventional oil, many attempts have been made to use unconventional oil such as oil sand bitumen and extra heavy oil, which are difficult to extract from a reservoir. Unconventional oil requires upgrading processes because its asphaltene compounds containing a lot of sulfur and heavy metals are much higher than those of conventional oil. Among various upgrading processes, solvent deasphalting (SDA) process has the advantage of lower operating cost than other oil conversion processes and high quality deasphalted oil (DAO) can be obtained by selectively rejecting asphaltene in the SDA process. In the SDA process, about 3â€“5 times more solvent than feed is used to extract DAO and the solvent recycling is the most energy consuming part. In this study, we propose a novel process that consumes less energy than the conventional SDA process. In the proposed new process, the temperature of the first separation column is adjusted to efficiently operate a heat exchanger and carbon dioxide is used as a stripping agent to separate the solvent from DAO in the last separation column. To compare energy consumption and cost of the conventional and proposed SDA process, a commercial software was used for numerical simulation. The energy consumption was calculated by a graphical method using composite and grand composite curves. After calculating minimal energy consumption, operating and capital costs were obtained with changing the type of hot and cold utilities. The proposed new SDA process can produce the same quality and quantity of DAO as the conventional SDA process, but its energy consumption and total cost are lower than those of the conventional process.

484549 Evaluation of Aromatics Formation in Thermal Cracking of Triglycerides

Abstract:

Liquid fuels produced by biomass conversion offer an alternative to our reliance on fossil sources. Among the process available in the literature, thermal cracking results on a liquid product (bio-oil) with similar characteristics to petroleum crude oil, especially when performed with triglyceride sources. However, undesired compounds, such as carboxylic acids and benzene might be formed as well. Although aromatics may have positive aspects (e.g. increasing the octane number on light fuels), they are toxic and, particularly benzene has its content restricted by local regulations. Moreover, benzene removal involves complex processes, being wiser to avoid its formation.

The aim of this work was to investigate one of the hypothesis for benzene formation, regarding a relationship between the degree of unsaturation of the biomass with the benzene content in bio-oil. Thermal cracking experiments were performed using triglyceride sources with three different degrees of unsaturation, in the same operational conditions. Commercial soybean oil and blends with 10% and 20% of hydrogenated were constituted as samples to be analysed. Density, acid index and iodine index were measured to evaluate the blending effect. The thermal cracking experiments were performed in triplicate at 525 °C of temperature and feed mass flow of 300 g/h in a continuous screw reactor operating over steady state conditions. The mass balance shows the liquid yield increased and coke formation decreased with for higher blends.

Benzene content decreased by 14% in the bio-oil produced by the biomass with the lowest degree of unsaturation. Also, the reactions for carboxyl group removal were favorable to the saturated compounds, reducing the acidity index in the biomass with the lowest unsaturation content. Notwithstanding, long-chain paraffinic compounds remained in the products, which suggests that saturated compounds does not crack easily.

500032 Tuning the Structure and Electronic Properties of Manganese Nitrides for Ammonia Production from First-Principles Modeling

Abstract:

The conventional Haber-Bosch ammonia synthesis process plays a significant role in the global food chain by feeding approximately 40% of worldâ€™s population, but is also extremely energy intensive, consuming 1-2% of the worldâ€™s energy supply. Ideally, simple technology enabling intermittent operation using renewable energy will be made available through step catalysis by exploiting manganese (Mn) based alloy materials to facilitate both nitrogen activation and ammonia production.

In this talk, a first-principles investigation based on periodic density functional theory (DFT) calculations has been carried out by focusing on the elucidation of molecular-level mechanisms for ammonia formation using a model based on the Mn_4N nitride phase, which can be readily obtained through the nitridation of metallic manganese. It is hypothesized that the reduction will follow a Mars-van Krevelan-type mechanism. Under a reducing environment at 1 atmospheric H_2 pressure, this computational study has shown that it is thermodynamically favorable to form NH utilizing the surface N species. Afterwards, the reaction becomes endothermic for ammonia evolution from NH. The diffusion of N in the nitride crystal lattice can influence the overall kinetics, and has been considered in this study as well. It turns out that both the overall thermodynamics and the N diffusion kinetics are the key for ammonia production and should be considered as the parameters to tune the materials performance. Also in this talk, the simple strategy to dope and alloy manganese with other transition metals to validate the research hypothesis has been explored computationally, and will be presented and discussed.

496330 Rapid and Extended Analysis of Syngas Using Micro GC Fusion

Abstract:

In 2012, the World Bank estimated that 140 billion cubic meters (bcm) of the 3370 bcm of natural gas produced worldwide was flared. Flaring burns surplus gas and is detrimental to the environment, contributing over 350 million tons of CO2 in annual emissions. An alternative to flaring is to convert the associated natural gas into synthesis gas (syngas), which is an intermediate gas used to produce synthetic liquid fuel. Converting syngas to liquid chemicals through the Fischer-Tropsch (FT) process presents an economical and greener alternative to flaring. Syngas is composed predominantly of hydrogen and carbon monoxide, though it may also contain carbon dioxide, nitrogen, methane, ethane, ethylene, or H2S in the high ppm to percent level. After catalytic conversion, determining the syngas composition is a critical step before using the gas in any downstream process. To ensure efficiency, the analysis of syngas must be analyzed quickly and accurately. Building on proven microelectromechanical systems (MEMS) technology, Micro GC Fusion is capable of providing rapid analysis of typical syngas components in a compact instrument. A method equipped with temperature programming provides an area %RSD of less than 1.0% and retention time %RSD of less than 0.05% repeatability. This accurate and repeatable analysis of syngas is performed in less than 50 seconds. Additionally, Micro GC Fusion is capable of performing analyses on more complex samples including C4-C5 olefins, and C1-C12 alkanes. This extended analysis of syngas is performed in less than 90 seconds.

490981 Dual-Alcohol Blending Effects on Gasoline Properties

Abstract:

A pure alcohol cannot provide the distillation temperature range required by spark ignition engines for smooth driveability. The front-end volatility in a distillation curve is of importance for cold start, engine warm-up, evaporative emissions, and vapor lock. Mid-range volatility is vital for warming up, acceleration, and cold weather performance. Finally, tail-end volatility is essential for running when the engine is hot, and it also indicates the propensity for combustion deposits and oil dilution. In addition, the high enthalpy of vaporization of alcohols is a prohibitive factor for good cold-start behavior. Therefore, both light and heavy hydrocarbon fractions of gasoline are needed to supply an ignitable mixture under these challenging conditions. In general, blends of gasoline with a single alcohol also have shortfalls. If lower alcohols (methanol and ethanol) are used, azeotropic behaviour due to the high polarity of these molecules, as well as their low calorific value, high enthalpy of vaporization and low stability would be problematic. If higher alcohols are used, achieving very low volatilities is challenging and can significantly limit the maximum volume that can be blended with gasoline. Thus, in order to increase the portion of bio-alcohols in the gasoline blend, another approach is needed. Using multi-alcohol-gasoline blends is an approach that might achieve a fuel with high portion of biofuel while meeting all standard specifications with no need for major changes in existing infrastructures and engines. Mixing a lower and a higher alcohol (e.g., ethanol and 1-butanol) in gasoline can provide a mixture that has the same vapor pressure as that of gasoline while reducing the evaporative emission on the fuel delivery system. However, other physicochemical and combustion related properties of dual alcoholâ€“ gasoline blends have not determined to fully characterize these fuel blends for a satisfactory performance in existing spark ignition engines. This presentation will describe the development of a database of all normal and cyclic C1-C10 alcohols and a procedure to downselect for a pair of alcohols that can be blended with gasoline in different target volumes from midrange (20 vol. %) to high range (40 vol. %). After selecting the desired pair of alcohols, a range of properties were measured to evaluate the fuel blends: distillation curve, calorific value, vapor pressure, vapor lock index, ignition quality, water tolerance and phase separation, density, and viscosity.

499675 NO Uptake and Desorption on Fe-Modified Pd/ZSM-5: Impact of Pd and Fe Loading

Abstract:

NO Uptake and Desorption on Fe-Modified Pd/ZSM-5:

Impact of Pd and Fe Loading

Kyle Karinshak*, Michael P. Harold*, and Kiran Premchand+

*Dept. of Chemical & Biomolecular Engineering, University of Houston, Houston, Texas

+FCA US LLC Auburn Hills, MI 48236

* corresponding author: M.P. Harold (mpharold@central.uh.edu)

Reducing cold-start emissions, emissions under 200 °C, will play a significant role in meeting future EPA and EU emissions standards. By adsorbing NO and NO_2 at low temperatures and releasing these gases in temperatures regimes where traditional emission control catalysts can operate, Passive NOx Adsorbers (PNAs) augment existing emission control systems and are a promising new field of catalysis. The scant published literature has focused on Pd supported on various materials, including ceria or alumina. In contrast, the patent literature suggests the efficacy of ZSM-5 as a support as well as the addition of Fe to PNA compositions [1-6]. Our preliminary investigations into Palladium-exchanged MFI zeolites show significant NOx uptake capabilities and complex desorption behavior.

Catalyst samples were prepared through a wet impregnation protocol before being deposited onto monolith cores. Palladium(II) nitrate was used as the precursor and dissolved into water equal to the pore volume of the ZSM-5 zeolite (SAR 30). Alumina solution was added to the washcoating solution to improve binding. Four catalysts – 0 wt% Pd, 1wt% Pd, 1 wt% Pd/1 wt% Fe, and 2 wt% Pd, were washcoated onto a 0.25in diameter, 2in length, 600 cspi monolith cores at a fixed loading of 1.5 g/in^3. A bench-scale quartz reactor was utilized to simulate emissions and study uptake and desorption characteristics. DRIFTS (diffuse reflectance infrared Fourier transform spectroscopy) experiments of catalysts samples were performed to examine surface species formation and the mechanism through which NOx uptake occurs on Pd/ZSM catalysts.

Figure 1 shows uptake characteristics of NOx uptake at 50 °C over a five minute duration for the four catalyst. 400 ppm of NO was flowed through the catalyst accompanied by 2% O2 and balance Ar. Similar trends were observed at uptake temperatures of 80 °C and 150 °C with decreasing uptake quantities and breakthrough periods. NO adsorption was accompanied by noticeable NO_2 slip from the catalyst. The addition of Fe to the catalyst composition significantly improved the performance of NOx uptake.

Figure 2 shows the NOx desorption peaks from the same experiments. Over half of desorbed NOx occurred in the form of NO_2. As NO adsorbs as a mono nitrosyl palladium complex, this shows that NO is being converted into NO_2 through a zeolite-surface reaction and likely storing on the Pd-less acid sites of the zeolite and binder. Unlike Murata et al, who reported a single NOx desorption peak during TPD experiments [7], our results reveal multiple desorption peaks. Three distinct peaks are identified in Pd-containing samples whereas only two exist in the Pd-less sample. A preliminary assignment of these peaks include acid sites of the binder, the acid sites of the zeolite, and the Pd nitrosyl complexes.

Figure 3 shows DRIFTS measurements of three samples: 0 wt% Pd, 1 wt% Pd/1 wt% Fe, and 2 wt% Pd. Experiments were conducted in the presence of oxygen. Peak analysis matches that reported by Descorme and Gelin, confirmed the NO adsorption mechanism [8]. The Pd/Fe sample did not show the peaks associated with NO adsorption onto Fe complexes, confirming that the role of Fe in the catalyst composition is ensure a high dispersion of Pd [9]. Experiments conducted without oxygen (not shown) display peaks associated with NO adsorption onto Pd but do not show peaks associated with NO_2 formation and storage.

Additional TPD experiments (not shown) have shown that the presence of water significantly suppresses NOx uptake and desorption. The introduction of water inhibits the surface-reaction formation of NO_2, likely through the suppression of zeolitic acid sites. Ongoing experiments include powdered catalyst TPD experiments as well as additional DRIFTS experiments to better understand and quantify the effects the addition of binder.

Ongoing work will include modifications to the catalyst compositions as well as the addition of a modelling component for both uptake and desorption. Additionally, concurrent experiments are being carried out to understand the effects of adding hydrocarbons to the gas environment.

Figure 3: Evolution of DRIFTS data over time from three samples.

488343 Insights into Influencing Factors on Oxygen Storage Characteristics for Ceria-Zirconia Catalysts

Abstract:

Ceria systems have been investigated extensively in the field of automotive catalysis as oxygen storage materials in order to compensate for deviating stoichiometric operation during exhaust aftertreatment. The ability of ceria to cycle between Ce^{4+} and Ce^{+3} provides an attractive characteristic desirable for reactions such as water-gas shift, hydrocarbon oxidation, and hydrocarbon reforming [1]. Often, an additional dopant is added into the ceria lattice in order to increase oxygen storage capacity (OSC) and oxygen mobility, with ceria zirconia (CZO) being the most common system used [2]. The oxygen release rate (ORR) by these materials can be further enhanced by depositing platinum group metal (PGM) such as Pd or Rh, which provides beneficial metal-support interface in order for rapid oxygen transfer to occur [3]. While these catalytic materials have been in use, additional fundamental insight in the mechanism of OSC and the effect of PGM dispersion, CZO surface area, material pre-treatment, and varying testing conditions on catalyst performance are explored in this study.

A series of Pd/CZO catalysts was prepared by incipient wetness impregnation with different loadings and aged in redox conditions at high temperatures. Characterization by BET, H_2 chemisorption, and XRD will be discussed. OSC measurements were conducted via lambda probes before and after the catalyst as a function of temperature, while ORR were evaluated by the production of CO_2 as measured by FTIR as a function of time. Structure-activity relationships were developed to determine what factors were most important in achieving high OSC and fast ORR, particularly after high temperature aging. In addition, catalysts were evaluated for CO oxidation and water-gas shift to establish the importance of OSC in either reaction. Results from this study will help provide guidelines in the design of future oxygen storage materials in order to meet future emission standards.

[1] R. Gorte, AIChE Journal 56 (2010) 1126-1135

[2] E. Mamontov, et al., J. Phys. Chem. B 104 (2000) 11110-11116

[3] H.W. Jen, et al., Catal. Today 50 (1999) 309-328

RTI INTERNATIONAL

AIChE

Pd doped $CaCo_xZr_{1-x}O_{3-\delta}$ perovskites for automotive emissions control

Nov 01, 2017, AIChE Annual Meeting

Qinghe (Angela) Zheng and Marty Lail*, PhDs
Energy Technology Division, RTI International
Research Triangle Park, NC 27709
qizheng@ri.org

Conventional supported PGM-based TWC *vs.* Perovskite-TWC

RTI INTERNATIONAL

Conventional PGM-based TWCs
($PdRh/\gamma Al_2O_3$ (CeO_2-ZrO_2))

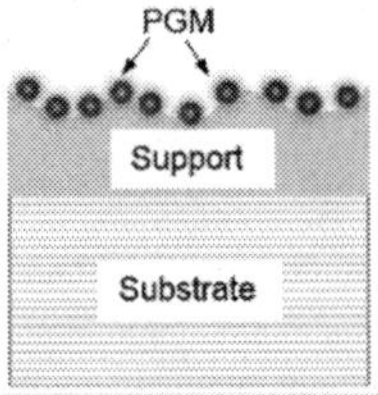

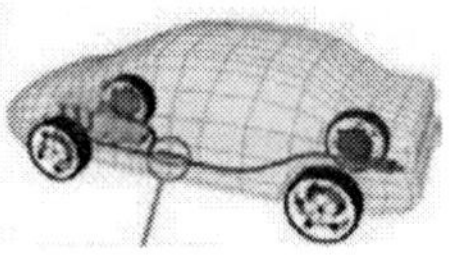

Three Way Catalytic (TWC) Converter

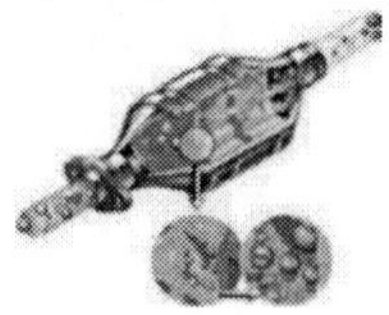

$ABO_{3-\delta}$ Perovskites

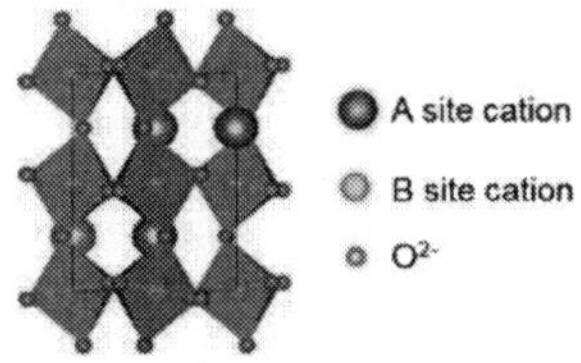

$$ABO_{3-\delta} + \frac{1}{2}\Delta\delta\, CO \rightarrow ABO_{3-\delta-\Delta\delta} + \frac{1}{2}\Delta\delta\, CO_2$$

$$ABO_{3-\delta} + \Delta\delta\, C_xH_y \rightarrow ABO_{3-\delta-\left(2x+\frac{1}{2}y\right)\Delta\delta} + \Delta\delta\,(x\, CO_2 + \frac{1}{2}y\, H_2O)$$

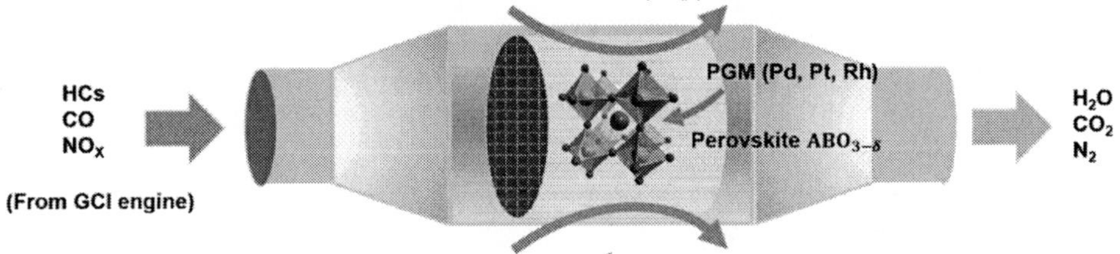

$$ABO_{3-\delta-\Delta\delta} + \frac{1}{2}\Delta\delta\, O_2 \rightarrow ABO_{3-\delta}$$

$$ABO_{3-\delta-\Delta\delta} + \Delta\delta\, NO \left(\text{or} \frac{1}{2}\Delta\delta\, NO_2\right) \rightarrow ABO_{3-\delta} + \frac{1}{2}\Delta\delta\, N_2 \ (\text{or} \frac{1}{4}\Delta\delta\, N_2)$$

2

Conventional supported PGM-based TWC *vs.* Perovskite-TWC

RTI INTERNATIONAL

Conventional PGM-based TWCs
($PdRh/\gamma Al_2O_3$ (CeO_2-ZrO_2))

- ✓ High activity (@ stoichiometric)
- ✓ Good performance with CeO_2 added
- ✓ High surface area support
- o Low thermal stability (@ fuel cutoff)
- o High cost of PGM
- o Market control of Ce resources

$ABO_{3-\delta}$ Perovskites

- ✓ Improved thermal stability
- ✓ Low cost
- ✓ Tunable compositions
- ✓ Large availability
- o Low surface area
- o Less activity

$$ABO_{3-\delta} + \frac{1}{2}\Delta\delta\, CO \rightarrow ABO_{3-\delta-\Delta\delta} + \frac{1}{2}\Delta\delta\, CO_2$$

$$ABO_{3-\delta} + \Delta\delta\, C_xH_y \rightarrow ABO_{3-\delta-(2x+\frac{1}{2}y)\Delta\delta} + \Delta\delta\,(x\, CO_2 + \frac{1}{2} y\, H_2O)$$

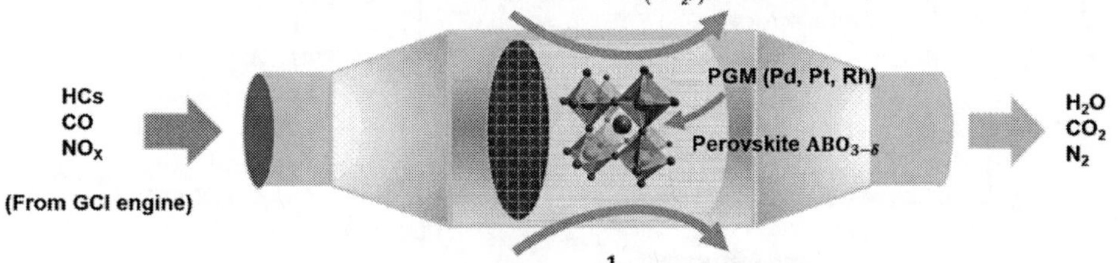

$$ABO_{3-\delta-\Delta\delta} + \frac{1}{2}\Delta\delta\, O_2 \rightarrow ABO_{3-\delta}$$

$$ABO_{3-\delta-\Delta\delta} + \Delta\delta\, NO\left(or \frac{1}{2}\Delta\delta\, NO_2\right) \rightarrow ABO_{3-\delta} + \frac{1}{2}\Delta\delta\, N_2\ (or \frac{1}{4}\Delta\delta\, N_2)$$

3

$ABO_{3-\delta}$ perovskite structure

RTI INTERNATIONAL

$ABO_{3-\delta}$ perovskite structure

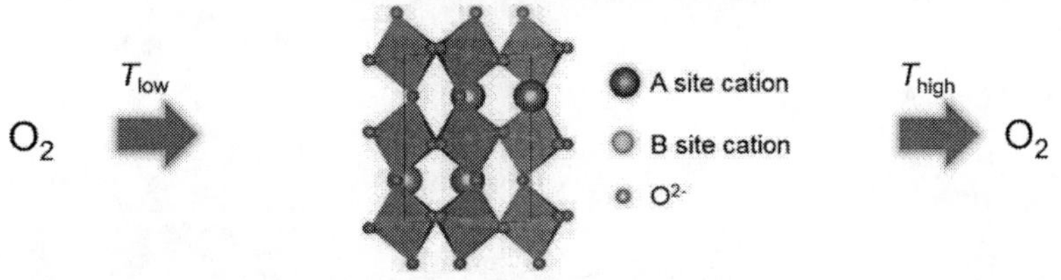

A site cation: alkaline earth, alkaline, and lanthanide ions in a 12-fold coordination (e.g. A= La, Sr, Ca, Ba)

- Important in perovskite structural formation
- Influence oxygen vacancy and structural defects density
- Influence charge-valence balance

B site cation: smaller transition metal ions in a 6-fold coordination with oxygen ions (e.g. B=Co, Mn, Fe, Cu)

- Perovskite oxygen storage capacity depends on the change of oxidation state of the B-site cation
- Perovskites with Co at B site was reported to show best activity in HC and CO oxidation
- Co-based perovskites have low thermal stability due to decomposition at high T
- Improved perovskite properties can be achieved by B-site doping

Oxygen non-stoichiometry *δ*: oxygen anion deficiency or excess depending on the total valence amount of A-site and B-site cations

4

Oxygen mobility within $ABO_{3-\delta}$ perovskite lattice

- $ABO_{3-\delta}$ perovskite can exhibit significant swings in δ through the reaction with gas phase O_2.
- δ can be tailored by doping A- or B-sites with appropriately selected cations
- δ may vary with T and P_{O2}
- Adsorption of superficial [O] at T<600 °C
  Activation of interfacial lattice [O] at T>600 °C } Mars-van-Kreveken mechanism

RTI developed novel $CaCo_xZr_{1-x}O_{3-\delta}$ type perovskites wi or w/o PGM doping

(Patent application RTI 16P027-PROV)

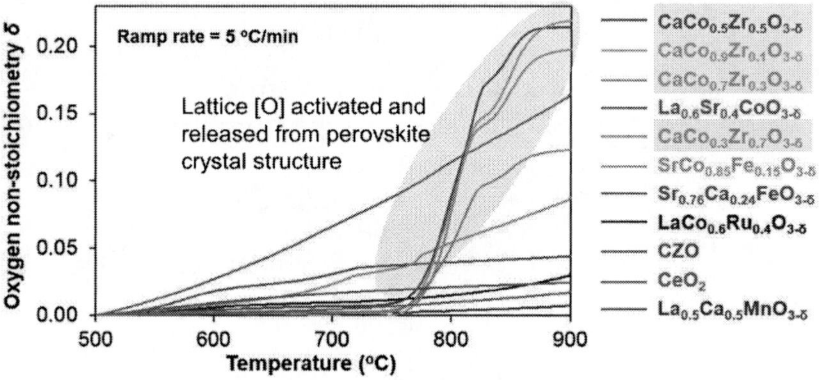

Fig. 1. O_2 non-stoichiometry as a function of sample temperature (500 ~ 900 °C) during TGA temperature programmed study of $ABO_{3-\delta}$ perovskite materials.

5

Perovskite with high OSC as potential TWC

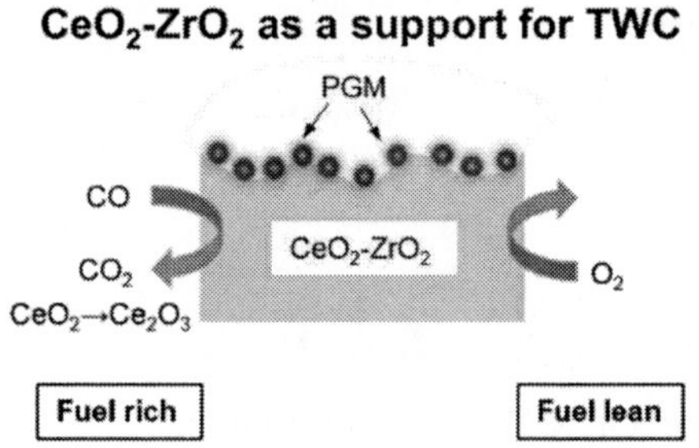

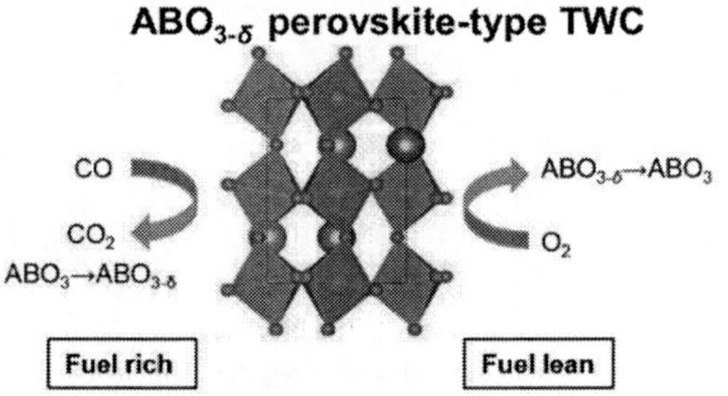

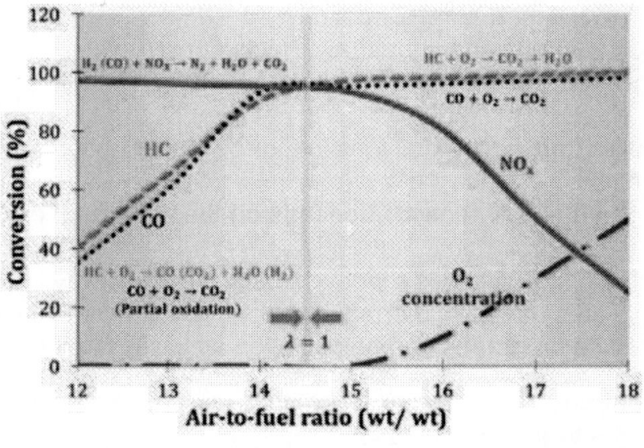

Figure 2. The TWC conversion profile as a function of air-to-fuel ratio (λ).

6

Experimental methodology

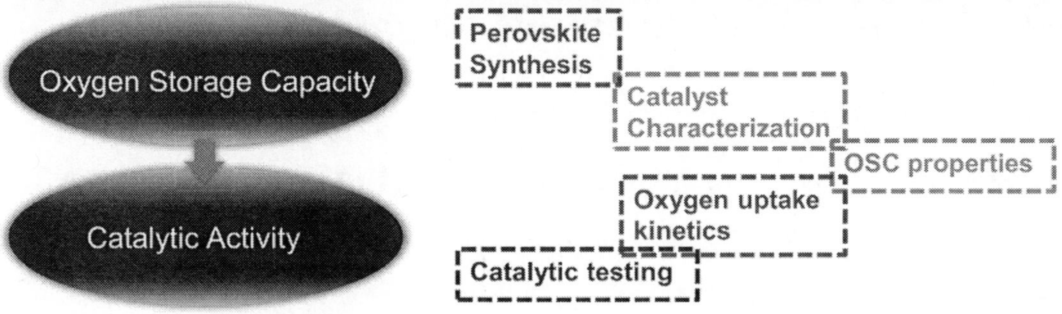

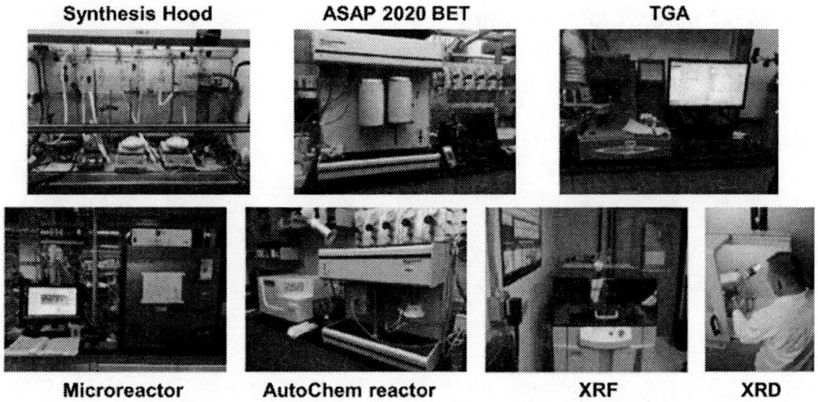

7

Synthesis and characterization of $CaCo_xZr_{1-x}(Pd)O_3$ perovskites

Table 1. Analytical result of as-synthesized (Pd-doped) $CaCo_xZr_{1-x}O_3$ perovskites.

#	Theoretical composition	Actual composition (XRF)	Tolerance factor, τ
CCZPd2	$CaCo_{0.4}Zr_{0.4}Pd_{0.2}O_3$	$CaCo_{0.4}Zr_{0.3}Pd_{0.2}O_3$	0.81
CCZPd1	$CaCo_{0.5}Zr_{0.4}Pd_{0.1}O_3$	$CaCo_{0.5}Zr_{0.3}Pd_{0.1}O_3$	0.82
CCZPd05	$CaCo_{0.55}Zr_{0.4}Pd_{0.05}O_3$	$CaCo_{0.55}Zr_{0.3}Pd_{0.06}O_3$	0.83
CCZ91	$CaCo_{0.9}Zr_{0.1}O_3$	$CaCo_{1.0}Zr_{0.1}O_3$	0.86
CCZ73	$CaCo_{0.7}Zr_{0.3}O_3$	$CaCo_{0.8}Zr_{0.2}O_3$	0.85
CCZ55	$CaCo_{0.5}Zr_{0.5}O_3$	$CaCo_{0.5}Zr_{0.5}O_3$	0.84
CCZ37	$CaCo_{0.3}Zr_{0.7}O_3$	$CaCo_{0.3}Zr_{0.6}O_3$	0.82
CZ10	$CaZrO_3$	$CaZrO_3$	0.80

- **Perovskite synthesis:** Pechini method, calcination @ 1200 °C;
- **Perovskite target compositions:**
  $CaCo_xZr_{1-x}O_{3-\delta}$ (x=0, 0.3, 0.5, 0.7, and 0.9),
  Pd doped $CaCo_{0.6-y}Zr_{0.4}Pd_yO_{3-\delta}$ (y=0.05, 0.1 and 0.2);

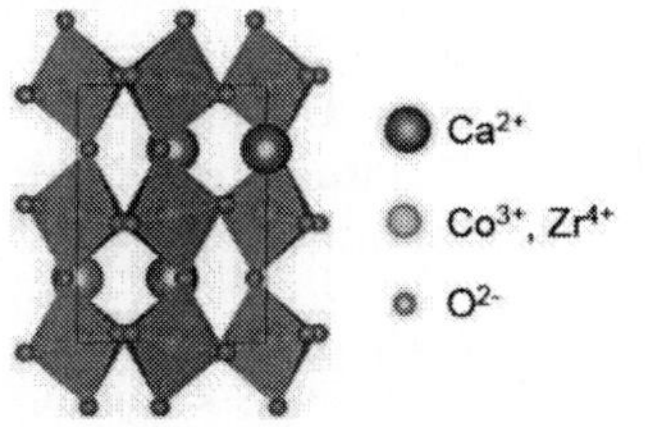

8

Synthesis and characterization of $CaCo_xZr_{1-x}(Pd)O_3$ perovskites

Table 1. Analytical result of as-synthesized (Pd-doped) $CaCo_xZr_{1-x}O_3$ perovskites.

#	Theoretical composition	Actual composition (XRF)	Tolerance factor, τ
CCZPd2	$CaCo_{0.4}Zr_{0.4}Pd_{0.2}O_3$	$CaCo_{0.4}Zr_{0.3}Pd_{0.2}O_3$	0.81
CCZPd1	$CaCo_{0.5}Zr_{0.4}Pd_{0.1}O_3$	$CaCo_{0.5}Zr_{0.3}Pd_{0.1}O_3$	0.82
CCZPd05	$CaCo_{0.55}Zr_{0.4}Pd_{0.05}O_3$	$CaCo_{0.55}Zr_{0.3}Pd_{0.06}O_3$	0.83
CCZ91	$CaCo_{0.9}Zr_{0.1}O_3$	$CaCo_{1.0}Zr_{0.1}O_3$	0.86
CCZ73	$CaCo_{0.7}Zr_{0.3}O_3$	$CaCo_{0.8}Zr_{0.2}O_3$	0.85
CCZ55	$CaCo_{0.5}Zr_{0.5}O_3$	$CaCo_{0.5}Zr_{0.5}O_3$	0.84
CCZ37	$CaCo_{0.3}Zr_{0.7}O_3$	$CaCo_{0.3}Zr_{0.6}O_3$	0.82
CZ10	$CaZrO_3$	$CaZrO_3$	0.80

- **Perovskite synthesis:** Pechini method, calcination @ 1200 °C;
- **Perovskite target compositions:**
  $CaCo_xZr_{1-x}O_{3-\delta}$ (x=0, 0.3, 0.5, 0.7, and 0.9),
  Pd doped $CaCo_{0.6-y}Zr_{0.4}Pd_yO_{3-\delta}$ (y=0.05, 0.1 and 0.2);
- **Perovskite actual compositions** confirmed by XRF;

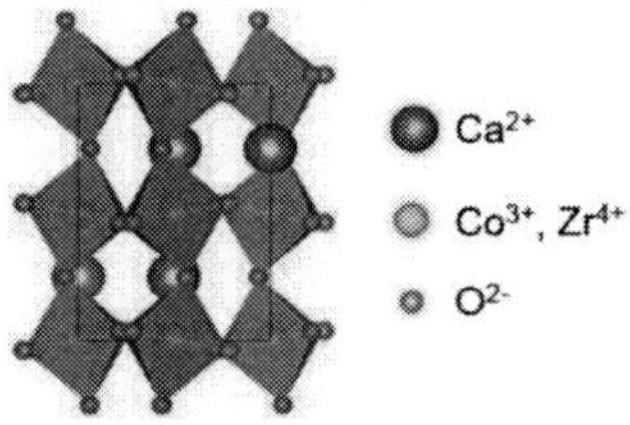

9

Synthesis and characterization of $CaCo_xZr_{1-x}(Pd)O_3$ perovskites

Table 1. Analytical result of as-synthesized (Pd-doped) $CaCo_xZr_{1-x}O_3$ perovskites.

#	Theoretical composition	Actual composition (XRF)	Tolerance factor, τ
CCZPd2	$CaCo_{0.4}Zr_{0.4}Pd_{0.2}O_3$	$CaCo_{0.4}Zr_{0.3}Pd_{0.2}O_3$	0.81
CCZPd1	$CaCo_{0.5}Zr_{0.4}Pd_{0.1}O_3$	$CaCo_{0.5}Zr_{0.3}Pd_{0.1}O_3$	0.82
CCZPd05	$CaCo_{0.55}Zr_{0.4}Pd_{0.05}O_3$	$CaCo_{0.55}Zr_{0.3}Pd_{0.06}O_3$	0.83
CCZ91	$CaCo_{0.9}Zr_{0.1}O_3$	$CaCo_{1.0}Zr_{0.1}O_3$	0.86
CCZ73	$CaCo_{0.7}Zr_{0.3}O_3$	$CaCo_{0.8}Zr_{0.2}O_3$	0.85
CCZ55	$CaCo_{0.5}Zr_{0.5}O_3$	$CaCo_{0.5}Zr_{0.5}O_3$	0.84
CCZ37	$CaCo_{0.3}Zr_{0.7}O_3$	$CaCo_{0.3}Zr_{0.6}O_3$	0.82
CZ10	$CaZrO_3$	$CaZrO_3$	0.80

- **Perovskite synthesis:** Pechini method, calcination @ 1200 °C;
- **Perovskite target compositions:**
  $CaCo_xZr_{1-x}O_{3-\delta}$ (x=0, 0.3, 0.5, 0.7, and 0.9),
  Pd doped $CaCo_{0.6-y}Zr_{0.4}Pd_yO_{3-\delta}$ (y=0.05, 0.1 and 0.2);
- **Perovskite actual compositions** confirmed by XRF;
- **Goldschmidt tolerance factor:**
  $\tau = \frac{r_A + r_O}{\sqrt{2}(r_B + r_O)}$
  (r_A, r_B, and r_O are the ionic radii for the A^{n+}, B^{m+}, and O^{2-});
- All the τ values are within the range of 0.8 ~0.9:
  Suggest orthorhombic structures
- **BET surface areas: <10 m^2/g**

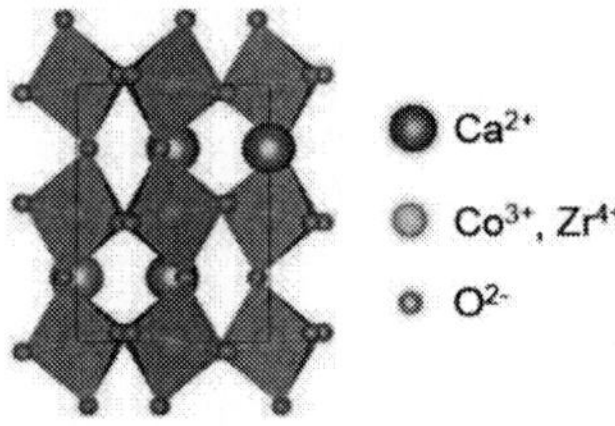

10

XRD: Partial B-site substitution by Zr enhances crystallinity Pd dopant type includes surface Pd^0 and bulk Pd^{2+}

RTI INTERNATIONAL

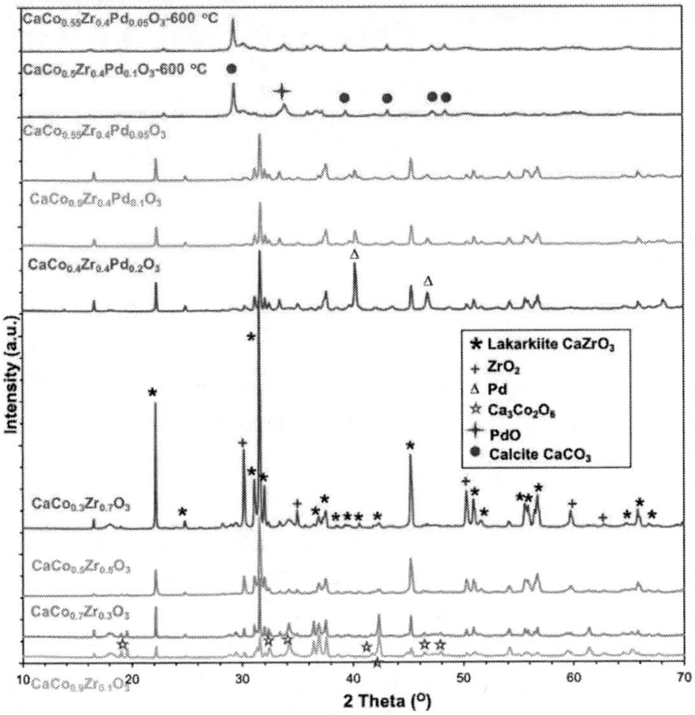

- For $CaCo_{0.6-y}Zr_{0.4}Pd_yO_3$ (calcined at low temperature 600 °C):
  - Perovskite structure not formed
  - Palladium presents as PdO
- For $CaCo_{0.6-y}Zr_{0.4}Pd_yO_3$:
  - Main phase remained $CaCo_{0.5}Zr_{0.5}O_3$ perovskite;
  - Doped Pd includes bulk Pd^{2+} and surface Pd^0.
- For $CaCo_xZr_{1-x}O_3$:
  - Retain Lakarkiite $CaZrO_3$ main structure;
  - Partial substitution of Co by Zr enhances the structural crystallinity of $CaCoO_3$;
  - When x=0.5, optimum crystallinity and minimum impurity.

Fig. 3. XRD patterns of the $CaCo_xZr_{1-x}O_3$ (x = 0, 0.3, 0.5, 0.7 and 0.9) and Pd-doped $CaCo_{0.6-y}Zr_{0.4}Pd_yO_3$ (y = 0.05, 0.1 and 0.2) perovskite samples.

11

H_2-TPR: Enhanced Co reduction by Zr substitution and Pd doping

RTI INTERNATIONAL

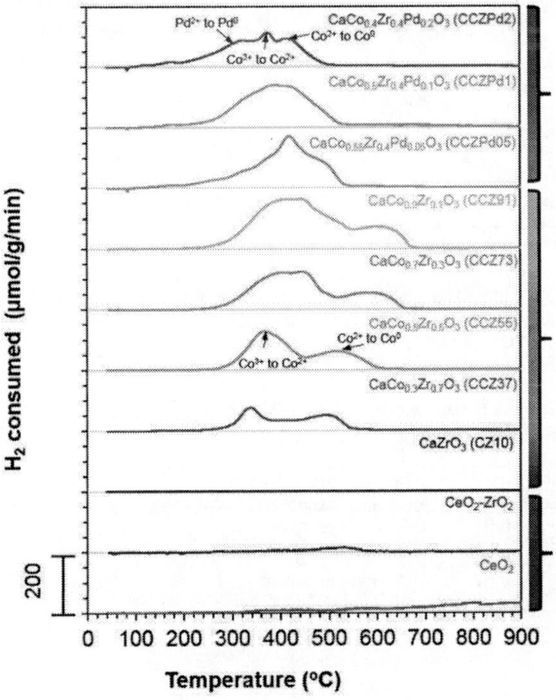

- For $CaCo_{0.6-y}Zr_{0.4}Pd_yO_3$:
  - Pd doping also facilitates the reduction of Co species;
  - $CaCo_{0.4}Zr_{0.4}Pd_{0.2}O_3$:
    Bulk $Pd^{2+} \rightarrow Pd^0$ (T_{max}=320 °C)
    $Co^{3+} \rightarrow Co^{2+}$ (T_{max}=390 °C)
    $Co^{2+} \rightarrow Co^0$ (T_{max}=420 °C);
  - Possible H_2 dissociation on surface Pd^0 particles followed by the successive spill-over of dissociated H atoms to the Co species.
- For $CaCo_xZr_{1-x}O_3$:
  - $Co^{3+} \rightarrow Co^{2+}$ (280-500 °C)
    $Co^{2+} \rightarrow Co^0$ (450-680 °C);
  - B-site substitution by Zr enhances the reduction of Co species by lowering the reduction T;
  - No reduction peak observed with $CaZrO_3$.
- Perovskite vs. CeO_2 and CeO_2-ZrO_2 (CZO):
  $CaCo_xZr_{1-x}(Pd)O_3$ perovskites have significantly higher reducibility/OSC than CeO_2 and CZO.

Fig. 4. H_2 consumption as a function of reaction time on stream and temperature of studied samples during H_2-TPR measurements (temperature ramp rate at 5 °C/min).

12

Redox cyclic tests: $CaCo_{0.5}Zr_{0.5}O_3$ showed optimum OSC and stability

RTI INTERNATIONAL

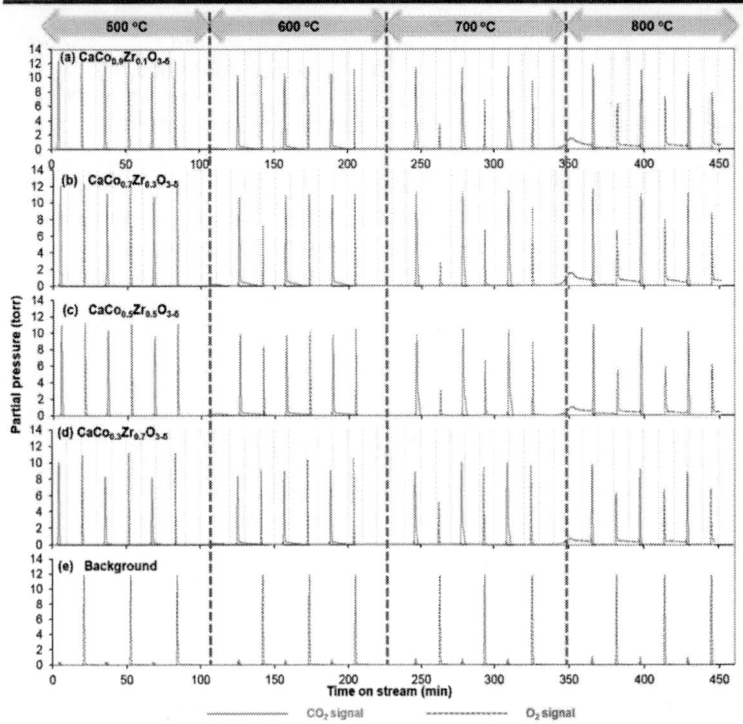

Reduction-Oxidation (3 cycles)

Reduction by 20% CO/He
Oxidation by air
Temperature: 500, 600, 700 and 800 °C

- Perovskite OSC increases with increasing T;
- Perovskite OSC increases with increasing *x* (Co content);
- Stable oxygen storage performance by Zr substitution;
- Optimum perovskite OSC and stability with $CaCo_{0.5}Zr_{0.5}O_3$.

Fig. 5. CO_2 production and O_2 intake vs. reaction time on stream (TOS) of $CaCo_xZr_{1-x}O_{3-\delta}$ (x=0.3, 0.5, 0.7, or 0.9) perovskite samples, as represented by MS partial pressure signals during isothermal CO reduction-air oxidation cyclic tests at 500, 600, 700, and 800 °C.

13

Oxygen intake kinetics measurements

RTI INTERNATIONAL

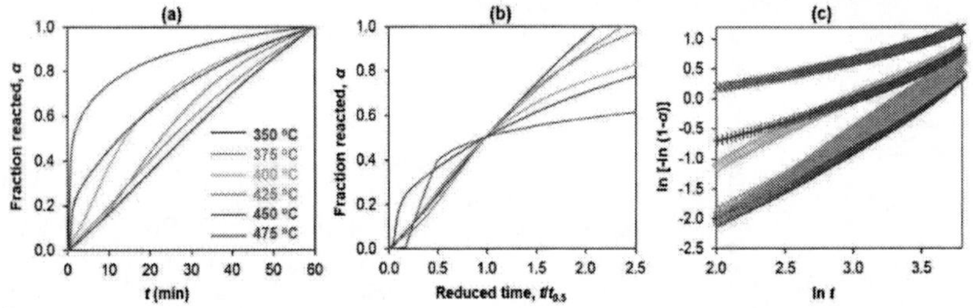

$$\alpha = \frac{m_0 - m_t}{m_0 - m_{final}}$$

$$\alpha \in [0, 1]$$

Fig. 6. Plots for oxygen-intake kinetic calculation of $CaCo_{0.5}Zr_{0.5}O_{3-\delta}$ perovskite showing (a) Fraction reacted α with time variation, (b) Fraction reacted α against reduced time $t/t_{0.5}$, and (c) $\ln[-\ln(1-\alpha)]$ as a function of $\ln t$ (with t in min) for the determination of reaction mechanisms at temperatures varied from 350 °C to 475 °C.

Table 2. List of solid-state rate equations with suggested *m* values studied for the oxygen-intake reaction for $CaCo_{0.5}Zr_{0.5}O_{3-\delta}$ and $CaCo_{0.9}Zr_{0.1}O_{3-\delta}$ perovskite samples. *m* values: slopes in Fig.6(c) at each temperature.

Eq. #	Function	Equation	m	Reaction mechanism
1	$D_1(\alpha)$	$\alpha^2 = kt$	0.62	Diffusion controlled
2	$D_2(\alpha)$	$(1-\alpha)\ln(1-\alpha) + \alpha = kt$	0.57	Diffusion controlled
3	$D_3(\alpha)$	$[1-(1-\alpha)^{1/3}]^2 = kt$	0.54	Diffusion controlled
4	$D_4(\alpha)$	$1 - 2\alpha/3 - (1-\alpha)^{2/3} = kt$	0.57	Diffusion controlled
5	$F_1(\alpha)$	$-\ln(1-\alpha) = kt$	1.00	First-order kinetics
6	$R_2(\alpha)$	$1-(1-\alpha)^{1/2} = kt$	1.11	Surface reaction controlled
7	$R_3(\alpha)$	$1-(1-\alpha)^{1/3} = kt$	1.07	Surface reaction controlled
8	Zero order	$\alpha = kt$	1.24	Zero-order kinetics

14

Oxygen intake kinetics measurements: First order kinetic model

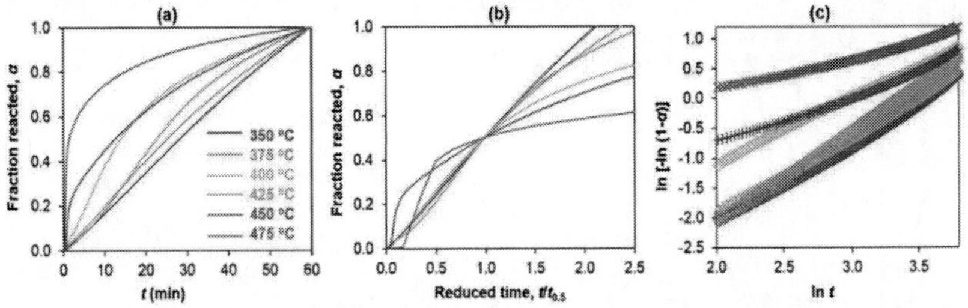

$$\alpha = \frac{m_0 - m_t}{m_0 - m_{final}}$$

$$\alpha \in [0, 1]$$

Fig. 6. Plots for oxygen-intake kinetic calculation of $CaCo_{0.5}Zr_{0.5}O_{3-\delta}$ perovskite showing (a) Fraction reacted α with time variation, (b) Fraction reacted α against reduced time $t/t_{0.5}$, and (c) $\ln[-\ln(1-\alpha)]$ as a function of $\ln t$ (with t in min) for the determination of reaction mechanisms at temperatures varied from 350 °C to 475 °C.

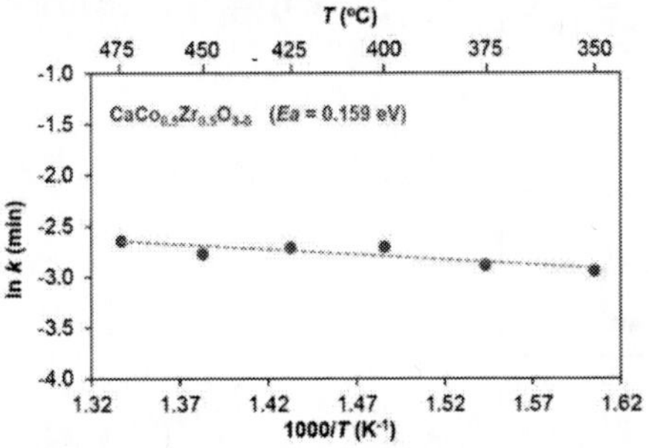

$-\ln(1-\alpha) = kt$ m = 0.6~1.3 First-order kinetics

Fig. 7. Temperature dependent-reaction rate constants k in natural log scale as a function of inverse temperature for the calculation of reaction activation energy for $CaCo_{0.5}Zr_{0.5}O_{3-\delta}$ perovskite. The k values were achieved by fitting reaction data with various kinetic models, and first-order kinetic model was selected.

15

Catalytic oxidation activity of Pd doped $CaCo_{0.6-y}Zr_{0.4}Pd_yO_3$

Simulated dry exhaust feed: 0.060-0.066% C_3H_6, 0.750-0.825% CO, 0.675-0.750% O_2, 19.5% CO_2, N_2 bal.

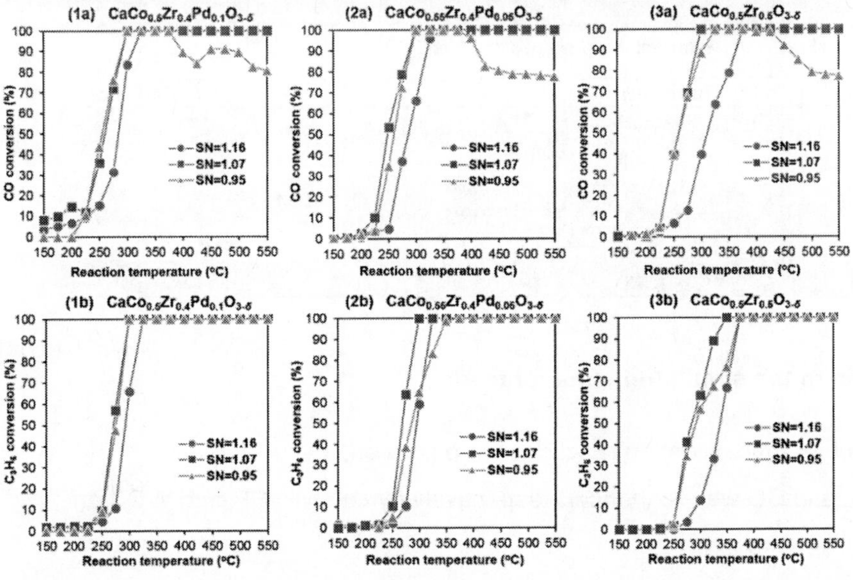

- Complete conversions of CO and C_3H_6 below 350 °C;
- Pd-doped samples showed higher oxidation activities than Pd-free one at all three conditions;
- Higher conversions at richer condition (SN=0.95) for Pd samples.
- "Pd regeneraetion" effect at rich condition (Tanaka *et al*)

Fig. 8. (1) CO and (2) C_3H_6 conversions during catalytic oxidation tests over (a) $CaCo_{0.5}Zr_{0.4}Pd_{0.1}O_{3-\delta}$, (b) $CaCo_{0.55}Zr_{0.4}Pd_{0.05}O_{3-\delta}$ and (c) $CaCo_{0.5}Zr_{0.5}O_{3-\delta}$ perovskite with simulated exhaust feed **at stoichiometric numbers (SNs) of 1.16, 1.07 and 0.95. SN=2[O_2]/([CO]+9[C_3H_6])**

16

Catalytic oxidation activity with addition of H_2O in the feed

Simulated wet exhaust feed: 0.056% C_3H_6, 0.702% CO, 0.648% O_2, 17.55% CO_2, 10% steam, N_2 bal.

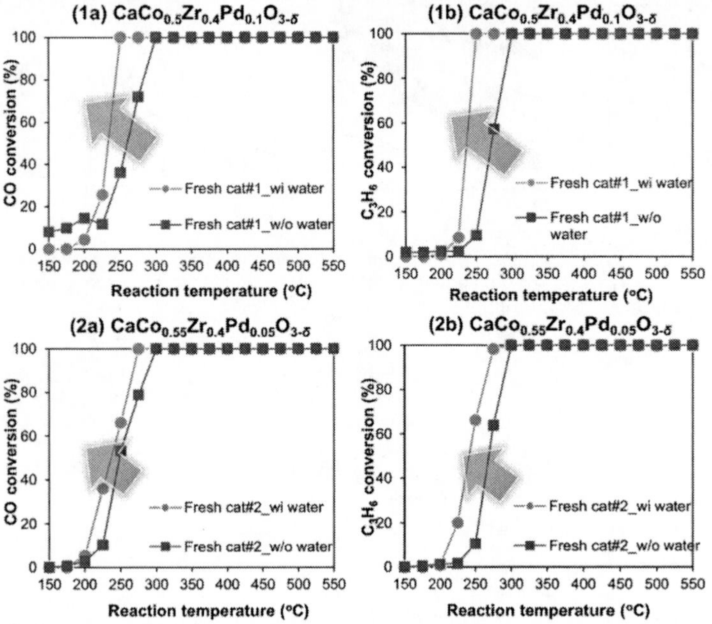

Fig. 9. (1) CO and (2) C_3H_6 conversions during catalytic oxidation tests over fresh and aged (a) $CaCo_{0.5}Zr_{0.4}Pd_{0.1}O_{3-\delta}$, (b) $CaCo_{0.55}Zr_{0.4}Pd_{0.05}O_{3-\delta}$ perovskites **at SN of 1.07, with presence of 10 vol-% H_2O in the feed.**

Enhancements in both CO and C_3H_6 conversions when water (steam) was added to the feed;

Catalytic oxidation activity of Pd doped $CaCo_{0.6-y}Zr_{0.4}Pd_yO_3$

Table 3. Main reaction pathways during catalytic conversions of CO and C_3H_6 over Pd doped $CaCo_xZr_{1-x}O_3$ perovskites by feeding simulated engine exhaust at SN=1.07, when water is present in the feed.

Eq. #	Reaction pathway	Reaction	ΔH^0_{298} (kJ/mol)
(1)	Oxidation of C_3H_6	$C_3H_6 + 4.5\ O_2 \xrightarrow{\text{Perovskite}} 3CO_2 + 3H_2O$	-2058
(2)	Oxidation of CO	$CO + 0.5\ O_2 \xrightarrow{\text{Perovskite}} CO_2$	-283
(3)	Steam reforming of C_3H_6	$C_3H_6 + 6\ H_2O \xrightarrow{\text{Perovskite}} 3CO_2 + 9H_2$	+110
(4)	Water gas shift	$CO + H_2O \xrightarrow{\text{Perovskite}} CO_2 + H_2$	-41
(5)	Combustion of H_2	$H_2 + 0.5\ O_2 \xrightarrow{\text{Perovskite}} H_2O$	-286

When water was added to the simulated exhaust feed:

- Reaction pathways other than oxidation reactions were favored;
- Conversions of C_3H_6 and CO were enhanced respectively by catalytic SR and WGS on the same catalyst ;
- H_2 generated by the two reactions was readily consumed by stoichiometric amount of O_2 in the feed.

Catalytic oxidation activity of perovskites after accelerated aging

Simulated wet exhaust feed: 0.056% C_3H_6, 0.702% CO, 0.648% O_2, 17.55% CO_2, 10% steam, N_2 bal.

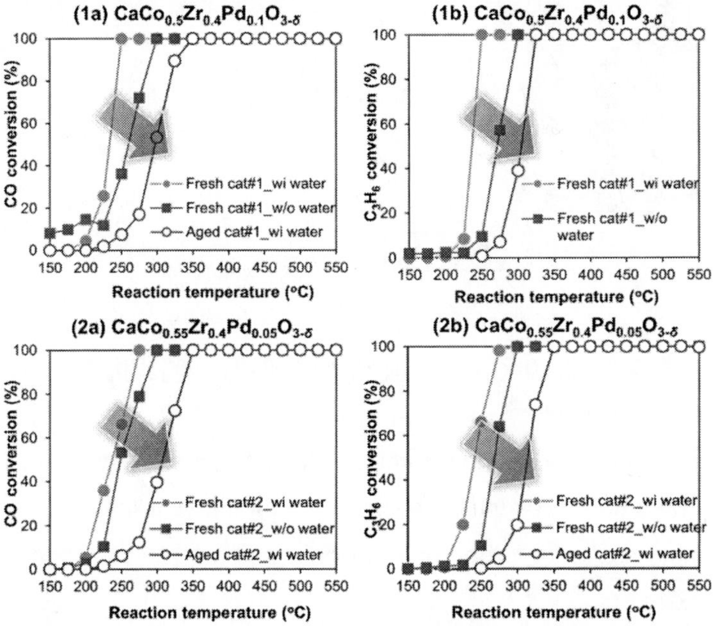

Fig. 10. (1) CO and (2) C_3H_6 conversions during catalytic oxidation tests over fresh and aged (a) $CaCo_{0.5}Zr_{0.4}Pd_{0.1}O_{3-\delta}$, (b) $CaCo_{0.55}Zr_{0.4}Pd_{0.05}O_{3-\delta}$ perovskites with at SN of 1.07, with presence of 10 vol-% H_2O in the feed. **Aging condition: 1000 °C in air for 50 hrs.**

Certain degree of catalyst deactivation observed after accelerated aging in air at 1000 °C for 50 hrs, due to extensive catalyst sintering.

19

Summary

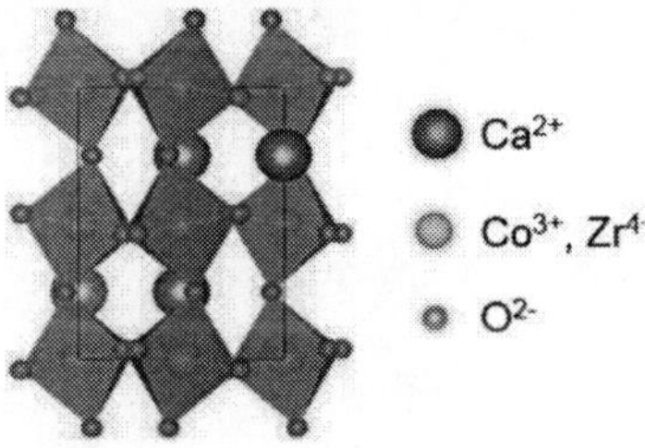

- Novel $CaCo_xZr_{1-x}O_{3-\delta}$ (x = 0, 0.3, 0.5, 0.7 and 0.9) and Pd-doped $CaCo_{0.6-y}Zr_{0.4}Pd_yO_3$ (y = 0.05, 0.1 and 0.2) perovskites were synthesized;
- The studied perovskites retain their main phase of orthorhombic Lakargiite $CaZrO_3$ structure, partial substitution of Co by Zr at B sites enhances the perovskite structural crystallinity;

20

Summary

RTI INTERNATIONAL

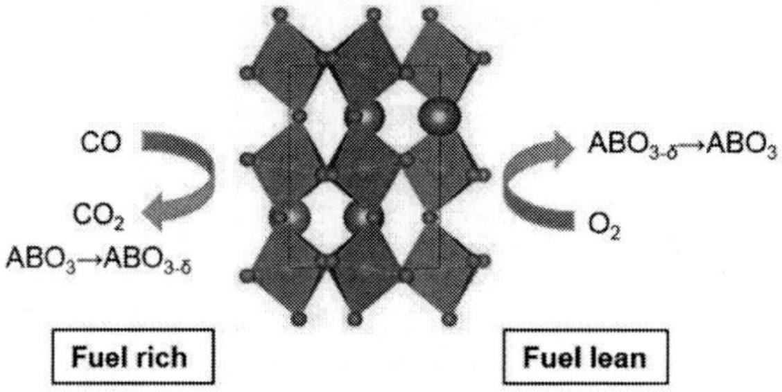

- Novel $CaCo_xZr_{1-x}O_{3-\delta}$ (x = 0, 0.3, 0.5, 0.7 and 0.9) and Pd-doped $CaCo_{0.6-y}Zr_{0.4}Pd_yO_3$ (y = 0.05, 0.1 and 0.2) perovskites were synthesized;
- The studied perovskites retain their main phase of orthorhombic Lakargiite $CaZrO_3$ structure, partial substitution of Co by Zr at B sites enhances the perovskite structural crystallinity;
- The studied perovskites showed improved redox property and OSC compared to the ceria-based oxygen storage materials (OSMs).

21

Summary

RTI INTERNATIONAL

$$ABO_{3-\delta} + \frac{1}{2}\Delta\delta\, CO \rightarrow ABO_{3-\delta-\Delta\delta} + \frac{1}{2}\Delta\delta\, CO_2$$

$$ABO_{3-\delta} + \Delta\delta\, C_xH_y \rightarrow ABO_{3-\delta-\left(2x+\frac{1}{2}y\right)\Delta\delta} + \Delta\delta\,(x\, CO_2 + \frac{1}{2} y\, H_2O)$$

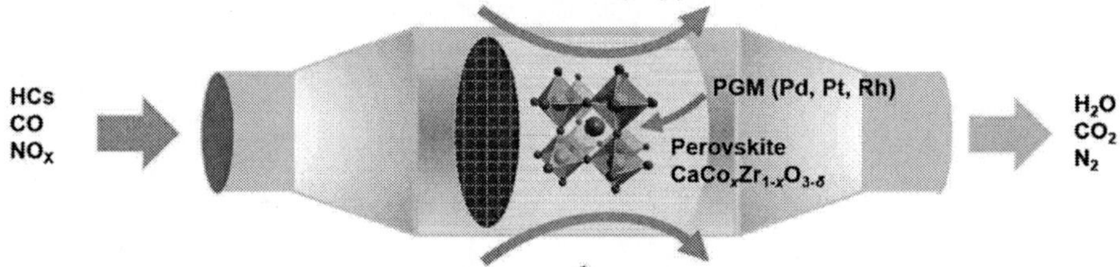

$$ABO_{3-\delta-\Delta\delta} + \frac{1}{2}\Delta\delta\, O_2 \rightarrow ABO_{3-\delta}$$

- Novel $CaCo_xZr_{1-x}O_{3-\delta}$ (x = 0, 0.3, 0.5, 0.7 and 0.9) and Pd-doped $CaCo_{0.6-y}Zr_{0.4}Pd_yO_3$ (y = 0.05, 0.1 and 0.2) perovskites were synthesized;
- The studied perovskites retain their main phase of orthorhombic Lakargiite $CaZrO_3$ structure, partial substitution of Co by Zr at B sites enhances the perovskite structural crystallinity;
- The studied perovskites showed improved redox property and OSC compared to the ceria-based oxygen storage materials (OSMs).
- Pd-doped samples showed promising catalytic activity towards C_3H_6 and CO oxidation under simulated exhaust conditions

22

Selected references

[1] A. Simson, K. Roark, R. Farrauto, A feasibility study of niobium-containing materials for oxygen storage in three way catalytic converters, Appl. Catal. B Environ. 158–159 (2014) 106–111. doi:10.1016/j.apcatb.2014.04.006.

[2] A. Schön, C. Dujardin, J.-P. Dacquin, P. Granger, Enhancing catalytic activity of perovskite-based catalysts in three-way catalysis by surface composition optimisation, Catal. Today. 258, Part 2 (2015) 543–548. doi:10.1016/j.cattod.2014.11.002.

[3] Q. Zheng, R. Farrauto, M. Deeba, I. Valsamakis, Part I: A Comparative Thermal Aging Study on the Regenerability of Rh/Al2O3 and Rh/CexOy-ZrO2 as Model Catalysts for Automotive Three Way Catalysts, Catalysts. 5 (2015) 1770–1796. doi:10.3390/catal5041770.

[4] R. You, Y. Zhang, D. Liu, M. Meng, L. Zheng, J. Zhang, T. Hu, YCeZrO Ternary Oxide Solid Solution Supported Nonplatinic Lean-Burn NOx Trap Catalysts Using LaCoO3 Perovskite as Active Phase, J. Phys. Chem. C. 118 (2014) 25403–25420. doi:10.1021/jp505601x.

[5] Q. Zheng, R. Farrauto, M. Deeba, Part II: Oxidative Thermal Aging of Pd/Al2O3 and Pd/CexOy-ZrO2 in Automotive Three Way Catalysts: The Effects of Fuel Shutoff and Attempted Fuel Rich Regeneration, Catalysts. 5 (2015) 1797–1814. doi:10.3390/catal5041797.

[6] S. Royer, H. Alamdari, D. Duprez, S. Kaliaguine, Oxygen storage capacity of La1−xBO3 perovskites (with A′ = Sr, Ce; B = Co, Mn)—relation with catalytic activity in the CH4 oxidation reaction, Appl. Catal. B Environ. 58 (2005) 273–288. doi:10.1016/j.apcatb.2004.12.010.

[7] A. Klimkowicz, K. Świerczek, A. Takasaki, B. Dabrowski, Oxygen storage capability in Co- and Fe-containing perovskite-type oxides, Solid State Ion. 257 (2014) 23–28. doi:10.1016/j.ssi.2014.01.018.

[8] A. Mishra, N. Galinsky, F. He, E.E. Santiso, F. Li, Perovskite-structured AMnxB1−xO3 (A = Ca or Ba; B = Fe or Ni) redox catalysts for partial oxidation of methane, Catal. Sci. Technol. 6 (2016) 4535–4544. doi:10.1039/C5CY02186C.

[9] B. Levasseur, S. Kaliaguine, Effects of iron and cerium in La1−yCeyCo1−xFexO3 perovskites as catalysts for VOC oxidation, Appl. Catal. B Environ. 88 (2009) 305–314. doi:10.1016/j.apcatb.2008.11.007.

[10] G. Perin, J. Fabro, M. Guiotto, Q. Xin, M.M. Natile, P. Cool, P. Canu, A. Glisenti, Cu@LaNiO3 based nanocomposites in TWC applications, Appl. Catal. B Environ. 209 (2017) 214–227. doi:10.1016/j.apcatb.2017.02.064.

[11] S. Sartipi, A.A. Khodadadi, Y. Mortazavi, Pd-doped LaCoO3 regenerative catalyst for automotive emissions control, Appl. Catal. B Environ. 83 (2008) 214–220. doi:10.1016/j.apcatb.2008.02.014.

[12] Z. Say, M. Dogac, E.I. Vovk, Y.E. Kalay, C.H. Kim, W. Li, E. Ozensoy, Palladium doped perovskite-based NO oxidation catalysts: The role of Pd and B-sites for NOx adsorption behavior via in-situ spectroscopy, Appl. Catal. B Environ. 154–155 (2014) 51–61. doi:10.1016/j.apcatb.2014.01.038.

23

Acknowledgement

Material Chemistry Manager: Dr. Marty Lail

Material Group: Dr. Mustapha Soukri
Dr. Jianping Shen
Dr. Ignacio Luz
Dr. John Thompson
Kelly Amato
Timothy Bellamy
Devaughn Body

U.S. DEPARTMENT OF ENERGY

RTI INTERNATIONAL

Thank you and please feel free to ask any question!

One patent application (RTI 16P027-PROV) on perovskite material development and applications.

One manuscript just accepted by *Catalysis Today*.

25

500801 Probing the Oxidation and Reduction Halves of the Cu Redox Cycle during Standard NH_3-Selective Catalytic Reduction of NO_x on Cu-SSZ-13

Abstract:

The selective catalytic reduction of NO_x (x=1,2) using NH_3 as a reductant (NH_3-SCR) is a leading strategy for NO_x abatement in oxygen-rich automotive exhaust to meet increasingly stringent emissions constraints. It was shown that low temperature (473 K) NOx SCR on Cu-SSZ-13 occurs via a Cu(II)/Cu(I) redox cycle [1] on two types of isolated Cu sites, both divalent Cu^{2+} cations compensating paired framework Al atoms and monovalent $[CuOH]^+$ complexes at isolated framework Al atoms, which are solvated by NH_3 during reaction [2]. Two different kinetic regimes that are characterized by different apparent orders, activation energies, and steady-state Cu(I)/Cu(II) distributions in the Cu(I)/Cu(II) redox cycle have been identified under standard SCR conditions, in which O_2 is used as an oxidant. The rate-limiting steps during standard SCR depends on the Cu density (determined by the Si/Al and Cu/Al) of the Cu-SSZ-13 catalysts. One regime corresponds to SCR cycles limited by Cu(I)->Cu(II) oxidation with O_2 and the other regime corresponds to cycles limited by Cu(II)->Cu(I) reduction, which we probe in *operando* (varying Cu density, gas pressures and temperature) and in transient XAS experiments. O_2-oxidation during low temperature SCR requires two Cu(I) sites to be oxidized, in contrast to reduction occurring at single, isolated Cu(II) sites. Therefore, standard SCR rates depend on Cu cation density and distribution in Cu-SSZ-13 zeolites when Cu(I) oxidation steps are kinetically-relevant, but are independent of Cu density otherwise. We probe the nature of oxidation and reduction half-cycles independently by measuring their intrinsic kinetics. In-situ monitoring of Cu(I)/Cu(II) fractions under oxidation and reduction-limited conditions showed the existence of two different pools of Cu sites, those that are within diffusion distances of another Cu cation and can be oxidized with O_2 in standard SCR cycles, and others that are stranded and isolated from other Cu.

[1] C. Paolucci, A.A. Verma, S.A. Bates, V.F. Kispersky, J.T. Miller, R. Gounder, et al., Isolation of the Copper Redox Steps in the Standard Selective Catalytic Reduction on Cu-SSZ-13, (2014) 12022â€"12027. doi:10.1002/ange.201407030.

[2] C. Paolucci, A.A. Parekh, I. Khurana, J.R. Di Iorio, J. Albarracin, A. Shih, et al., Catalysis in a Cage: Condition-Dependent Speciation and Dynamics of Exchanged Cu Cations In SSZ-13 Zeolites, J. Am. Chem. Soc. (2016) 1â€"63. doi:10.1021/jacs.6b02651.

493807 Sulfur Oxidation Studies with Bimetallic Pd/Pt Catalysts

Abstract:

Bimetallic Pd/Pt catalysts are good automotive-engine oxidation aftertreatment catalyst candidates due to their improved activity and potential sintering resistance relative to the monometallic formulations. Any of these catalyst systems have to endure extended times on stream and tolerate exposure to high temperatures, water, and trace sulfur. Of note, the literature states that the Pd/Pt bimetallic benefits only exist in the absence of sulfur. Previous studies with bimetallic Pd/Pt catalysts show that sulfur sorption and oxidation characteristics are influenced by both Pd:Pt mole ratio and precious metal particle size. Here, studies were conducted with mono- and bimetallic Pd/Pt/Al_2O_3 catalysts to investigate more specifically how SO_2 oxidation activity is influenced by the Pd:Pt mole ratio. Diffuse reflectance infrared Fourier transform spectroscopy (DRIFTS), temperature-programmed oxidation, and steady-state oxidation studies were conducted on catalysts with different precious metal composition but similar precious metal crystallite particle sizes. During surface studies, it was found that catalysts with a higher Pd content tended to have greater activity for oxidizing sulfur species at low temperatures and as a result formed more SO_3 and SO_4 species. However, reactor studies showed that catalysts with a lower Pd content appeared to have greater activity for oxidizing sulfur species at a low temperature because more conversion of SO_2 to SO_3 species was detected at the reactor outlet. Due to this obvious disconnect, studies are now focused on investigating how the oxidation products formed on the catalyst surface influence the catalystâ€™s apparent activity for the SO_2 oxidation reaction. This work provides evidence that the SO_2 oxidation characteristics, as well as the potential for sulfur decay, for mono- and bimetallic Pd/Pt catalysts are heavily influenced by the Pd:Pt mole ratio.

495480 Nature of SO_2 Poisoned Cu-SSZ-13 Catalysts Under Ammonia Selective Catalytic Reduction (NH_3-SCR) Conditions

Abstract:

Nature of SO_2 poisoned Cu-SSZ-13 Catalysts under Ammonia Selective Catalytic Reduction (NH_3-SCR) Conditions

Arthur J. Shih[1], Hui Li[2], Ashok Kumar[3], Juan M. Gonzalez[1,4], Ishant Khurana[1], Christopher Paolucci[2], Jonatan D. Albarracin Caballero[1], Atish A. Parekh[1], Aída Luz Villa[4], W. Nicholas Delgass[1], Rajamani Gounder[1], Aleksey Yezerets[3], William F. Schneider[2], Jeffrey T. Miller[1], and Fabio H. Ribeiro[1]

[1]Davidson School of Chemical Engineering, Purdue University, West Lafayette, IN

[2]Department of Chemical and Biomolecular Engineering, University of Notre Dame, IN

[3]Cummins Incorporated, Columbus, IN

[4]Chemical Engineering Department, Universidad de Antioquia, Medellín, Colombia

Reduction of NO_x emissions from diesel engine exhaust is an environmental issue driven by increasingly stringent emission regulations. Cu-SSZ-13 catalysts have been shown to be promising for this application due to their hydrothermal stability and selectivity in reducing NO_x to N_2 using NH_3 compared to other zeolite catalysts. SO_x formed from the combustion of ppm levels of sulfur in diesel fuel deactivates the Cu-SSZ-13 catalyst. However, its effect on Cu^{2+} and CuOH active sites and the SCR mechanism is unclear and under debate.[1-4]

Two Cu-SSZ-13 catalysts, one with only Cu^{2+} (3.8 Cu wt%, 100% Cu^{2+}) and another with primarily $[CuOH]^{1+}$ (1.5 Cu wt%, 80% CuOH) active sites were synthesized, hereby denoted as Z_2Cu and ZCuOH, respectively.[5] Each catalyst was sulfated with dry SO_2 at 200°C and 400°C, resulting in sulfur contents of 0.7 wt% and 1.0 wt%, respectively. Sulfation at 400°C decreased the standard SCR rate by 90% for the ZCuOH catalyst but by only 60% for the Z_2Cu catalyst. In addition, the apparent activation energy for ZCuOH sites decreased from 70 kJ mol^{-1} to 11 kJ mol^{-1} with increasing extents of sulfur poisoning but remained constant at 70 kJ mol^{-1} for the Z_2Cu catalyst. These kinetic results suggest two different deactivation mechanisms for the Z_2Cu and ZCuOH sites. The decrease in apparent activation energy for the ZCuOH catalyst is attributed to changes in the binding energy of NH_3 to the Cu active sites, corroborated by evidence from UV-Visible where sulfur-Cu ligands were observed on ZCuOH but not on Z_2Cu. The unchanged apparent activation energy and reactant orders for the Z_2Cu catalyst after sulfation are attributed to pore-filling by ammonium sulfate salts; this reduces the total number of accessible Cu, as supported by argon micropore measurements.

Z_2Cu sites are preferred over ZCuOH sites due to the lower impact of sulfur poisoning on the rate and apparent activation energy. These results predict that synthesizing catalysts with higher fractions of Z_2Cu sites will lead to improved emission control catalysts for commercial application.

References:

1. Jangjou, Y.; Wang D.;Kumar, A.; Li J.; Epling, W.S. *ACS Catal.* **2016**, *6*, 10.
2. Shen, M.; Wen, H.; Hao, T.; Yu, T.; Fan, D.; Wang, J.; Li, W.; Wang, J. *Catal. Sci. Technol.* **2015**, *5*, 1741.
3. Zhang, L.; Wang, D.; Liu, Y.; Kamasamudram, K.; Li, J.; Epling, W. *Appl. Catal. B: Environ.* **2014**, *156-157*, 371.
4. Luo, J.; Wang, D.; Kumar, A.; Li, J.; Kamasamudram, K.; Currier, N.; Yezerets, A. *Catal. Today* **2016**, *267*, 3.
5. Paolucci, C.; Parekh, A.A.; Khurana, I.; Di Iorio, J.R.; Li, H.; Albarracin Caballero, J.D.; Shih, A.J.; Anggara, T.; Delgass, W.N.; Miller, J.T.; Ribeiro, F.H.; Gounder, R.; Schneider, W.F. *J. Am. Chem. Soc.* **2016**, *138*, 18.

496535 Measurement of the Oxygen Storage Capacity and Its Role in the Catalytic Oxidation

Abstract:

CeO_2, thanks to its unique redox property of oxygen storage and release under lean and rich conditions, respectively, has been widely used as the catalyst support component. This property of oxygen storage and release can usually be quantitatively measured, called the oxygen storage capacity (OSC), via a reduction method using different techniques. The amount of OSC measured depends on the type of reducing agent and the technique used. Furthermore, the OSC is also influenced by the synthesis method of CeO_2, the modification by other elements like ZrO_2, the physical and geometric properties, and the metal supported on CeO_2-based materials. On the other hand, it is well accepted that the catalytic activity benefits from the unique characteristic of oxygen storage and release for reactions carried out over the CeO_2-based catalysts. Unfortunately, the significant enhancement of catalytic activity is still poorly correlated by the OSC, or the role of OSC in favoring the catalytic reaction is still poorly understood. We establish the measurement of OSC by comparing the use of different reducing agents (CO and H_2) and the use of different techniques (TCD and TGA). The effects of synthesis method, high temperature aging, precious metal loading, and the structure modification by ZrO_2 (i.e. CZO) and other stabilizer elements on CeO_2 have been also investigated. The OSC measured by different methods and the influencing factors are analyzed with emphasis on to the catalytic activity like CO oxidation for the objectives to establish a correlation between OSC and catalytic activity.

495203 Activation of Pd/SSZ-13 in Low Temperature NO Adsorption for Cold Start Application

Abstract:

As the regulation on NOx emission in exhaust gas has been globally issued and enforced, the purification of NOx from diesel engine is regarded as a great challenge. Although selective catalytic reduction (SCR) catalyst effectively reduce NOx in exhaust gas, this system cannot operate properly below 200 Â°C (i.e. during cold start period, initial ca. 100 â €“ 200 sec of engine operation). Therefore, during the warm-up of NOx reduction catalyst, the considerable amount of NOx is emitted into atmosphere without any treatment [1]. Recently, cold start NOx trap catalysts attract significant attention with a concept of storing NOx at low temperature (i.e. during cold start) and releasing them at elevated temperature, where SCR operates. In this work, we tried to apply various Pd/SSZ-13 catalysts with different Pd loading method and investigate the effects of hydrothermal aging (HTA) treatment on the NOx adsorption ability and the physicochemical change of Pd/SSZ-13.

NATURAL GAS LIQUIDS (NGL): FRACTIONATE OR NOT TO FRACTIONATE AT THE GAS PROCESSING PLANT

Bijal V. Gangar, Texas A&M University-Kingsville, Kingsville, TX, USA
Ali A. Pilehvari, Texas A&M University-Kingsville, Kingsville, TX, USA

Abstract

Natural Gas Liquid (NGL) is the liquid product obtained from the bottom of the demethanizer in a gas processing plant. It consists of ethane, propane, butanes and natural gasoline (heavier hydrocarbons). Here in USA, the gas processors in the planning stages of a new gas plant, are in dilemma whether they should include the fractionation of NGL at the plant and sell the individual products or send the NGL to a central processing facility such as the one at Mont Belvieu, Texas. Overseas, most of the large gas plants fractionate the NGL at the plant site. However, in USA the gas processors must consider the economic feasibility before making such decisions, which is determined by supply and demand relationship and availability of pipelines to transport the NGL and/or NGL products. In fact, most of the old large gas plants in USA have the NGL fractionation facility but many recent gas plants have been designed without the NGL fractionation units. This paper investigates the economic feasibility or unfeasibility of having the NGL fractionation train at gas plants. The gas plants can operate in either ethane recovery or ethane rejection mode.

In case of ethane recovery, most of ethane is recovered from the bottom of the dmethanizer with NGL. However, in case of ethane rejection some ethane is allowed to escape with the residue gas by warming up the demethanizer bottom. Decision whether to recover or reject ethane is made based on the prices of sales gas and ethane. If the price of sales gas is high and the price of ethane is low, then ethane rejection would be more economical.

There are many parameters that affect the economics of whether to fractionate at the plant site or not. Amongst the important parameters are price of sales gas, NGL and NGL products, cost of fractionation, transportation and storage of NGL and fractionation cost of NGL at the hub. However, in this work only sales prices, and costs of fractionation and transportation were considered.

HYSYS simulation of NGL fractionation units was carried out for typical shale gas composition. Three different flowrates (150, 200 and 250 MMSCFD) were considered and the cases were analyzed in both ethane recovery and ethane rejection modes. The estimation of total module cost, operating cost and the resulting economic feasibility was analyzed for both current product prices and other assumed reasonable prices of NGL, sales gas and NGL products.

Incremental economic analysis approach was utilized meaning, NGL market value was considered as a raw material cost for the case of fractionation. For the gas composition considered and the current range of prices, the preliminary results are as follows. For both ethane recovery and ethane rejection modes at current prices, NGL fractionation at the plant site is not economical.

Introduction

Natural Gas Liquids (NGL) refers to ethane, propane, n-butane, i-butane and C_5^+ (natural gasoline). Demethanizer column separates its feed into residue gas and NGL. Residue gas, which is mainly methane, is further compressed and sold as sales gas based on its heating value. However, NGL derived from the bottom of the demethanizer column is pumped to deethanizer column where it is fractionated to ethane and C_3^+. C_3^+ is fed to the depropanizer column and split into propane and C_4^+. C_4^+ is then fed to the debutanizer column where it is separated into butanes (n-butane and i-butane) and C_5^+ (natural gasoline).

The butanes can be further fractionated to i-butane and n-butane but this is generally not done at gas processing plants.

Besides, gas processing plants are operated in either ethane recovery or ethane rejection mode based on the season, demand of ethane and price of the residue gas. Several NGL recovery process schemes have been described in the literature based on their inlet conditions, product requirements, mechanical constraints and operational requirements. Most common schemes among them are ISS (Industrial Standard Single-stage process), GSP (gas sub-cooled), CRR (cold recycle residue), RSV (recycle-split vapor), IPSI-1 (enhanced NGL recovery process) and IPSI-2 (internal refrigeration for enhanced NGL recovery process) [1]. Additional ethane recovery schemes and a new process scheme have been proposed [2] to perform a techno-economic analysis of each scheme.

The main focus of this work is to decide if one should fractionate NGL at the gas processing plant or send it to a central fractionating facility such as Mont Belvieu. However, in resolving such issues one must also consider a few factors. Besides, in this study only costs of fractionation and transportation have been considered. NGL can be transported via pipelines to the fractionation hub. NGL products can also be transported via pipelines from fractionators to storage and from storage facilities to end users or secondary storage [3]. Transportation fees of NGL from various basins to Mont Belvieu, ranging from 5 cpg for Permian basin to 21 cpg (cents per gallon) for Bakken field have been delineated [4]. Further, transportation and fractionation (T&F) costs of NGL has ranged from 35 cpg for old contracts to 55 cpg for newer contracts [5].

The objective of this work is

- To simulate the NGL fractionation units for different flowrates using both ethane recovery and ethane rejection modes.
- To determine the economic feasibility of NGL fractionation at the gas processing plant site.
  The variables considered in achieving these objectives are flowrate of gas, current and other assumed reasonable price of residue gas, NGL and NGL products.

Process Description

The process scheme used in this work is similar to sub-cooled process (GSP) scheme, where maximum heat is recovered from the process streams and need for external refrigeration streams is eliminated. The cold stream from the bottom of the tower are withdrawn and warmed up by exchanging heat with the stream going to the cold separator, before being recycled. Both JT (Joule-Thomson) valve and turbo-expander have been used to provide the necessary chilling to the rectifying section of the demethanizer. The same scheme can be used for both ethane recovery and ethane rejection mode by just turning off and on the demethanizer trim reboiler respectively.

Feed gas is sweetened and dehydrated and the gas flowing from the molecular sieve unit is split into two parts. One stream is fed to the HX-2 where it is cooled by overhead stream from the demethanizer column. While, the other stream is first cooled in the HX-1 and then further cooled in HX-3. These two split streams are then mixed together and fed to the cold separator as depicted in Fig. 1.

The cold separator flashes this stream into vapor and liquid streams. The vapor stream is further split into two parts. One part is cooled by pressure reduction from turbo-expander and fed to the demethanizer. While, the other part is cooled in LNG exchanger-1, passes through JT valve where its pressure is dropped and fed to the demethanizer as top feed. The liquid stream coming out from the cold separator also passes through JT valve where its pressure and temperature are dropped before being fed to the demethanizer column.

The overhead stream of the demethanizer provides necessary cooling for the HX-2, compressed, air cooled and sent to the sales gas pipeline. The two streams from the stripping section of the demethanizer provide the cooling for HX-3, warmed up and recycled back to the column. The second

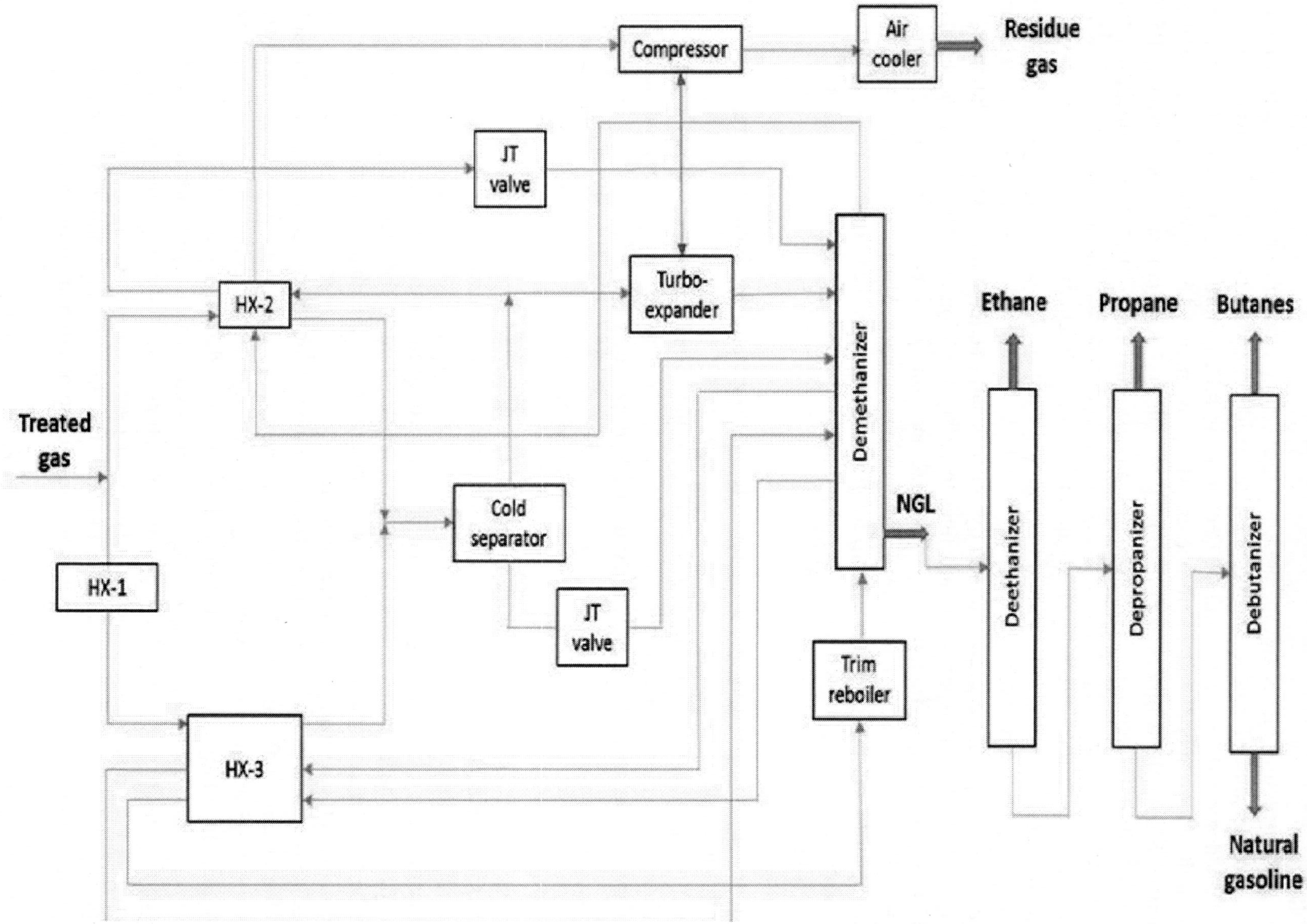

Fig. 1: Block diagram of NGL fractionation in ethane recovery mode

stream passes through the trim reboiler (which is turned off for ethane recovery mode) before being recycled back to the column. NGL is withdrawn as a liquid product from the bottom of the demethanizer. This way maximum ethane is recovered as demethanizer trim reboiler is turned off and column bottom is cold. Also, about 3% of methane is allowed in NGL stream [6]. This option is preferred when demand for ethane is high and the price of ethane is higher than in sales gas (ethane recovery mode). All fractionation column conditions are chosen according to the GPSA specifications [6].

The only difference in ethane rejection mode being cooling for HX-3 is provided by liquid stream from the cold separator which is then fed to the column. No side streams are withdrawn from the column and trim reboiler is turned on. This way column bottom section is hotter and NGL stream doesn't contain any methane. However, in this case some of the ethane is allowed to escape with the overhead stream (within specifications). In ethane rejection mode, the flowrate of NGL also reduces because it contains less ethane.

HYSYS Simulation

HYSYS process simulations were developed for 150, 200 and 250 MMSCFD of gas in both ethane recovery and ethane rejection modes. The feed conditions and composition are rendered in Table 1. The feed compositions used are from an existing gas plant in South Texas. The composition of the gas represents a typical Eagle Ford Shale Plates. The process description for each scheme is described below.

NGL fractionation in ethane recovery mode

The process description for ethane recovery mode is described in earlier section. The turbo-expander (K-100) provides the power for compressor (K-101). The flowrate of side-streams withdrawn from the de-methanizer column T-501 are controlled to achieve desired top (residue gas) and bottom product (NGL) specifications. The NGL fractionation train includes de-ethanizer, de-propanizer and de-butanizer column. The product specifications are adopted from GPSA data book.

NGL fractionation in ethane rejection mode

In ethane rejection mode, the trim reboiler (E-205) is turned on to warm up the de-methanizer column bottom section and let some of the ethane escape with the top product. The flowrate of stream 35 going to the trim reboiler and then recycled as stream 36 to the de-methanizer column is controlled to achieve the desired product specifications. The air cooler is added before the de-ethanizer column to cool the incoming NGL stream.

Table 1: Feed conditions and composition for both ethane recovery and ethane rejection

Stream specifications	**Feed**		
Flowrate (MMSCFD)	150	200	250
Pressure (psia)	950	950	950
Temperature (°F)	120	120	120
Components	**Composition (mol %)**	**Components**	**Composition (mol %)**
N_2	0.0030	n-C_4	0.0057
CO_2	0.0045	i-C_5	0.0019
C_1	0.8940	n-C_5	0.0013
C_2	0.0591	n-C_6	0.0013
C_3	0.0234	n-C_7^+	0.0011
i-C_4	0.0047		

Economic Analysis

The decision whether to fractionate or not to fractionate the NGL is made based on profitability of selling NGL vs NGL products. The prices for NGL, sales gas and NGL products are based on the values at Mont Belvieu, Texas. Transportation costs have been deducted from the prices at Mont Belvieu to determine the prices at the plant site area considered, i.e. Permian and Marcellus basin. The recent prices (February 2017) and price from 2012 (when the price of ethane and other NGL products were very high) of NGL and NGL products were considered [7].

Capital Cost Estimation

Capital cost was estimated using specifications of major equipment from HYSYS simulation. The process units considered in the capital cost estimation contain only the NGL fractionation train including heat exchangers, pumps, distillation towers used in de-ethanizer, de-propanizer and de-butanizer columns. The sizing information for all these units including pressure ranges, area, number of trays, diameter and height of column were obtained from HYSYS simulations. The heat exchanger duties were also found from the HYSYS simulations.

The capital cost estimation is made using total module cost technique, which is also called as InSide Battery Limit (ISBL) plant. It includes direct and indirect costs, contingency and fee costs. But unlike grass root costs, it does not include auxiliary costs which means NGL fractionation units are considered as an expansion to the existing facility. Direct and indirect costs were considered. The direct costs include equipments, material, utilities and operating labor while, the indirect costs consider freight, overhead, administration, distribution and selling expenses and others.

The contingency and fee costs are considered to be 15% and 3% of the bare module cost respectively [8]. The cost module is evaluated for year 2017 based on the Chemical Engineering Plant and Cost Index (CEPCI=553, [9]). The construction period of one year and project life of 15 years was assumed. The annualized interest rate of 10% was assumed with taxation rate of 30%. The total manufacturing cost is calculated by equation 4.1 formulated based on Turton et al. [8] and adopted to present operation.

$$COM = 2.48 * C_{OL} + 0.236 * FCI + 1.11(C_{RM} + C_{UT}) \quad (4.1)$$

where COM is the cost of manufacturing (M$), C_{OL} is the cost of operating labor (M$), FCI is the fixed capital investment (M$), C_{RM} is the cost of raw material (M$) and C_{UT} is the cost of utilities (M$).

Capcost was used to perform the economic analysis. The prices of sales gas, NGL and NGL products used for this study are portrayed in Table 2 [7]. The transportation costs from Marcellus and Permian basin to Mont Belvieu were considered to be 0.15 and 0.05 $/gal respectively [4]. Operating cost for each case was calculated per the procedure outlined by Turton et al. [8].

Table 2: Current and 2012 prices of NGL, sales gas and NGL products

Component	Current prices (Feb 2017)	2012 prices
Sales gas ($/MMBtu)	3.05	3.5
NGL ($/MMBtu)	7.5	11
Ethane ($/MMBtu)	3.37	11.5
Propane ($/MMBtu)	9.6	16
Butanes ($/MMBtu)	11.38	21
Natural gasoline ($/MMBtu)	12.5	22

Economic analysis at Marcellus basin location and current prices

The estimate of discounted Net Present Value (NPV) Discounted Cash Flow Rate Of Return (DCFROR) and Discounted Pay-Back Period (DPBP) was obtained for NGL fractionation in ethane recovery and ethane rejection mode for 150, 200 and 250 MMscfd gas flowrate. Fig. 2 represents the discounted NPV for all three cases is highly negative and with increase in flowrate, revenue losses increase indicating that NGL fractionation in these conditions is not economical. Fig. 3 depicts the economic analysis at current prices and Marcellus basin location for 150, 200 and 250 MMscfd gas flowrate in ethane rejection mode. It can be observed that NPVs for all the three flowrates are highly negative. However, from Fig. 2 and 3 it can be deduced that more losses are incurred in ethane recovery mode as compared to ethane rejection mode. The reason Discounted NPV is better in case of ethane rejection mode compared to ethane recovery mode is the fact that ethane as residue gas has higher value than ethane as pure. Hence, at current prices ethane rejection mode is preferred.

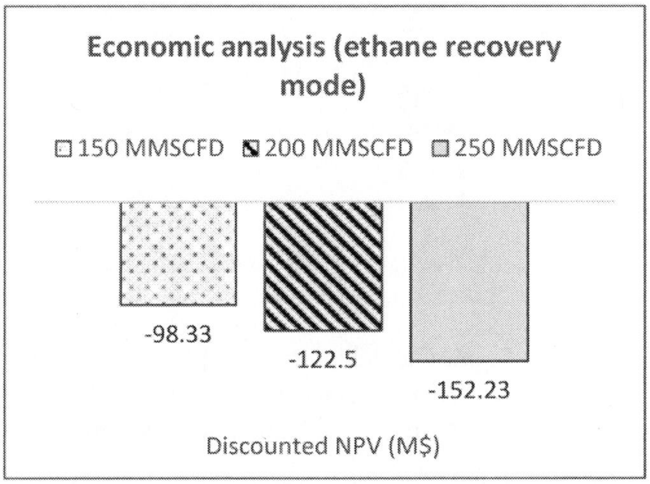

Fig. 2: Economic analysis at current prices and Marcellus basin location for 150, 200 and 250 MMSCFD gas flowrate in ethane recovery mode

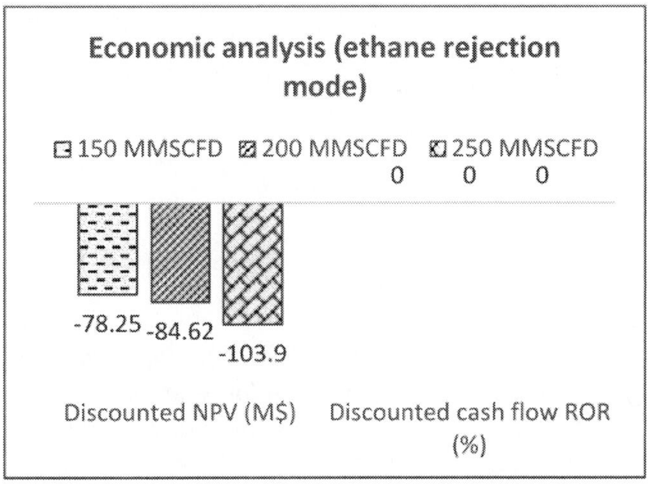

Fig. 3: Economic analysis at current prices and Marcellus basin location for 150, 200 and 250 MMSCFD gas flowrate in ethane rejection mode

Economic analysis at Permian basin location and current prices

The economic analysis at current prices and three different flowrates was also performed at Permian basin location in both ethane recovery and ethane rejection modes. Even if the distance and subsequently transportation costs decreases, NPVs are still highly negative for both ethane recovery and ethane rejection modes. Which implies that, it is uneconomical to fractionate NGL on plant site in these cases.

Effect of variation in location, price of sales gas, NGL and NGL products

The economic analysis was also performed at 2012 prices. Since, there was a boom in oil and gas market at that time and it is important to check if it was economical to fractionate the NGL at the gas processing plant at such higher prices.

a. Economic analysis at Marcellus basin location and 2012 prices

Fig. 4 and 5 represents the economic analysis for 2012 vs current prices and Marcellus basin location for 200 MMscfd gas flowrate in ethane recovery and ethane rejection mode. It is evident from Fig. 4 that it was highly profitable to fractionate the NGL on plant site in ethane recovery mode with DCFROR of 154% and DPBP of 0.1 year. Nevertheless, it was uneconomical to fractionate the NGL on plant site even for 2012 prices in ethane rejection mode (Fig. 5). As ethane price was very high (Table 2) compared to residue gas and by adopting ethane rejection mode, some of the ethane is allowed to escape with the residue gas, reducing the flowrate of pure ethane.

b. Economic analysis at Permian basin location and 2012 prices

Further, similar analysis was performed for Permian basin location for ethane recovery and ethane rejection modes respectively and the results are similar to that for Marcellus basin location. Indicating, at 2012 prices it was economical to fractionate the NGL on plant site for ethane recovery mode but not for ethane rejection mode.

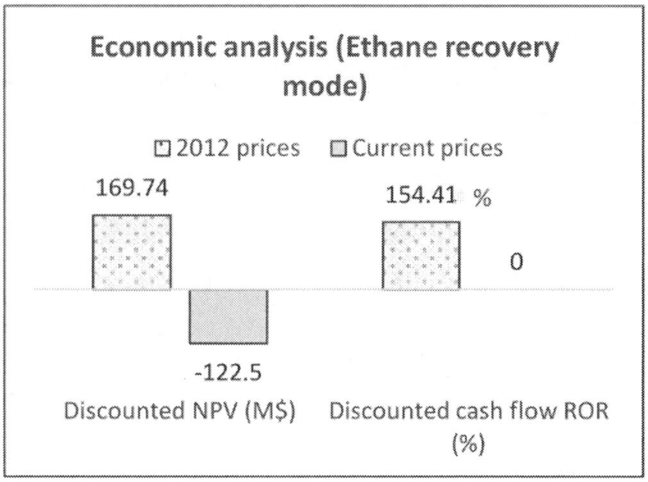

Fig. 4: Economic analysis for 2012 vs current prices and Marcellus basin location for 200 MMSCFD gas flowrate in ethane recovery mode

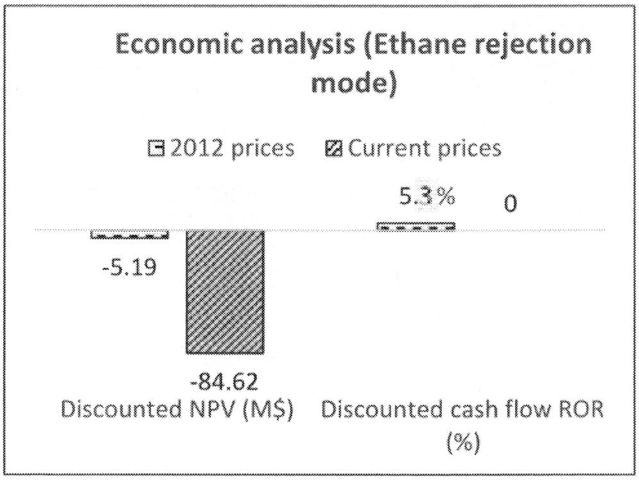

Fig. 5: Economic analysis for 2012 vs current prices and Marcellus basin location for 200 MMSCFD gas flowrate in ethane rejection mode

Product revenue distribution at Marcellus basin location for current and 2012 prices

The share of each product in the revenues was analyzed at both current and 2012 prices for 200 MMscfd gas flowrate and Marcellus basin location in ethane recovery and ethane rejection modes. This analysis helps to understand the importance of each product in revenue distribution and also identifies which product affects the revenues most significantly.

Besides, since at current prices, ethane price is significantly low (Table 2) compared to other NGL products, propane plays a major role in product revenue distribution for both ethane recovery and ethane rejection modes.

However, in 2012 ethane prices were higher than sales gas and comparable to other products. Thus, for 2012 prices in ethane recovery mode, ethane plays a dominant role in product revenue distribution. In ethane rejection mode at 2012 prices, again propane becomes the prevailing component as ethane flowrate decreases due to rejection of ethane in sales gas.

Conclusions

- For both ethane recovery and ethane rejection modes at current prices, NGL fractionation at the plant site is not economical. However, ethane rejection mode yields better results than ethane recovery mode.
- In contrast, NGL fractionation for 2012 prices in ethane recovery mode was highly profitable at the plant site but it was not economical in ethane rejection mode.
- At current prices, fractionation of NGL at the gas processing plant is not recommended.

- Further, between the prices of 2012 and 2017 there were a set of prices, during which fractionating at the plant became more economical than not fractionating it.

References

[1] Risk based optimization for representative natural gas liquid (NGL) recovery processes by considering uncertainty from the plant inlet, M. Getu, S. Mahadzir, Y. Samyudia, M.S. Khan, A. Bahadori, M. Lee, J. Nat. Gas Sci. Eng. 27 (2015) 42-54

[2] Techno-economic evaluation of a novel NGL recovery scheme with nine patented schemes for offshore applications, J.H. Park, M.S. Khan, R. Andika, M. Getu, A. Bahadori, M. Lee, J. Nat. Gas Sci. Eng. 27 (2015) 2-17

[3] Natural gas liquids (NGLs), D. Brockett, Penn state extension (2015)

[4] Consider GTL as an economic alternative for stranded natural gas and ethane, J. Cantrell, J.A. Bullin, G. Mcintyre, C. Butts, B. Cheatham, Hydrocarb. Process. 92(10) (2013) 77-80

[5] Can Mont Belvieu Handle the NGL Supply Surge? - Part II, C. Mitchell (2012)

[6] Engineering data book, Gas processors Suppliers Association, 12th edition (2004)

[7] https://www.eia.gov/energyexplained/index.cfm?page=hgls_prices

[8] Analysis, synthesis and design of chemical processes, R. Turton, R.C. Bailie, W.B. Whiting, J.A. Shaeiwitz, 3rd edition, (2009)

[9] CEPCI: http://www.chemengonline.com/current-economic-trends-cepci-january-prelim-and-december-final/

LNG MARKET AND TECHNOLOGY

Jeffrey Zhang, CH-IV International Inc., Houston, TX

Introduction

U.S. LNG Outlook

According to the latest projections by the United States Energy Information Agency (EIA)[1], U.S. liquefied natural gas (LNG) exports are predicted to increase to 88 million metric tons per annum (MTPA) in 2040 from 9 MTPA in 2015 (Figure 1). This paper seeks to assess these projections and provide other insights into the U.S. LNG market addressing the following points:

- The availability of current and projected U.S. domestic gas supply to support gas volume requirements of the export projections;
- The alignment of projected export figures with existing and planned U.S. LNG liquefaction facilities;
- The dominant liquefaction technologies selected by industry to support the increase in liquefaction volumes
- Overview of the identified dominant technologies as well as high-level project evaluation guidelines are not covered in this paper but are contained within the accompanying conference presentation.

This paper does not seek to analyze the global natural gas market nor provide insight into the "Shale Revolution". The effective date of this paper is April 5th, 2017, all collected data and actual developments which occur beyond this date is not covered in this paper.

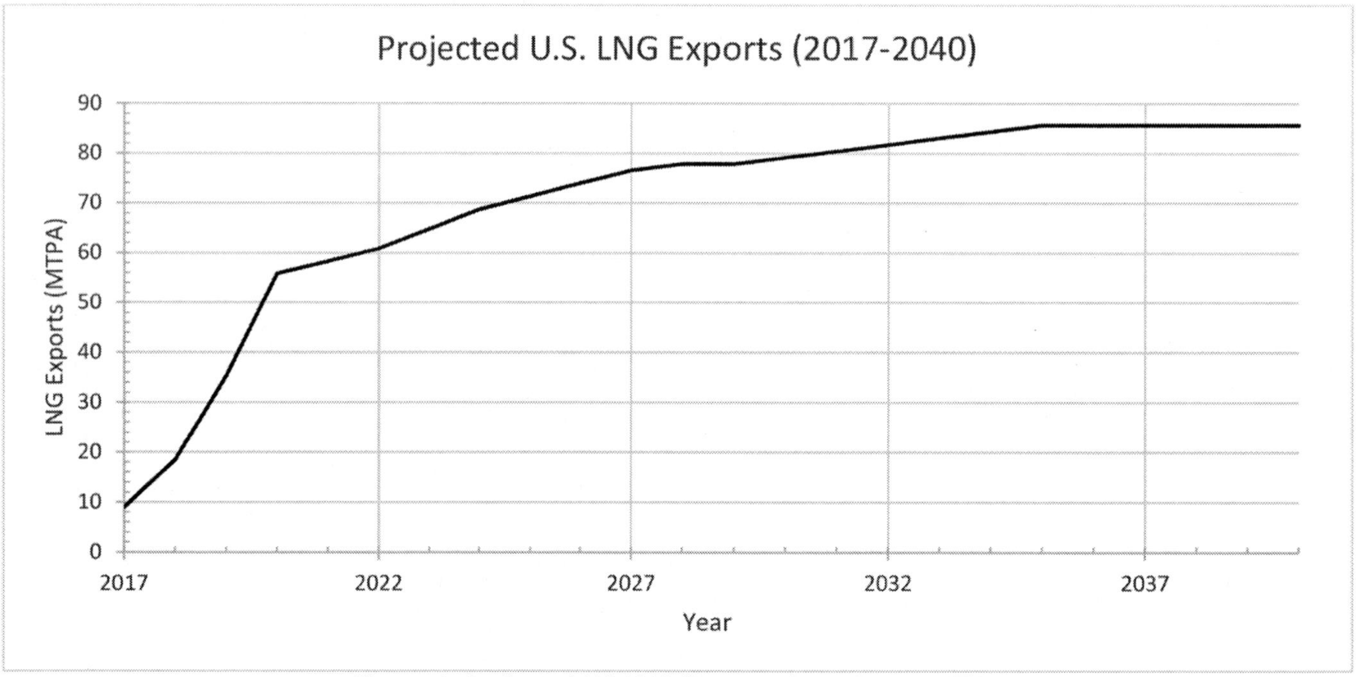

Figure 1: Projected US LNG Exports. Data Source: EIA.

Method

U.S. Natural Gas Market – Past, Present, and Future Projections

The general linkage between the availability of U.S. LNG for export and the existence of surplus natural gas in the US market (defined as marketed natural gas production less natural gas consumption less gas exports by non-LNG means) is established through a high level comparison of the U.S. domestic natural gas marketed production[2] and consumption[3] from 1949 to 2016 to actual LNG exports U.S. LNG exports 1985 to 2016[4]. The comparison of historical production and consumption is also used to help illustrate the dramatic impact of the Shale Revolution upon total U.S. Marketed Natural Gas Production[5]. The EIA's domestic natural gas production and consumption projections to 2040 is presented and is used to obtain the maximum projected available LNG for export. This is used to place the EIA's LNG export projections 2017-2040 into context with projected surplus natural gas. The production projections was also checked against 2017 contingent resources[6,7] reported by the United States Geological Survey (USGS)[8] to place the LNG exports in further context.

U.S. LNG Export Capacity – Projects Based Analysis

A comprehensive tally of existing, under construction, and otherwise proposed (U.S. facilities only) LNG import and export facilities was compiled using available government[9] and industry[10] data. The information includes construction location, import/export capacity, liquefaction technology (export only), onshore/floating (import only), and (projected if under construction or in the proposal phase) start up dates for these facilities. Only projects with announced start dates are included in the analysis. The data was used to:

- Validate U.S. government projections of U.S. LNG exports 2017-2040 by comparing projected installed (existing and under construction) export capacity to 2040 as well as projected federally authorized (DOE and FERC approval for both FTA and non-FTA export) capacity with announced start dates to 2040
- Determine projected U.S. industry proposed (includes all installed) capacity to 2040 and as compared to surplus natural gas of the reference case to 2040 and as compared to projected federally authorized facilities capacity to 2040
- Determined the additional gas required to feed the industry proposed export capacity is added to the reference U.S. cumulative production to 2040 and a new required cumulative production to 2040 is obtained. This new required cumulative production is compared to the 2017 contingent resources to determine if there is enough gas available to support the industry proposed export capacities.
- Determined total world installed import and export capacity to 2040
- Place both federally authorized and proposed U.S. export capacity to 2040 in context with world non-U.S. import capacity and non-U.S. installed export capacity to 2040
- Determine the most utilized liquefaction technologies in the U.S. by export capacity

Results and Discussion

U.S. Natural Gas Market – Past, Present, and Future Projections

A high level comparison of the U.S. domestic natural gas marketed production and consumption from 1949 to 2016 (Figure 2) shows that the market history may be divided into four (4) distinct segments: (1) growth and market match between 1949 and 1970 with production and consumption growing in tandem and with production slightly outstripping demand for this period, (2) relative stability with domestic production slightly above consumption for the period between 1970 and 1984 with relatively no change in the market size, (3) production decline and leading to a yawning net production gap between 1984 and 2002, and (4) production increase after 2002 such that by 2016, the net production gap had closed.

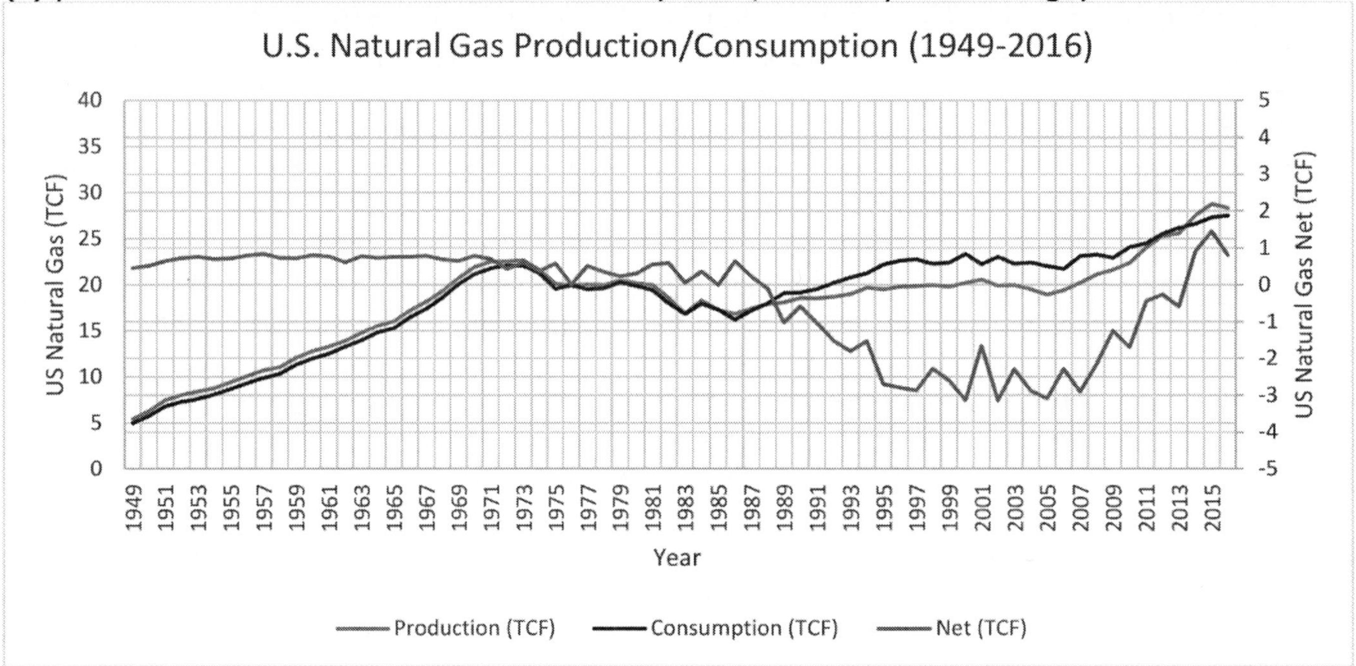

Figure 2: U.S. Natural Gas Production/Consumption (1949-2016) with Net Production. Data Source: EIA.

The linkage between surplus natural gas and LNG exports is demonstrated by the historical U.S. LNG export data from 1985 to 2016 (Figure 3). During the period of natural gas deficit (negative surplus), exports were negligible compared to projections for 2017 and beyond. It was only after the effect of the Shale Revolution became apparent in the 2010s that LNG exports began to increase.

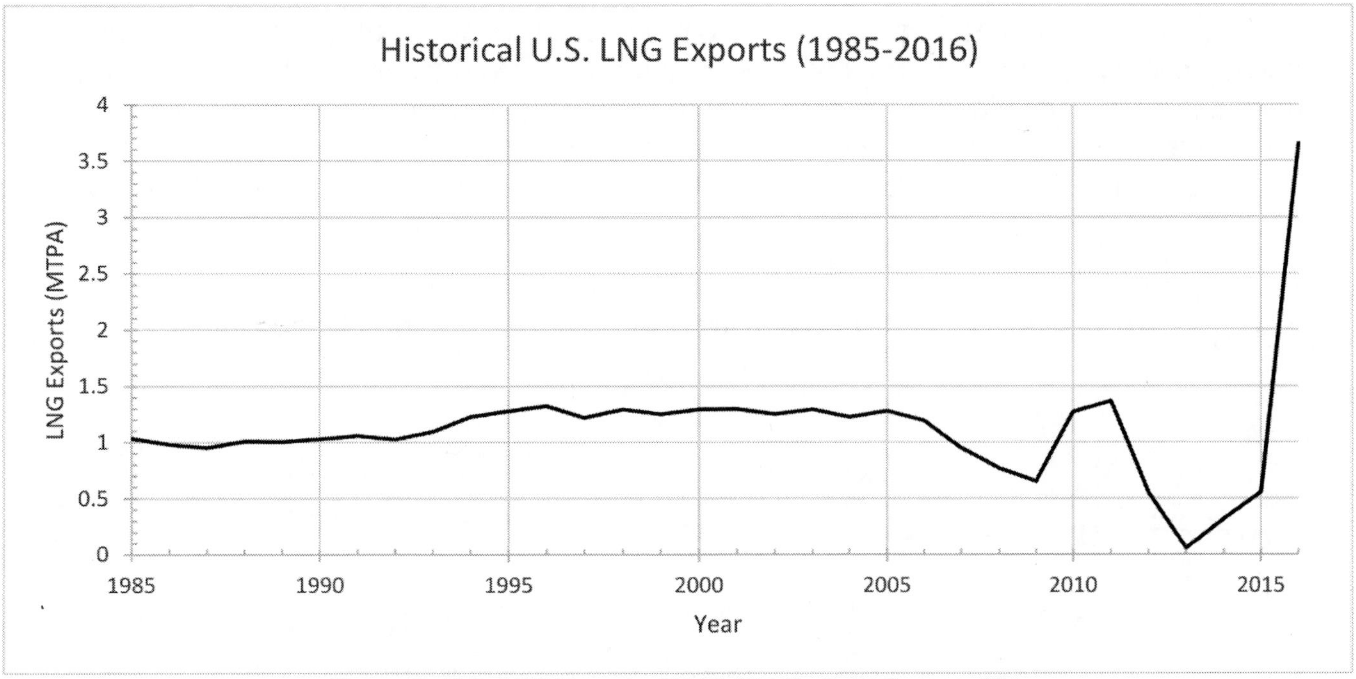

Figure 3: U.S. LNG Exports (1985-2016). Data Source: EIA.

The recovery of U.S. domestic natural gas production between 2002 and the present day is broadly as a result of increase in natural gas production from unconventional sources since 2002. The increase is so substantial that production from shale sources tripled between 2010 and 2015[5]. Constituting 6% of total US gas production in 2007 to 53% in 2015 (Figure 4).

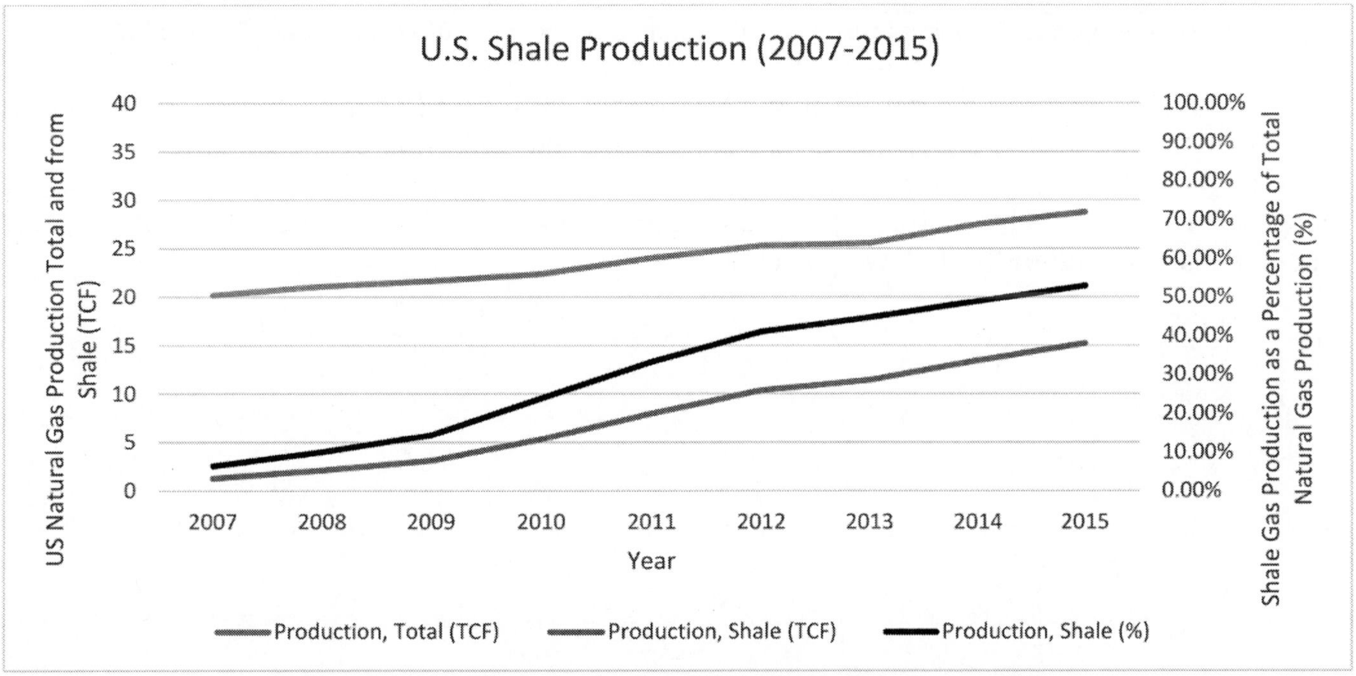

Figure 4: U.S. Shale Production (2007-2015). Data Source: EIA.

With the natural gas production from unconventional sources providing increased productive capability, the EIA projects greater natural gas production into the future (Figure 5) producing large potential gas surpluses which may be used in LNG exports. The EIA's reference yearly marketed gas production, consumption, and pipeline exports projections to 2040 was utilized to establish the reference yearly surplus natural gas available for LNG exports (Figure 5).

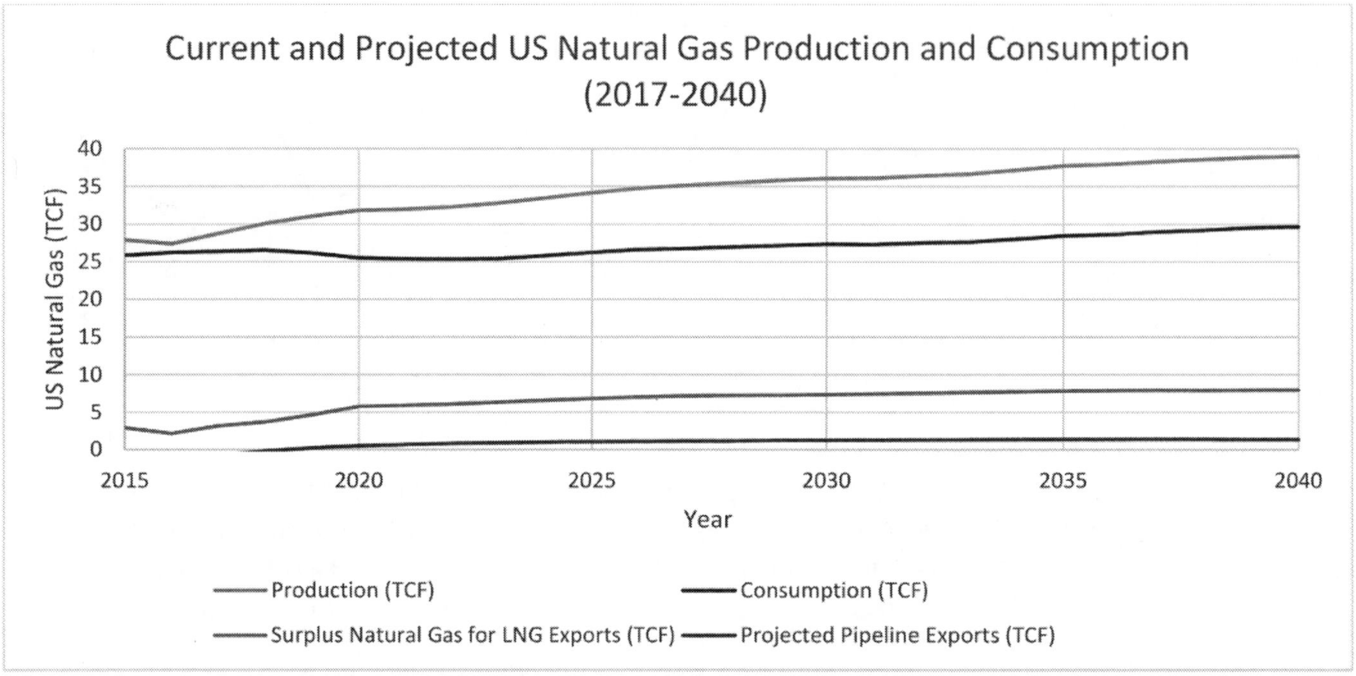

Figure 5: Projected US Natural Gas Production and Consumption (2017-2040). Data Source: EIA.

This reference surplus natural gas is compared with the EIA's LNG export projections to 2040 (Figure 6). The U.S. government reference projected LNG export case is found to compare favorably with the U.S. government projected surplus natural gas available for LNG exports. The comparison of cumulative natural gas production to the 2017 total assessed unconventional resources is deferred to the following section.

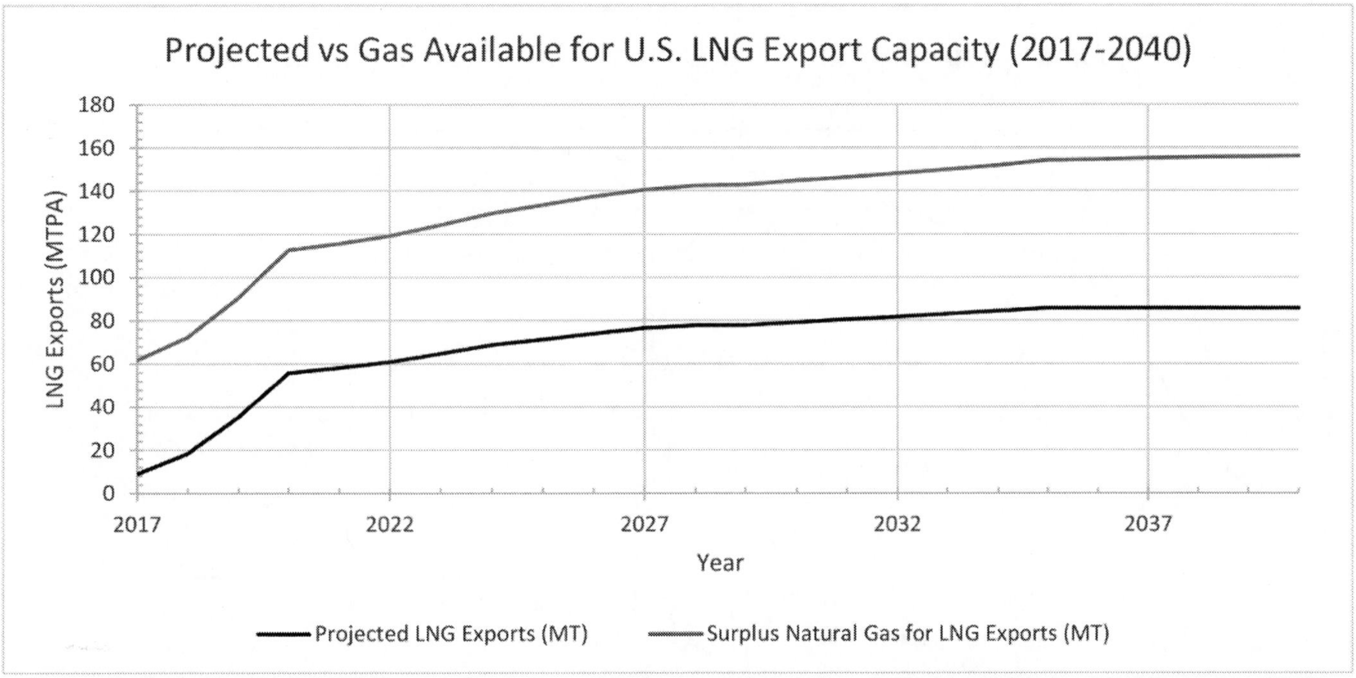

Figure 6: Projected U.S. LNG Export Capacity vs. Reference U.S. Natural Gas Available for LNG Export to 2040. Data Source: EIA.

U.S. LNG Export Capacity – Projects Based Analysis

This section focuses on placing the federal projections of U.S. LNG exports within the context of available natural gas supply and actual installed capacity, both in the United States and around the world. The current projected U.S. installed capacity falls short of the EIA's projected U.S. LNG exports to 2040, however, the EIA's projected LNG exports are largely in line with already announced federally authorized export capacity to 2040 (Figure 7). Thus, the government projected LNG export volumes may be largely fulfilled meet with already authorized capacity. There is also 24 MTPA of federally authorized export capacity whose project owners have not announced start dates. While these were not included in the formal data analysis of this paper, the existence of this authorized capacity means that the already authorized export capacity will most likely be able to provide sufficient export capacity to fully satisfy the EIA projected LNG export figures. The federally authorized capacity is compared to all U.S. proposed LNG export facilities with announced start dates. The results are presented in Figure 8.

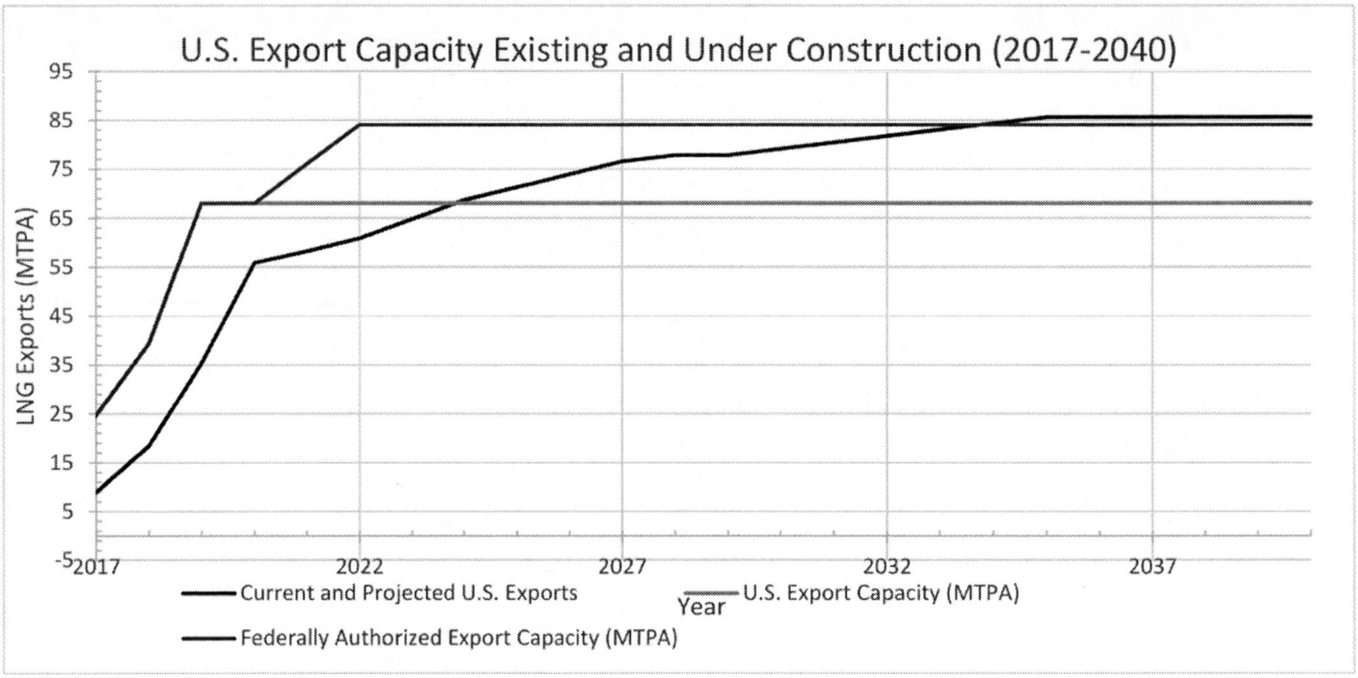

Figure 7: U.S. Export Capacity Existing and Under Construction (2017-2040). Data Source: FERC.

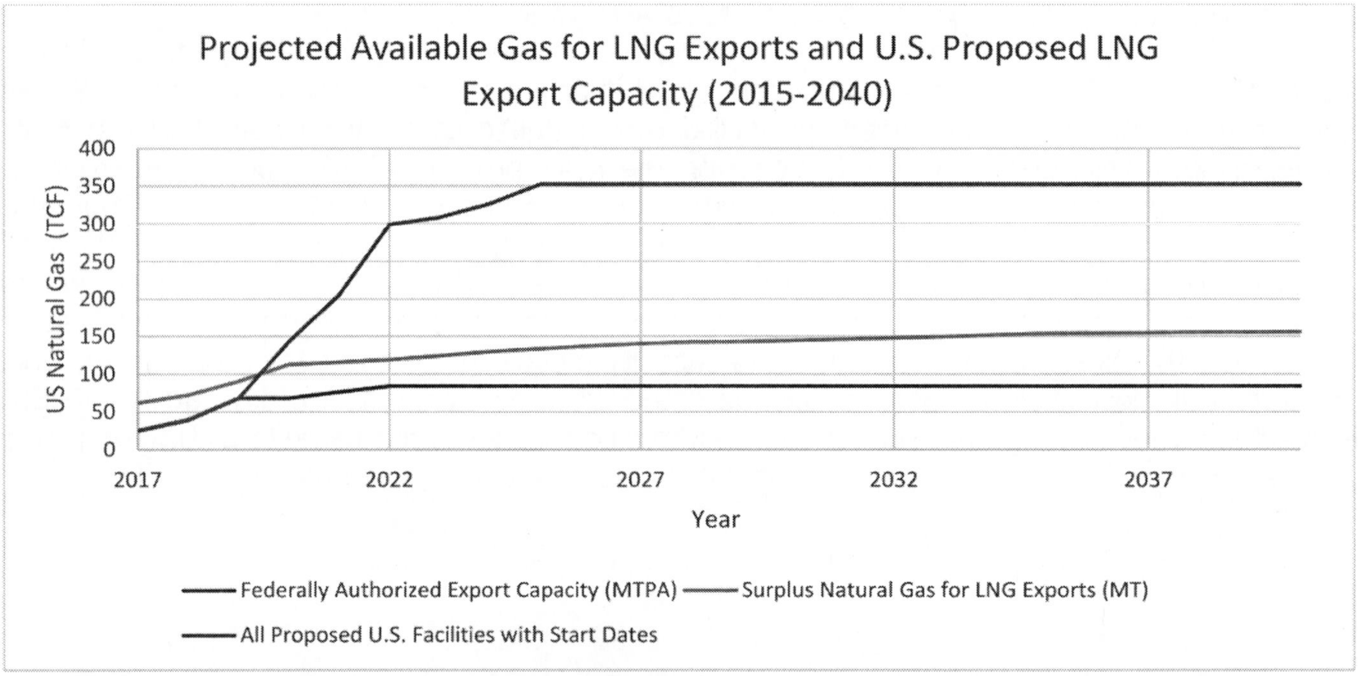

Figure 8: Projected Available Gas for LNG Exports and U.S. Proposed LNG Export Capacity (2017-2040) (2015-2040). Data Source: FERC, EIA, IGU.

From Figure 8, U.S. industry desired export capacity, defined here as the sum of all proposed export projects with announced start dates, far outstrips projected surplus natural gas supply available for export. Two questions are posed, does the United States possess enough natural gas resources to support the export capacity proposed by industry to 2040 and does the rest of the world have enough installed import capacity to accommodate the industry desired export capacity to 2040.

The natural gas volumes equivalent to industry desired export capacity to 2040 is added to projected baseline natural gas production to 2040. Cumulative production with and without industry desired export capacity as is compared to the assessed unconventional natural gas resources available in the U.S. as of 2017 (Figure 9). Unconventional resources are used as a stand-in for all contingent resources due to the already large role that unconventional reserves play in overall U.S. natural gas production (Figure 4). The results show that while the U.S. has enough domestic resources to support baseline LNG export projections (which is included in baseline production), additional resources may be required to between now and roughly 2035 in order to fully accommodate U.S. industry desired export capacity to 2040.

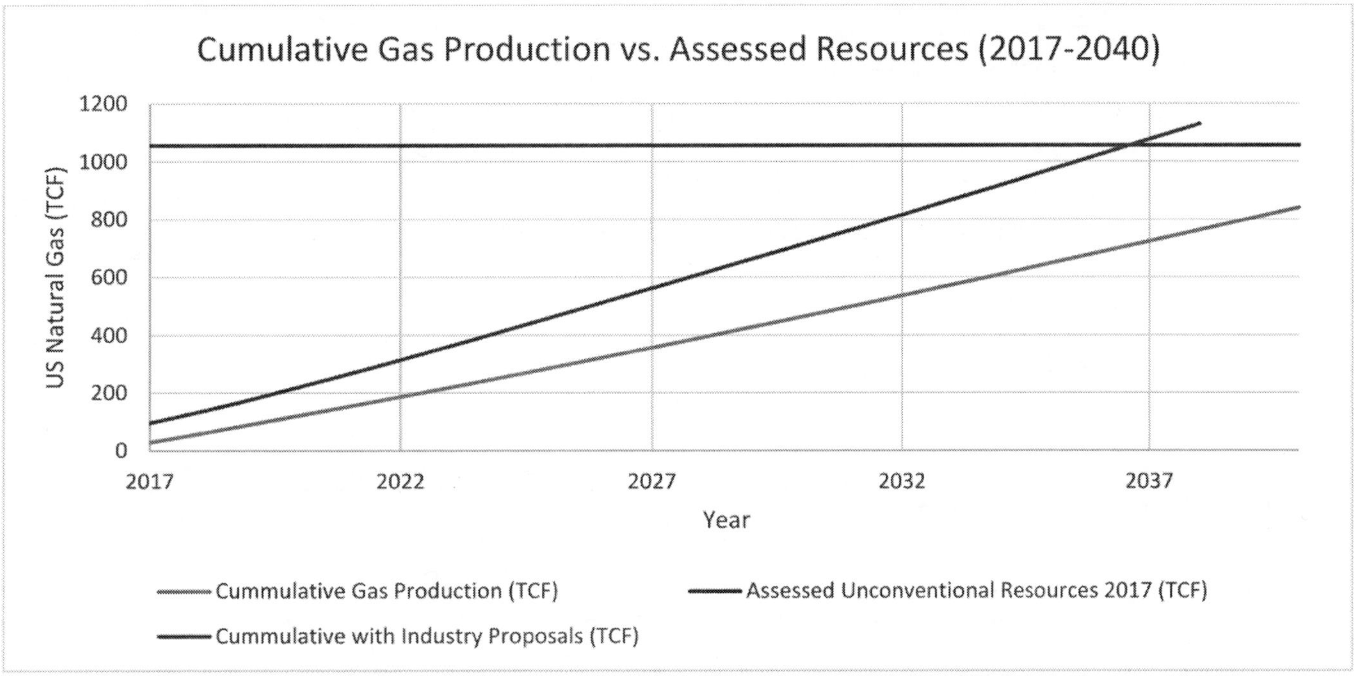

Figure 9: Cumulative Gas Production vs. Assessed Resources (2017-2040). Data Source: USGS.

The U.S. desired export capacity is placed within the context of global LNG installed import and export capacity. Historically, global import capacity has outstripped global export capacity by a wide margin (Figure 10). U.S. desired export capacity (which includes U.S. installed capacity) represents the entirety of the available import between non-U.S. installed export capacity and world installed import capacity to 2040 (Figure 11). The U.S. desired export capacity is shown to be theoretically feasible according to the preceding projected scenario.

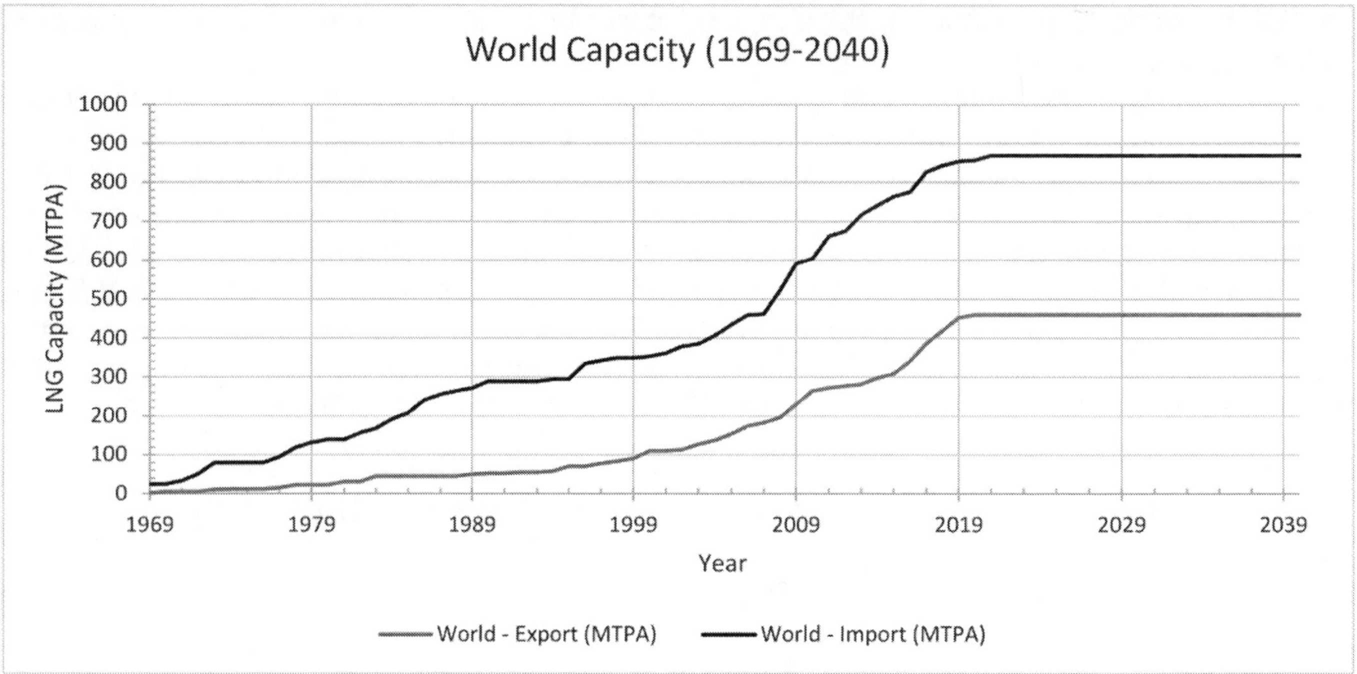

Figure 10: World Capacity (1969-2040). Data Source: IGU.

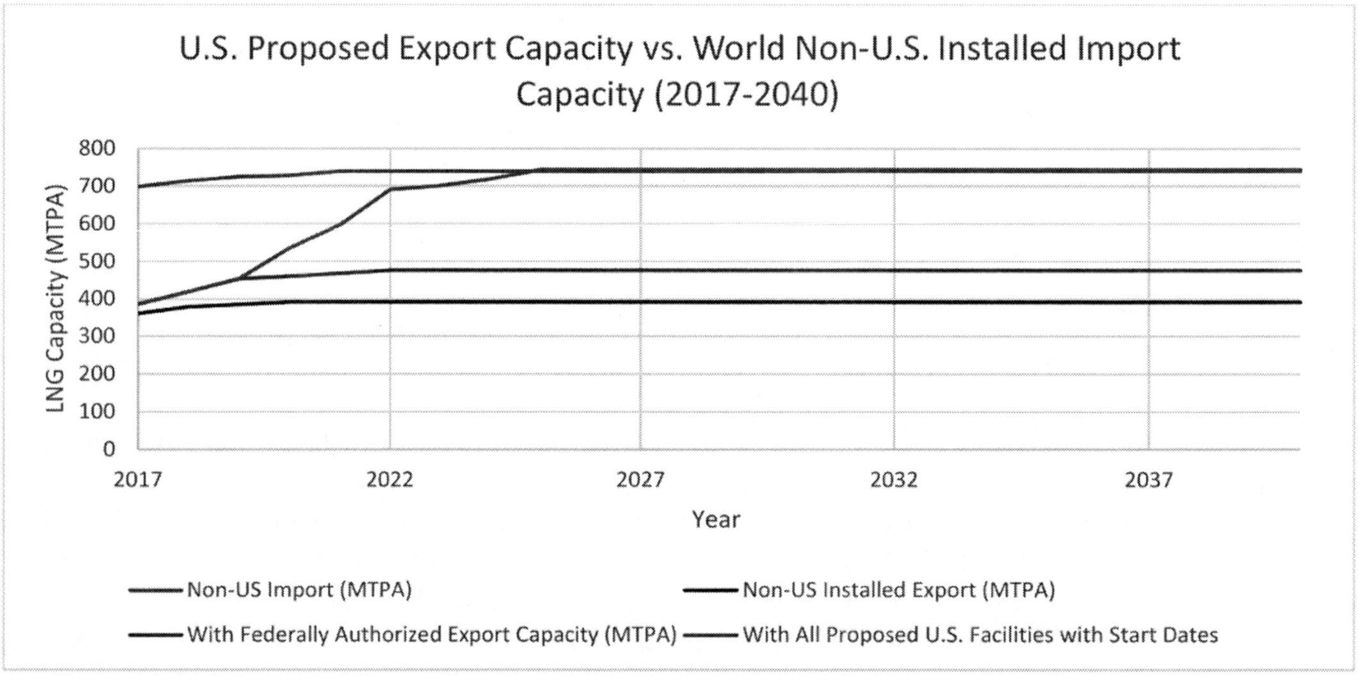

Figure 11: U.S. Proposed Export Capacity vs. World Non-U.S. Installed Import Capacity (2017-2040). Data Source: FERC, IGU.

Conclusion

From the analysis presented, the author concludes that the LNG projected exports by the EIA is possible and may be achieved with the reference gas production and consumption trend as well as the already authorized export capacity (if portions of authorized export capacity with no announced start dates are included). The author concludes that new gas resources may be required to support U.S. desired export capacity and that similarly, new global import capacity may be required to support the same U.S. desired export capacity.

Conversion Table

Table 1: Conversion Table

From		To
1 Trillion Cubic Feet of Natural Gas per Year (1 TCF/yr)	=	19.17 Million Tonnes Per Annum of LNG (MTPA)
1 Quadrillion BTU Equivalent of Gas per Year (1 Quad/yr)	=	19.55 MTPA LNG

References

1. (EIA) USEIA. *Annual Energy Outlook 2017 (AEO2017).* Washington D.C. 2017.
2. Administration USEI. Natural Gas Gross Withdrawals and Production. 2017; https://www.eia.gov/dnav/ng/NG_PROD_SUM_A_EPG0_VGM_MMCF_A.htm.
3. Administration USEI. Natural Gas Consumption by End Use. 2017; https://www.eia.gov/dnav/ng/hist/n9140us2A.htm.
4. Administration USEI. Liquefied U.S. Natural Gas Exports. 2017.
5. Administration USEI. Shale Gas Production. 2017; https://www.eia.gov/dnav/ng/ng_prod_shalegas_s1_a.htm.
6. SPE. *Petroleum Resources Management System*2007.
7. OGRC S. *Guidelines for Application of the Petroleum Resources Management System*2011.
8. Survey USG. Energy Resources Program - Assessment Summary Maps/Tables. 2017; https://energy.usgs.gov/OilGas/AssessmentsData/NationalOilGasAssessment/AssessmentUpdates.aspx.
9. Commission FER. FERC: Documents & Filing - FERC Online - eLibrary. 2017; https://www.ferc.gov/docs-filing/elibrary.asp.
10. (IGU) IGU. *2017 World LNG Report.* Barcelona, Spain 2017.

Visual Measurements of Hydrocarbon Freeze-out Temperature in LNG Mixtures

Arman Siavashi

Saif Al Ghafri and Eric May

2017 AIChE Annual Meeting
Minneapolis, USA
1st November 2017

ACLNGF

Vast Gas Resources: Australia's Northwest Shelf

Estimated $150B+ investment during this decade

Research Foci: Offshore production (flow assurance)

Gas processing & LNG Production

Carbon Capture & Sequestration

Project	Resource	(Estimated) Start-up	Proposed Capacity [mmtpa]	Concept
Pluto	5.0 TCF	2012	4.3	LNG Plant at Burrup Peninsula
Gorgon	47.3TCF 295MMbblc	2016	15	LNG Plant on Barrow Island
Prelude	2.8 TCF 120MMbblc	2017	3.6	Floating LNG
Wheatstone	4.5 TCF	2018	8.9	LNG Plant at Onslow
Ichthys	12.8 TCF 527MMbblc	2018	8.4	LNG Plant at Darwin
Browse	13.3 TCF 312MMbblc	2020+	7-15	FLNG

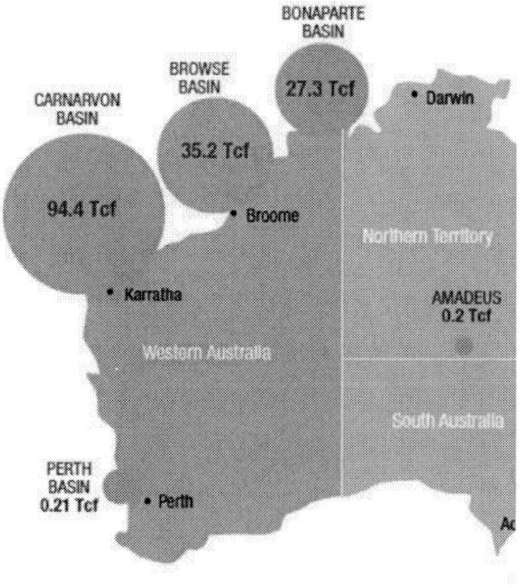

2

RasGas LNG Train 4: Scrub Column & Main Cryogenic Heat Exchanger (MCHE)

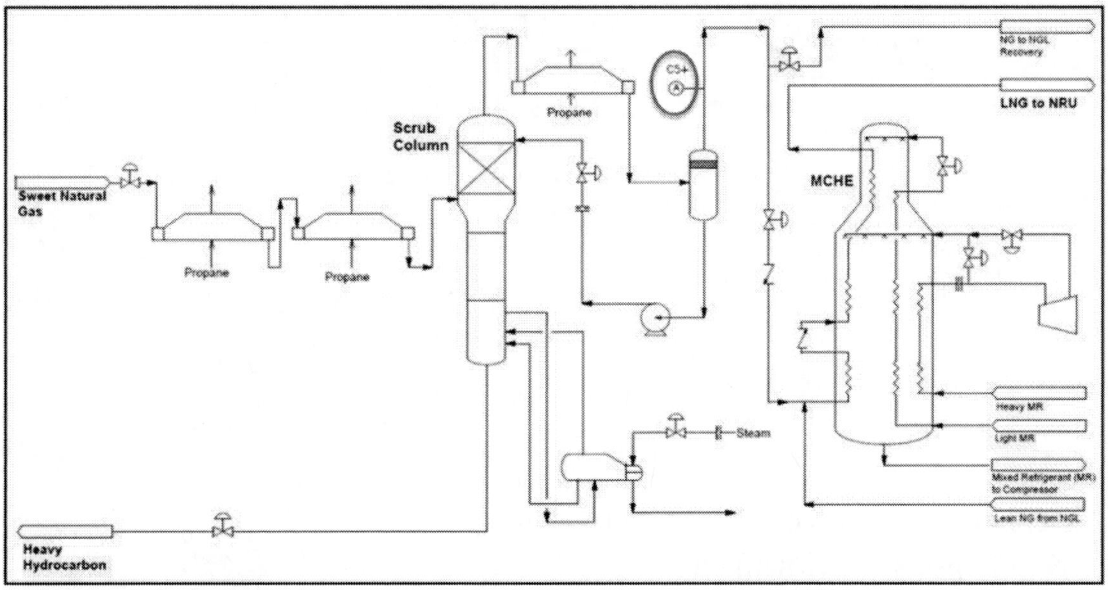

From Ismail & Al Thani, LNG-18, Perth 2016

3

C_{5+} content of MCHE feed gas

(adapted from Ismail & Al Thani, LNG-18, Perth 2016)

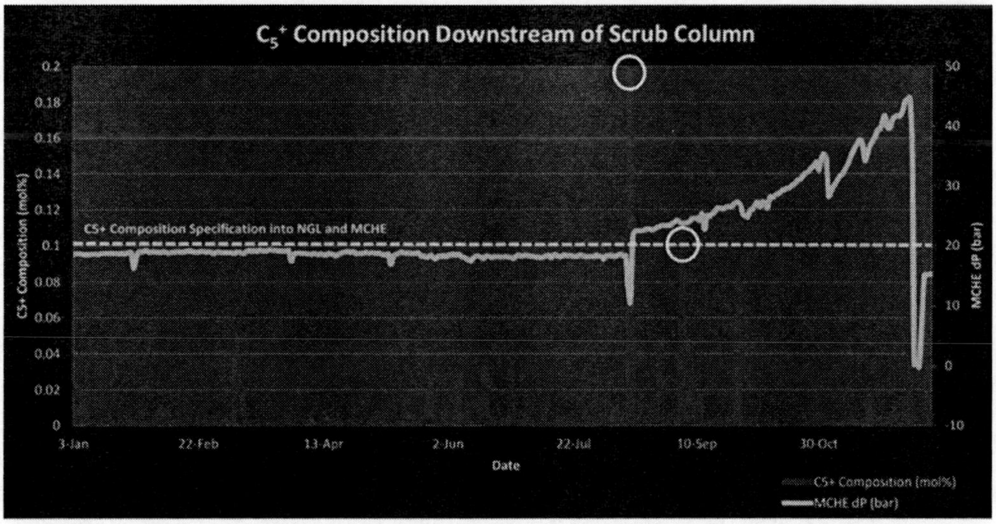

One clear event of off-spec gas (C_{5+} > 0.2 mol%) correlated with start of increasing Δp. (2nd event irrelevant as train in Lean LNG mode)

Operators ignored this event as considered negligible risk.

4

Methane + BTEX: Data

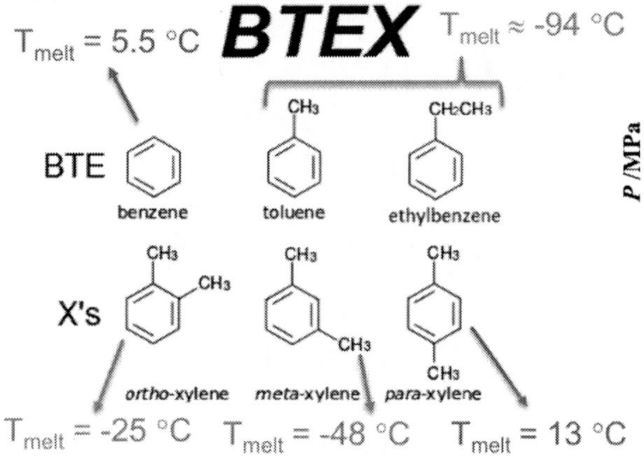

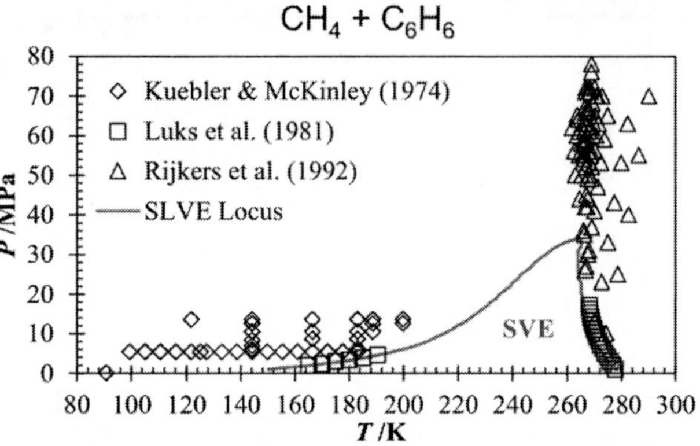

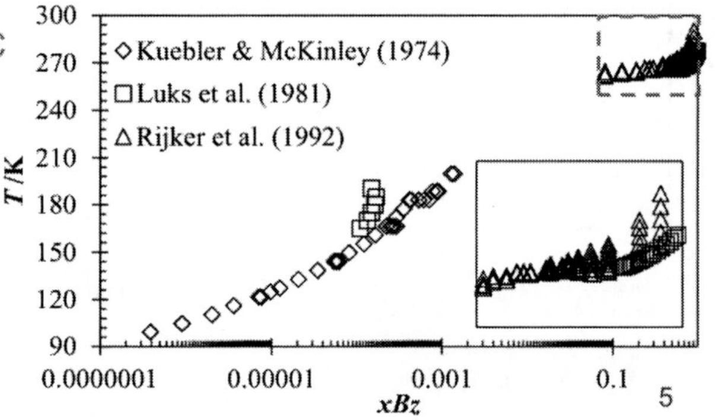

Limited SLE data for CH_4+C_6H_6 at LNG-relevant conditions

No SLE data for CH_4+p-C_8H_{10} at LNG-relevant conditions

5

Methane + BTEX: Models

Current industry-standard approach for calculating BTEX solubility in LNG is empirical activity coefficient model: **KLSSP from GPA**

No pressure dependence, limited concentration range

Alternative: use solid equation with cubic EOS:

$$\ln(f_i^s) = \ln(\varphi_{pure,i}^L P) - \frac{\Delta H_{f,i}}{RT_{m,i}}\left[\frac{T_{m,i}}{T} - 1\right] + \frac{\Delta c_{p,i}^{L\to S}}{R}\left[\frac{T_{m,i}}{T} - 1 + \ln\left(\frac{T}{T_{m,i}}\right)\right] - \frac{\Delta v_i^{L\to S}(P - P_m)}{RT}$$

$$x_i = \frac{f_i^s}{\varphi_i^L P}$$

From PR EOS with binary interaction parameter tuned to SLE data

Implement in freeware: *ThermoFAST*

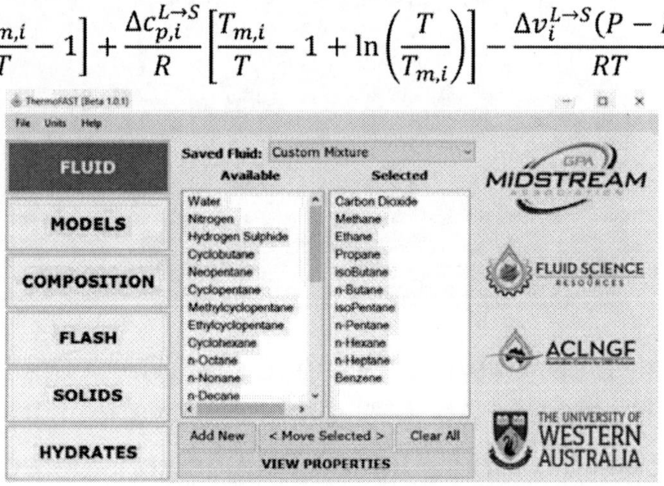

http://fsr.ecm.uwa.edu.au/

6

C

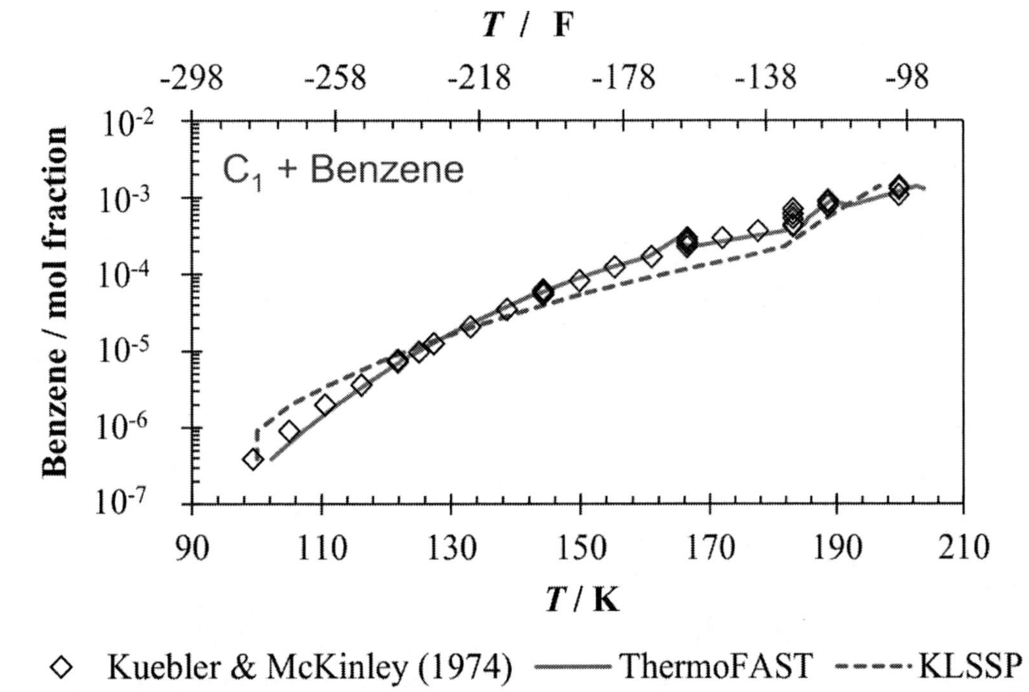

What about other binaries for which there is little or no SLE data?

Visual SLE Cell

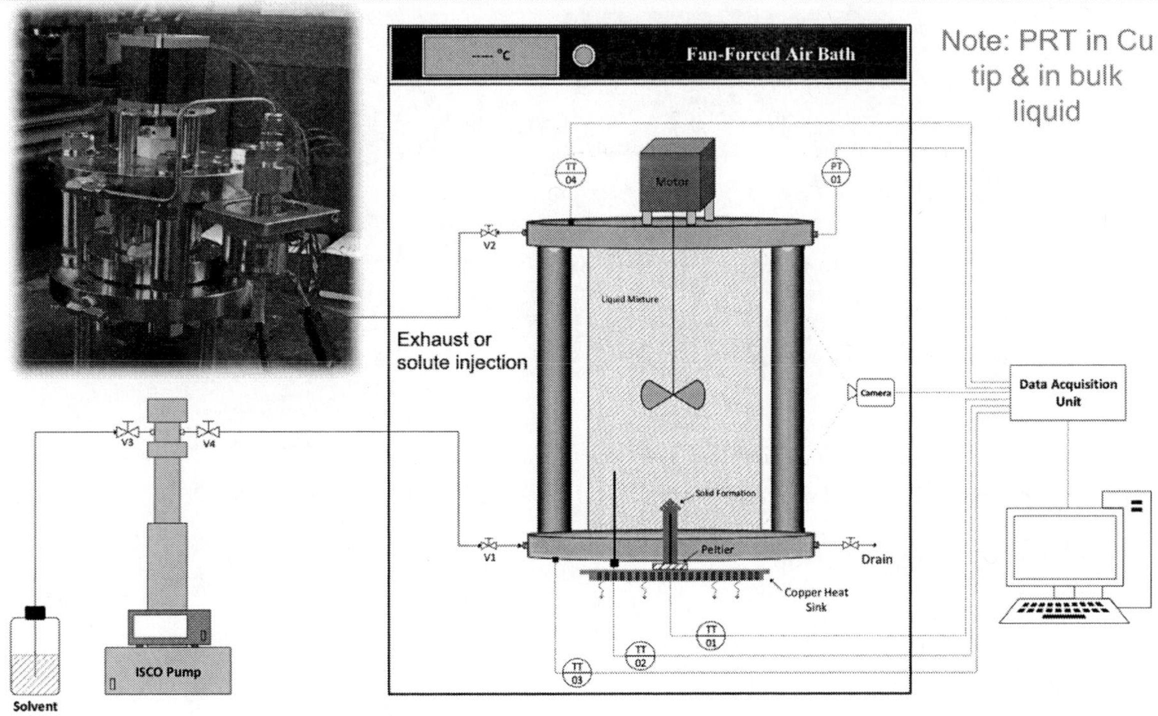

Siahvashi et al. (2017) *J. Chem. Eng. Data* DOI:10.1021/acs.jced.7b00171

Melting & Freezing Temperature Measurements

1. Cool bulk liquid down to 0.5 K above expected T_{melt}
2. Use Peltier to cool Cu tip at 0.01 K·min^{-1} until solids form.
3. Slowly increase Cu tip temperature until solid phase disappears. Repeat with tighter T bounds if required.

SNR: min detectable solid ≈ 10 mg in 50 mL ≈ 0.03% by mass

Standard uncertainties:

$u(x_{solute}) \approx 0.003$, $u(T_{melt}) \approx 0.2$ K, $u(p) \approx 20$ kPa

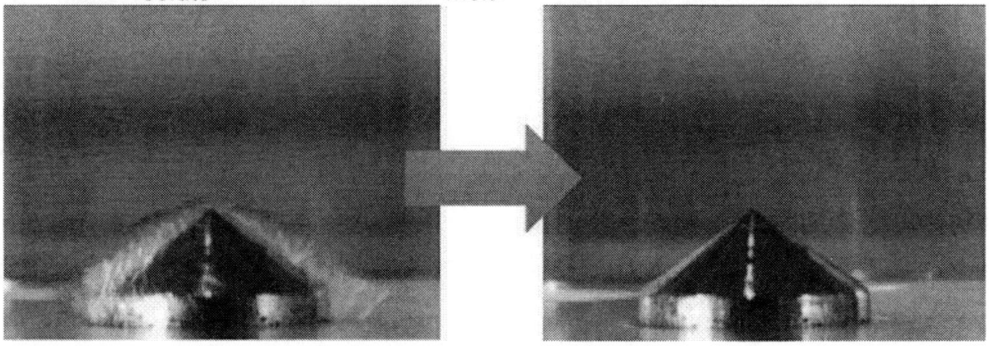

9

Solid-liquid-vapour eqbm: C_1+C_{pxy}

For gaseous components, need to describe VLE of binary to calculate x_{solute} at T_{melt}

Future upgrade to include liquid phase sampling & analysis

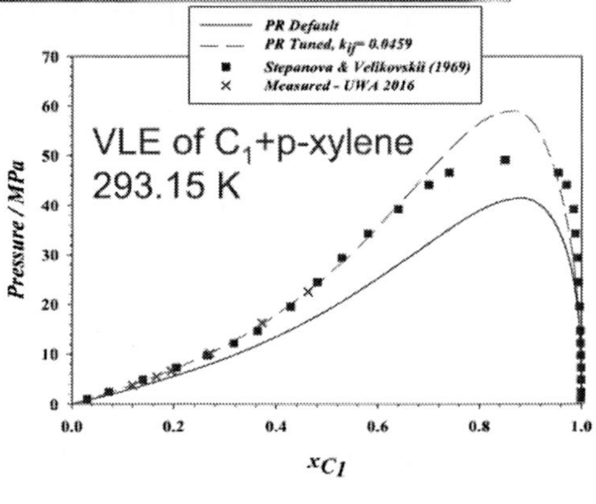

10

Apparatus Validation: C_{18} + cyclohexane

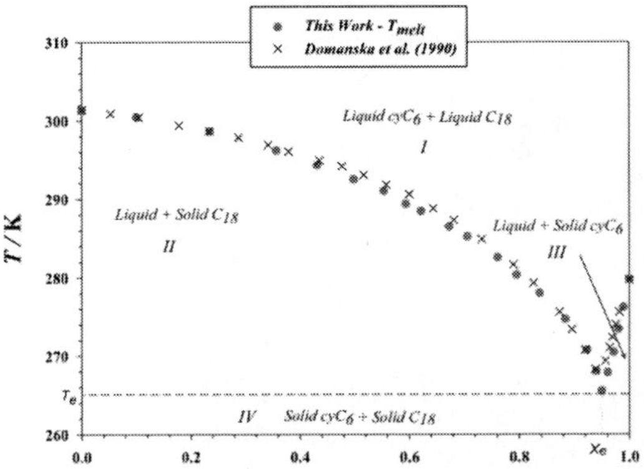

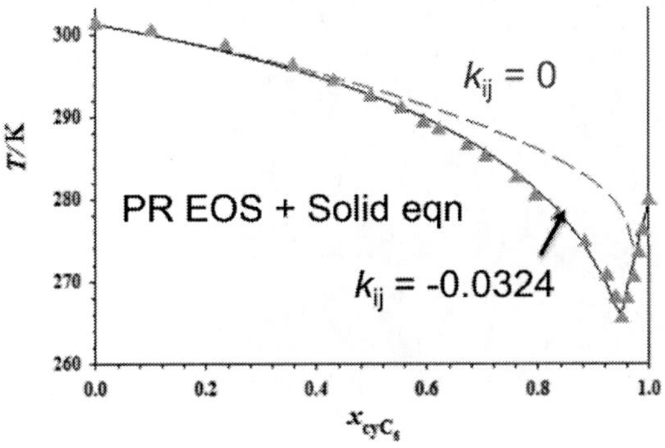

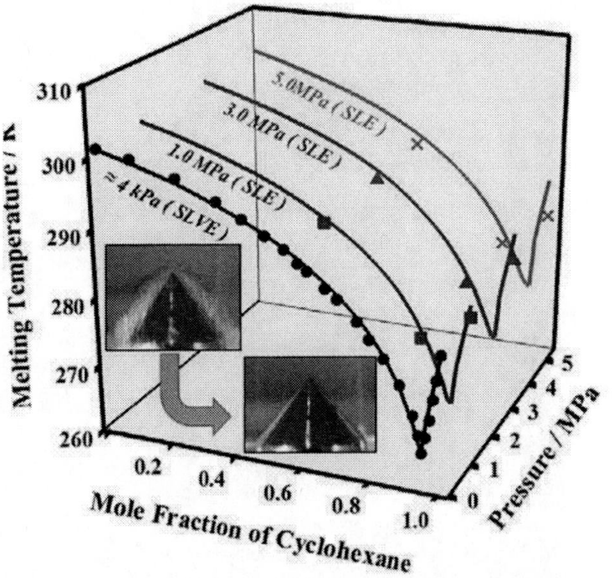

Cubic EOS tuned to SLVE data gives better high pressure SLE predictions than empirical models

11

Cryogenic Operation

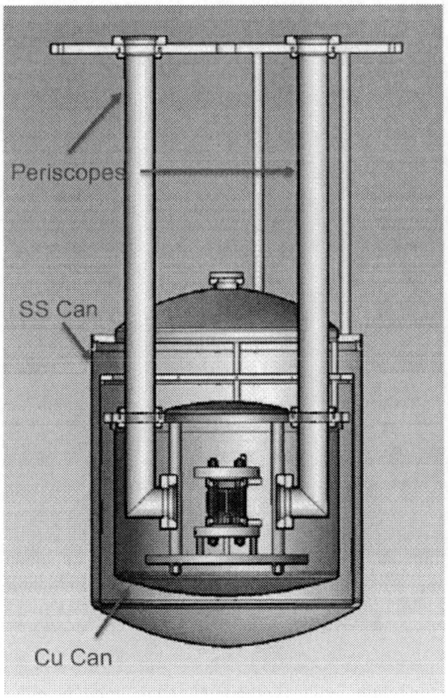

12

Second Phase of the Project: Dewar System T down to -190°C

13

0.8 ethane + 0.2 p-xylene: -35 °C, 3.5 MPa

Illustrative video only: Accurate T_{melt} determination involves mixing as Cu tip heated

14

First T_{melt} data for $p\text{-}C_8H_{10}$ in C_1 or C_2

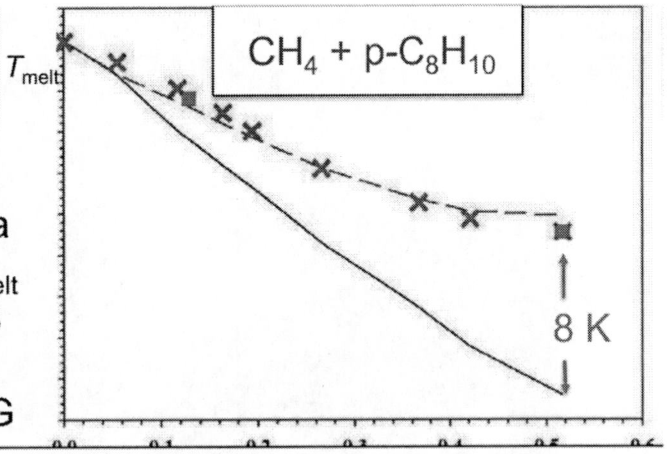

CH_4 + p-xylene binary data:

- $0 < x_{C1} < 0.52$, *p* to 28 MPa
- 8 K under-prediction of T_{melt} at $x_{C1} = 0.5$ by default EOS
- Tune BIP to T_{melt} data for improved prediction of LNG freeze-out risk

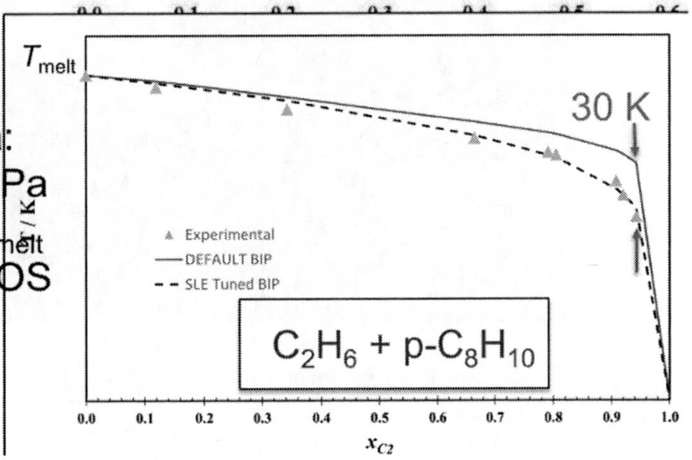

C_2H_6 + p-xylene binary data:

- $0 < x_{C2} < 0.95$, *p* to 5.9 MPa
- 30 K over-prediction of T_{melt} at $x_{C1} = 0.95$ by default EOS
- Tune BIP to T_{melt} data

15

Take-away message

- BTEX freeze-out is a significant operational issue in LNG production
- Developed robust SLE model tuned to limited available data
- Commissioned cryogenic visual SLE cell operating down to 90 K at pressures to 30 MPa

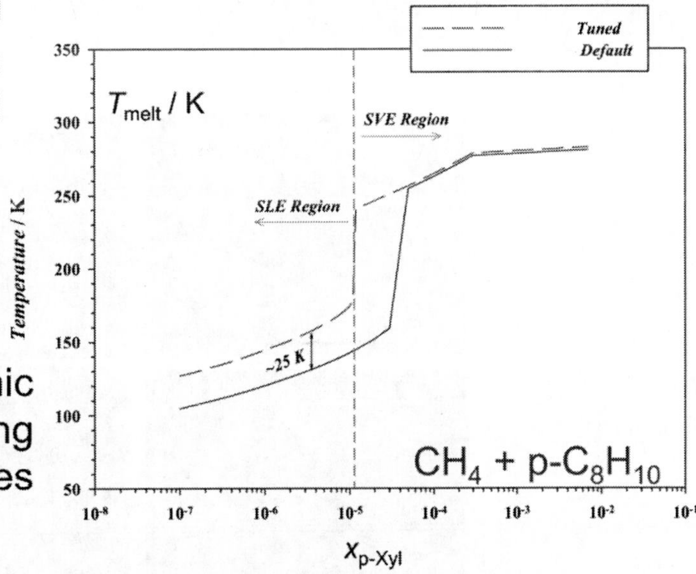

- Produced first SLE & SLVE data for p-Xylene in C_1 or C_2
- Un-tuned models for p-Xylene in C_1 under-predict T_{melt} at LNG conditions by 25 K compared to new, tuned-model

16

THE UNIVERSITY OF WESTERN AUSTRALIA

THANK YOU

Fluid Science & Resources
www.fsr.ecm.uwa.edu.au/

Australian Government
Australian Research Council

Chevron

GPA MIDSTREAM ASSOCIATION

ACLNGF

MDEA injection increased WAT by 6.5°C

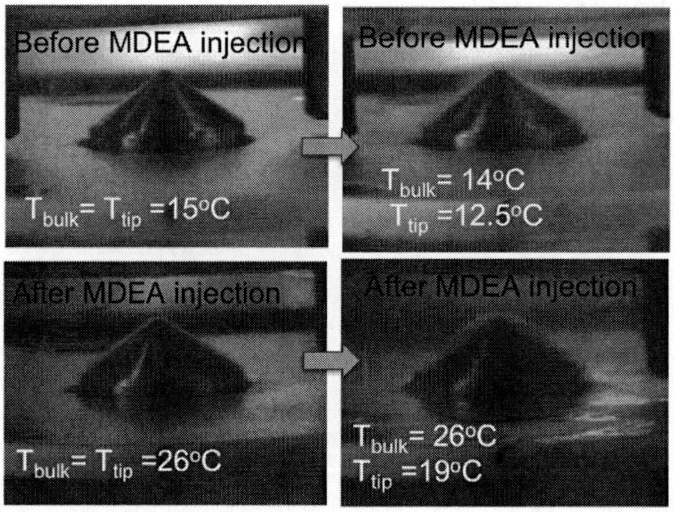

18

Comparisons with literature SLE data

Captures pressure dependence too (does not assume at bubble point)

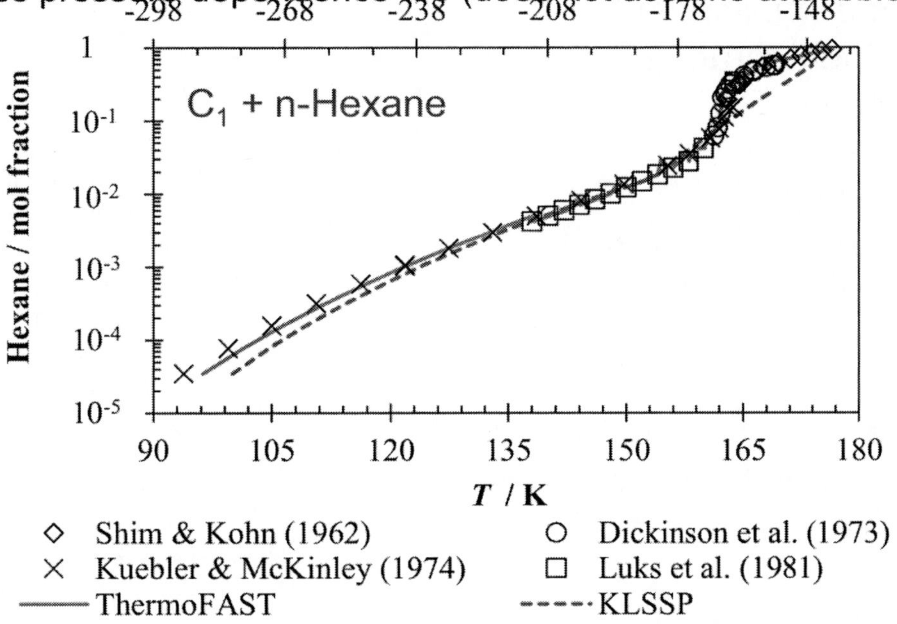

Captures pressure dependence too (does not assume at bubble point)

19

First Phase of Visual SLE Cell Development: Incubation System 0 < T /°C < 70

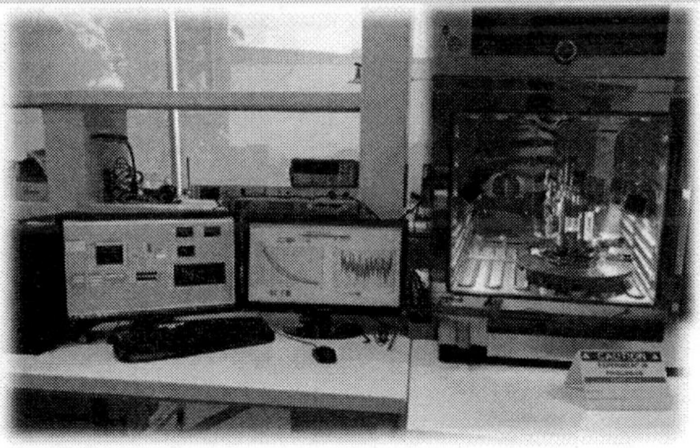

P_{max} = 300 bar

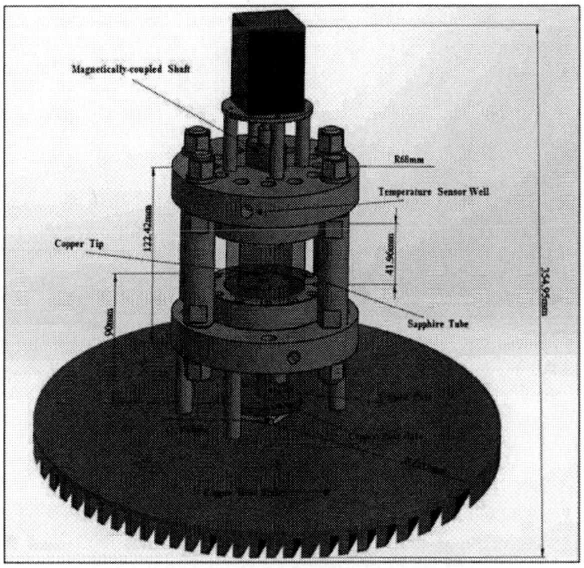

20

Induction Time Measurements

Determine equilibrium (melting) temperature then superheat

Set bulk liquid to equilibrium temperature & sub-cool copper tip to T_{freeze}

Measure induction time required for first solid to form on tip

Useful for future studies of nucleation rates

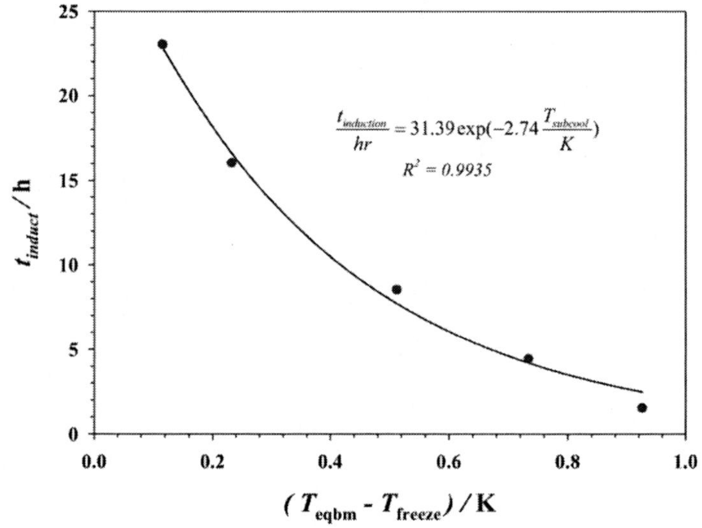

21

Cyclohexane + C_{18} SLE

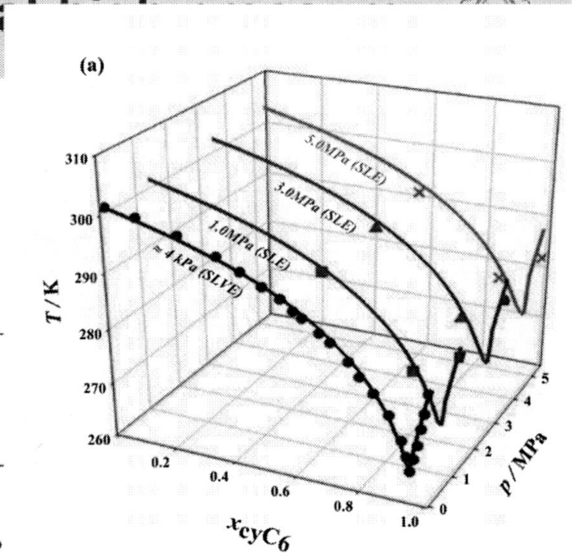

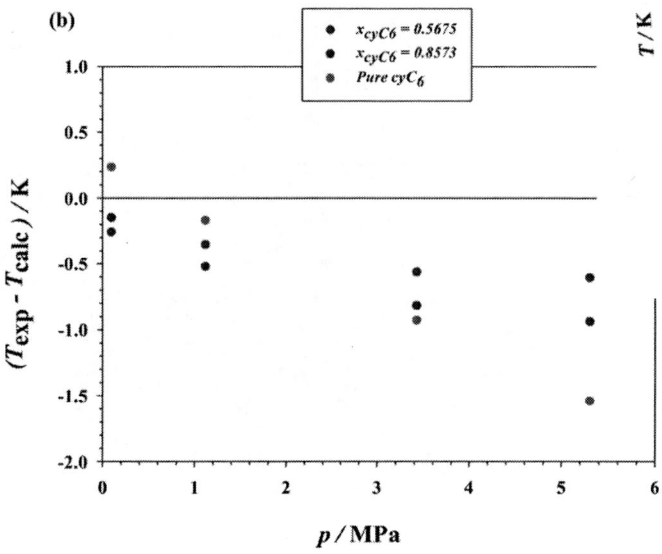

22

Level Measurement – Air Bath at 297 K

BEFORE methane injection (Pure pxy)- Photo taken after 1 hr for equilibration.

AFTER methane injection – Photo taken after 2 hr step-stirring for equilibration.

23

498907 Plant-Wide Modeling, Techno-Economic Analysis and Optimization of the Shale-Gas to Dimethyl Ether (DME) Process Via Direct and Indirect Synthesis Route

Abstract:

Due to the uncertainty in supply and prices of petroleum-based fuels, various alternative, clean fuels are being investigated. Dimethyl ether (DME) is a potential alternative to petroleum-based fuel because it is nonâ€"toxic, eco-friendly, has minimal particulate emissions, and has properties similar to LPG and diesel with a very high cetane number (55- 60)[1]. DME can be produced from a variety of feed stocks like coal, biomass, natural gas, and shale gas via direct or indirect synthesis route. Studies have shown that direct synthesis route has a higher DME production compared to indirect synthesis route. Currently, majority of the DME is produced commercially from coal via indirect synthesis route in the Asia-pacific region. However, in the United States, shale gas is available in abundance and seems to be a more attractive feedstock. Very limited studies are available in the open literature on techno-economic evaluation and optimization of shale gas-to-DME process via direct and indirect synthesis routes. The focus of this presentation will be on the process synthesis, economic evaluation, and optimization of the shale gas-to-DME process via direct and indirect synthesis in a multi-software environment.

In this study, plant-wide models for shale gas-to-DME synthesis processes via direct and indirect routes are first developed in Aspen Plus V8.4®. Kinetic models of the pre-reforming reactor, autothermal reforming reactor and DME synthesis reactors are developed and validated with the experimental data. For acid gas removal (AGR), two technologies, namely the Rectisol and MDEA/PZ technologies, are evaluated. A novel DME separation process that can efficiently separate DME, syngas and CO_2 has been developed. Binary interaction parameters for the vapor liquid equilibrium (VLE) model of the methanol-CO_2, DME-CO_2, DME-H_2O and DME-CO mixtures are regressed using the experimental data. Aspen Energy Analyzer V8.4® is used to design an optimal heat exchanger network by pinch analysis and the heat exchangers are sized rigorously using Aspen Exchanger Design and Rating V8.4®. The modeled equipment is mapped using the Icarus database in Aspen Process Economic Analyzer V8.4® (APEA) and economic evaluation is performed. An optimal design of the process is obtained by maximizing the Net Present Value (NPV) by using a multi-software platform comprising of Matlab, Microsoft Excel, Visual Basic, Aspen Plus, and APEA.

To summarize, the presentation will focus on the following aspects:

- Synthesis of the plant-wide models for DME synthesis following direct and indirect routes
- System level energy analysis of shale gas-to-DME production processes for varying H_2/CO ratio, CO_2 capture rate, and AGR technology
- Uncertainty analysis due to the raw materials price, product price and scale of the plant
- Optimal design of the process by maximizing NPV

References

Ogawa, T.; Inoue, N.; Shikada, T.; Inokoshi, O.; Ohno, Y. Direct Dimethyl Ether (DME) Synthesis from Natural Gas. Studies in Surface Science and Catalysis Natural Gas Conversion VII, Proceedings of the 7th Natural Gas Conversion Symposium. 2004, 379â€"384.

502790 New Thermophysical Properties Measurements of Complex Mixtures Relevant to Liquefied Natural Gas (LNG) Processing

Abstract:

Knowledge and understanding of the thermodynamic and thermophysical properties such as viscosity, surface tension, heat capacity and density have always been essential to LNG's (Liquefied Natural Gas) evolution because of the tremendous technical challenges present at all stages in the supply chain from production to liquefaction to transport. Each stage is designed using predictions of the mixture's thermophysical properties as a function of temperature, pressure and composition. Unfortunately, current available models lead to equipment over designs of 15% or more. To achieve better designs that work more effectively and/or over a wider range of conditions, new fundamental property data are essential, both to resolve discrepancies in our current predictive capabilities and to extend them to the higher-pressure conditions characteristic of many new gas fields. Due to the insufficient current available experimental data, new innovative techniques are required to measure different thermophysical properties at high pressures and wide range of temperatures including near the mixture's critical points (where gas and liquid become indistinguishable and the existing predictive fluid property models used breakdown).

In this work, we present a wide range of experimental measurements on different binaries and ternaries mixtures relevant to LNG processing's, with particular focus on viscosity, surface tension, heat capacity and density. For this purpose, customized and specialized high-pressure wide-temperature range apparatus were designed and constructed. The viscosity was measured using a vibrating wire viscometer apparatus consisting of centreless ground tungsten rod while mixtures densities were obtained by means of a single-Titanium sinker magnetic-suspension densimeter and an Isochoric cell apparatus. The surface tension was measured using a capillary rise method while the heat capacity was measured using a differential scanning calorimetry (DSC), that was also used to obtain the enthalpy of fusion and melting temperatures. Mixtures compositions were determined by knowing the mass of each component used to prepare the mixture. This was verified by means of sampling using a Gas Chromatography. Measurements were then made of (CH_4 + C_3H_8), (CH_4 + C_3H_8 + CO_2) and (CH_4 + C_3H_8 + C_7H_{16}) over the temperature range (203 to 423) K, pressures up to 35 MPa and different compositions, with a combined overall standard uncertainty of 0.5% for density, 3% for viscosity, 5% for heat capacity and 5% for surface tension. The extensive experimental data gathered in this work were compared with a variety of different advanced models frequently used for predicting thermophysical properties of LNG mixtures.

The current work presents a step forward towards accurate design of LNG processing's by providing essential data on complex mixtures that wasn't studied before. This work enabled new models, implemented in process simulation software, to predict reliably the fluid properties needed for equipment and process design. The current work also aided the community of scientists working to advance theoretical descriptions of fluid properties by allowing to identify deficiencies in theoretical descriptions and calculations.

492074 Numerical Investigation for Heat Transfer of Supercritical Methane Heated in a Horizontal Circular Tube

Abstract:

In the receiving terminal, LNG needs to go through the gasifier to absorb heat and transfer to natural gas in the normal temperature. The gasification pressure is generally between 5-9MPA. In this process, CH_4, the main component of LNG, is in supercritical state. Its physical properties will change greatly, especially when they are near the pseudo critical point. Unusual as its heat transfer law may be, there are few studies about supercritical methane at present. Most of the studies on other supercritical fluids show that the heat transfer becomes worse when the heat flux is very large in the vertical tube. However, in the LNG gasifier, the heat flux is not large, and there are a lot of horizontal tubes. In this case, will the heat transfer law be similar to what mentioned in literature? So the heat transfer of supercritical methane in a horizontal tube is simulated. Compared to the experiment, numerical simulation helps us to see more details of the flow and heat transfer process.

The simulation conditions are similar to the actual IFV (Intermediate Fluid Vaporizer) heat transfer conditions: the diameter of the tube is 12mm, the flow rate 0.03-0.07kg/s, the pressure 6-9MPA, and the temperature range 150-300K. Research findings:

In the low temperature region, with higher flux density, the influence of the buoyancy force is more obvious, and the temperature difference between the top and bottom of the tube is greater. Along the flow direction, the heat transfer is enhanced at the top of the tube, but at the bottom, the heat transfer deterioration occurs at first, and then the heat transfer coefficient increases. The greater the heat flux is, the more obvious the phenomenon of heat transfer enhancement and deterioration will be. Observe the internal flow and heat transfer of the tube, as the supercritical methane density varies greatly with temperature, the fluid in the middle which is far from the wall of the tube is in low temperature with big density. It sinks down gradually along the flow direction; The fluid near the wall is in higher temperature with lower density, it rises up to the top. The buoyancy leads to the great temperature difference and different heat transfer coefficients between top and bottom. In some literatures, the buoyancy is equivalent to the natural convection. However, further study shows that there is no strong natural convection in the circular tube, and the buoyancy force is only responsible for the uneven distribution of density. In addition, under the same heat flux, when the mass flow rate is higher, the heat transfer coefficient will become larger and the influence of the buoyancy force smaller. The influence of pressure can be ignored.

Near the pseudo critical point, the influence of the buoyancy force is more obvious, and the temperature difference is greater between the top and bottom of the tube than that in the low temperature region. Different from the low temperature region, the heat transfer coefficient at the bottom is very close under different heat flux, but it decreases greatly at the top along the flow direction, and it will not rise again. This phenomenon will be more obvious with greater heat flux. And we also studied the effects of density, heat capacity, thermal conductivity, viscosity change on the heat transfer. It is found that the heat transfer coefficient is positively related to the specific heat and thermal conductivity, and negatively related to the viscosity. The change of density will lead to the heat transfer deterioration at the top and heat transfer strengthen at the bottom. The heat transfer of supercritical methane is influenced not only by the turbulent kinetic energy mentioned in literature, but also by the changes of all the physical properties. Near the pseudo critical point, the specific heat and thermal conductivity change greatly. Their influence is very severe, which makes the heat transfer law different from that in the low temperature zone. In addition, under the same heat flux, if the mass flow rate gets higher, the heat transfer coefficient will be larger, and the influence of the buoyancy force smaller. The heat transfer law under different pressures is very close, but the temperature difference between the top and bottom becomes larger when the pressure is smaller, for the physical property changes more greatly under smaller pressure.

In the high temperature region, the influence of the buoyancy force is very small with more uniform density distribution, and the temperature and heat transfer coefficients of the top and bottom are very close, for the density does not change much in the temperature far from the pseudo critical point. Under the same mass flow rate, if the heat flow becomes higher, the heat transfer coefficient will be smaller, and it decreases along the flow direction. The overall heat transfer coefficient becomes larger with the increase of pressure, which should be attributed to the different physical properties under different pressures.

In addition, the criterion of floating force $Gr/Re^2<10^{-3}$ in literature is compared with the simulation results, and the criterion is proven accurate.

491179 Study on Ice Storage System Based on Cold Energy Utilization in Lcng Fueling Stations

Abstract:

As the increasing of the CNG vehicles for the advantages of low cost and environmental-friendly, the amount of the LCNG fueling station is also rapidly increasing. As we all know, the cold energy will be waste when the LNG (Liquefied Natural Gas) gasify into CNG (Compressed Natural Gas). So, it is necessary to design a cold energy recycle system in the LCNG fueling stations, which can achieve reasonable utilization of resources. The discontinuous flow makes it is hard to reuse the cold energy, so we design a heat exchanger and the ice storage facilities in the main system. In the heat exchangers, the LNG gasify into CNG, releasing the cold energy to the intermediate fluid. The ice storage facilities can be used to storage the cold energy into ice, and when we need this cold energy we can melt the ice into cold water for air-conditioning in the LCNG stations. Comparing to the previous ice storage systems, the new ice storage systems use the cold energy from the intermediate fluid at low temperature., so, the intermediate fluid can provide more cold energy for the air-conditions. As a result, the optimal operating point will be different from the normal systems.

There are many restrictions to the type of the intermediate fluid, such as low-freezing, high potential heat value, environmental and safety. The EGS (Ethylene glycol solution), which can meet the conditions perfectly, is one of the most common intermediate fluid. With the increasing of the mass concentration of the EGS, the freezing temperate decrease. As a result, in the LCNG station, we should choose the EGS with high concentration. Considering the viscosity, 50% mass concentration of EGS is suitable for the study.

In this paper, two-dimension model is performed to investigate the dynamic performance of the ice storage. Commercial CFD software ANSYS Fluent 15.0 and meshing software ICEM are used in this study. The model is a casing pipe with length of 1000 mm. The EGS is in the inner pipe with diameter of 20mm and the water is in the outer one with diameter of 100mm. A uniform velocity assumption is applied at the inlet boundary of the EGS. The outlet boundary condition of the EGS is outflow boundary condition. The water is static in a fixed temperature. According to the size of the mesh, we set the time step as 0.1s and the iterations as 30 times. Using the solidification & melting model in ANSYS Fluent 15.0, we can get the ice storage result in 6000 seconds.

Three influence factors are put forward through analyzing the actual situation——water temperature, EGS temperature, and EGS flow velocity. The range of EGS temperature is from 258K to 268K; the water is from 278K to 293K, and the EGS flow velocity range from0.2m/s to 1.5m/s. Based on the control variable method, we change different variables and calculate 11 different models in the Fluent 15.0. Then, we can obtain the temperature and the ice packing factor in the casing pipe.

Analyzing the three different factors, we can draw the conclusion: to some extent, the lower the EGS and the water temperature is, the faster the water freezes. However, the relationship between the EGS flow velocity and the freezing time is not linearity. There is an optimal velocity for the model. Comparing with the three different factors, we discovered that the EGS temperature have the most important influence to the freezing. So when we choose the optimal operating point, we should take it into consideration.

In order to verify the result of the software simulation, we also designed an experiment system. Three key section consist of the system. The first part is the precooling system. In consideration of safety in the lab, LNG is substituted by LN_2 (liquid nitrogen). PHE(Plate Heat Exchanger)is used to cooling the EGS from the cold energy of LN_2.This heat transfer process simulates the gasification process of LCNG station. The second part is the main part of the system. A visual casing pipe consisted of three different pipe is made in the experiment. The EGS is in the inner pipe made by 304 stainless steel, and the water is in the middle pipe made by. Polymethylmethacrylate The outer pipe is also the Polymethylmethacrylate pipe. A vacuum pump is used in the air between the middle pipe and the outer pipe to keep the vacuum. Through the visual design, we can not only observe the forming of the ice, but also satisfy the demand of thermal insulation of the system. The third part is the water cycle system composed with water chilling unit and flowmeter, which can adjust the initial temperature of the water.

On the one hand, temperature sensors, pressure sensors and flowmeters are used in the experiment to measure and collect the data of the parameters; on the other hand, a camera and image processing techniques are also used to observe the icing phenomenon with time in different working condition.

The influence of subcooled temperature and the gravity can be observed in the experiment. The ice is quite thick under the inner pipe than the ice above the inner pipe. Since the software simulate does not take the subcooled temperature into account when the water solidifies, we can find the subcooled temperature in the different condition in our experiment.

Based on the software simulate and the experiment, one can find out the theory and the influence factors of the icing. What's more, an optimal operating point can be found to fit for the status of LCNG fueling station, this will become useful advice for the development of cold energy utilization in LCNG station.

496550 The Structural Characteristics of Oil Shale Kerogens Changed with Their Humic Degree

Abstract:

The structural characteristics of oil shale kerogens changed with their humic degree

Qian Wang[1], Qing Liu[1], Fan Yang[1], Yucui Hou[2], Weize Wu[1,*]

[1]State Key Laboratory of Chemical Resource Engineering, Beijing University of Chemical Technology, Beijing 100029, China

[2]Department of Chemistry, Taiyuan Normal University, Taiyuan 030031, China

Oil shale is an important alternative energy resource, and the structure, especially the structure of kerogen is the foundation of its application. In this work, structural characteristics of Longkou, Huadian and Yilan kerogens with an increasing humic degree were studied. The relationship between structural characteristics and their humic degree were revealed. ^{13}C NMR, FTIR, XPS, XRD and the ion-exchange were used to obtain information on carbon skeletal structure, functional groups, the forms of heteroatom, the form of aliphatic carbon and the contents of carboxyl and phenolic hydroxyl groups. It was found that the kerogens are mainly composed of aliphatic structure, dominantly as CH_2 chain. Moreover, the ratio of aliphatic carbon (f_{al}, aliphaticity) and average methylene chain length (C_n) decrease with increasing humic degree. In these kerogens, most of their aromatic clusters are disorder. Furthermore, the ratio of aromatic carbon (f_a, aromaticity) and average aromatic cluster size increase while the substitutive degree of aromatic ring decreases with increasing humic degree. With increasing humic degree, the content of O=Câ˜'O (ester and carboxyl) decreases and that of Câ˜'O (alcohol, phenol, and ether) increase. The organic nitrogen in the kerogens exists as pyridinic, amine, pyrrolic and chemisorbed nitrogen oxides. The contents of pyridinic and chemisorbed nitrogen oxides increase while those of amino and pyrrolic decrease with increasing humic degree. In addition, organic sulfur in these kerogens exists as aromatic and aliphatic sulfur, sulfone and sulfoxide. The contents of aromatic sulfur and sulfoxide increase while that of aliphatic sulfur and sulfone decrease with increasing humic degree.

This work was supported by the National Basic Research Program of China (2014CB744301).

Keywords: Structural characteristics; Oil shale kerogen; Humic degree; Relationship

---

* Corresponding author; E-mail: wzwu@mail.buct.edu.cn; Tel./Fax: +86 10 64427603.

490992 Development and Application of a Fuel Property Database for Mono-Alcohols As Fuel Blend Components for Spark Ignition Engines

Abstract:

A wide spectrum of alcohol molecules could be considered for use as fuels or in fuel blends, but it is not possible to experimentally investigate the fuel potential of all of these molecules. Thus, experimental investigation must be done only on the most promising fuel candidates. The goal of this study was to establish a detailed list of properties of all possible molecular structures of normal and cyclic alcohols (C1-C10) with one hydroxyl group, and subsequently to narrow the potential fuel candidates to find the most auspicious alcohols for blending with gasoline in mid (less than 15 vol. %) and high (greater than 40 vol. %) range volumes for spark ignition engines. Where available, property data were obtained from literature reports. Property estimation methods were exploited for compounds without property data. An initial decision frame for screening out problematic compounds was devised and applied. Next, more stringent constraints were considered for the two scenarios of mid- and high-range blends of alcohol into gasoline. After this second stage, 46 alcohols were recognized as good candidates for blending with gasoline in low- and mid-levels; however, only 4 alcohols were accepted as potential fuels for blending with gasoline in high concentrations. Specifically, 1-butanol, iso-butanol, 2-methyl-2-pentanol, and (1-methylcyclopropyl)methanol were identified as most promising alcohol molecules suitable for blending with gasoline in a wide range of concentrations.

495640 Study on Crude Oil Extraction Method from Sludge Including Oil

Abstract:

Large amount of sludge including oil is discharged at oil production sites and is disposed as industrial waste at a high cost. In this work, we studied on crude oil extraction method from sludge and a configuration of a low cost plant that recovered crude oil from sludge. We selected naphtha as extractant. Naphtha is component of crude oil included in sludge and dissolves crude oil in sludge. At distillation tower, component of vaporized naphtha from extractant is approximately equal to that from sludge. Therefore we confirmed that our crude oil extraction method using naphtha as extractant stabilize in comparison with other extraction methods.

496482 Insight into Coal Structure Based on Benzene Carboxylic Acids from Coal Via Oxidation

Abstract:

Insight into coal structure based on benzene carboxylic acids from coal via oxidation

Fan Yang[1], Yucui Hou[2], Muge Niu[1], Shuhang Ren[1], Weize Wu[1,*]

[1]State Key Laboratory of Chemical Resource Engineering, Beijing University of Chemical Technology, Beijing 100029, China

[2]Department of Chemistry, Taiyuan Normal University, Taiyuan 030031, China

Coal structures, the basis of its applications, have been widely studied. Benzene carboxylic acids (BCAs, 12 types) can be generated from different aromatic clusters of coal via oxidation, and it has been found the 12 BCAs have different yield distributions for different coals. The results suggest that there is a certain relationship between BCAs and coal structure that provides a favorable basis for us to study the coal structures. In this work, based on the BCAs distributions and ^{13}C-NMR analyses, we investigated the structure characters of coal with different ranks. The results indicate that with the increase of coal rank, the yield of BCAs increase, and the structures of coal becomes more and more difficult to be degraded. There are great differences in the yield distributions of BCAs with the increase of coal rank, and more and more BCAs with a small number of carboxyl are obtained in high rank coal. Single-ring aromatic clusters and double-ring aromatic clusters are mainly aromatic structures in low rank coal. With an increase in the process of coalification, the aromatic clusters increase in size and the degree of condensed of aromatic rings also increases. When the carbon content of coal was approximate 87 %, the structure of coal had a mutation. At last, we proposed a structural model of Huolinhe lignite based on the yield distribution of BCAs. The work provides a new way to study the coal structures and construct the structural model of coal.

The project is financially supported by the National Natural Science Foundation of China (No. 21676019 and 21076138).

Keywords: Coal structure; Benzene carboxylic acid; Oxidation; ^{13}C-NMR

---

* Corresponding author; E-mail: wzwu@mail.buct.edu.cn; Tel./Fax: +86 10 64427603.

496953 Production of Carboxylic Acids from Lignite with Two-Stage Alkali-O_2 Oxidation

Abstract:

Alkali-O_2 oxidation of lignite can yield carboxylic acids (CAs). But the species with different reactivities in lignite are oxidized under identical conditions, resulting in a low CA yield. In this work, we proposed a two-stage oxidation process of lignite to increase the CA yield. First, lignite was extracted using a NaOH aqueous solution to divide the lignite into two parts: extract (Ex, 25.4 %) and residue (Re), which have different reactivities. Then, Ex and Re were oxidized separately with alkali-O_2 oxidation under different conditions. The extraction conditions and the alkali-oxygen oxidation conditions of Ex and Re were investigated. The results indicated that the carbon yield of CAs from Ex was 35.32 %, including 5.60 % formic acid, 6.15 % acetic acid, 11.95 % oxalic acid, 0.85 % succinic acid, and 11.13 % benzene carboxylic acids (BCAs) at 230 °C, an initial O_2 pressure of 4.5 MPa and a NaOH content of 1M for 40 min. The yield of carboxylic acids from Re was 36.69 %, including 0.35 % formic acid, 5.05 % acetic acid, 11.81 % oxalic acid, and 19.48 % BCAs at 260 °C, an initial O_2 pressure of 5.5 MPa and a NaOH content of 3 M for 70 min. The total CAs yield was 36.28 %, which is much higher than 26.7% with the traditional one-stage alkali-O_2 oxidation. The structural properties of Ex, Re and lignite were studied in detail with ^{13}C NMR and FT-IR to reveal the different reactivities and yields.

The project is financially supported by the National Natural Science Foundation of China (No. 21676019 and 21076138).

486553 Methane Production from Crude Solid Residue: A Minimal Organic Waste Strategy

Abstract:

International environmental concerns such as Global warming have forced the scientific community to look beyond the fossil fuels to meet the energy demand and security. To match up with ever increasing energy demand, new technology must be robust and environmentally safe while using the raw material which is inexpensive and easily available. Unlike conventional substrates like coal, municipal solid waste (MSW) is renewable, abundantly available and rich in organic acids for their easy transformation into clean energy. Existing disposal strategies of MSW are not efficient for sustainable energy production processes due to the release of biogas, volatile organic compounds, and leachates that may cause public health hazards. Contrary to this, Thermophilic anaerobic digestion (TAD) of MSW provides endless possibilities for clean, renewable energy production as well as waste management. TAD has endless potential to produce methane (CH_4) and other flammable gasses which can be derivatized to other renewable biofuels viz. methanol, butanol, etc. Classic anaerobic fermentation (AF) can also be combined with TAD to materialize the concept of the release of zero-organic-waste-to-environment.

MSW is mainly constituted by Food (21% w/w) and paper wastes (28% w/w) efficient management of which has become imperative today. Thus, we have combined TAD (using thermophilic methanogenic consortia, TMC) with AF, for efficient breakdown of complex, undefined food and paper waste mixture for methane production. The feedstock for TAD is termed as "secondary waste (SW)" as it is the residual of anaerobic hydrolysis carried out using CRUDE (Conversion of Raw and Untreated Disposal into Ethanol) process. Batch TAD experiments were carried out with SW using TMC at 60 °C, pH 7.5 up to 30 days. Methane yield using SW was 95±7 L kg^{-1} VS (volatile solid) after 17 days treatment (60°C; 50 rpm; pH 7.5) compared to 92±11 L kg^{-1} VS with primary waste. These data confirms the potentials of SW as feedstock for methane production. After optimizing the substrate to inoculum ratio, more concentrated waste streams could be used and methane yields increased up to ~200±5 L CH_4 Kg^{-1} VS. Gas chromatography confirmed the dominance of methane (80±6% v/v) in samples analyzed, with traces of CO_2 and H_2.

500574 Production of Carboxylic Acids from Lignite with Two-Stage Alkali-Oxygen Oxidation

Abstract:

Production of carboxylic acids from lignite with two-stage alkali-oxygen oxidation

Yucui Hou[1,*], Wenbin Li[2], Fan Yang[2], Shuhang Ren[2], Weize Wu[2]

[1] Department of Chemistry, Taiyuan Normal University, Taiyuan 030031, China

[2] State Key Laboratory of Chemical Resource Engineering, Beijing University of Chemical Technology, Beijing 100029, China

Alkali-O_2 oxidation of lignite can yield carboxylic acids (CAs). But the species with different reactivities in lignite are oxidized under identical conditions, resulting in a low CA yield. In this work, we proposed a two-stage oxidation process of lignite to increase the CA yield. First, lignite was extracted using a NaOH aqueous solution to divide the lignite into two parts: extract (Ex, 25.4 %) and residue (Re), which have different reactivities. Then, Ex and Re were oxidized separately with alkali-O_2 oxidation under different conditions. The extraction conditions and the alkali-oxygen oxidation conditions of Ex and Re were investigated. The results indicated that the carbon yield of CAs from Ex was 35.32 %, including 5.60 % formic acid, 6.15 % acetic acid, 11.95 % oxalic acid, 0.85 % succinic acid, and 11.13 % benzene carboxylic acids (BCAs) at 230 °C, an initial O_2 pressure of 4.5 MPa and a NaOH content of 1M for 40 min. The yield of carboxylic acids from Re was 36.69 %, including 0.35 % formic acid, 5.05 % acetic acid, 11.81 % oxalic acid, and 19.48 % BCAs at 260 °C, an initial O_2 pressure of 5.5 MPa and a NaOH content of 3 M for 70 min. The total CAs yield was 36.28 %, which is much higher than 26.7% with the traditional one-stage alkali-O_2 oxidation. The structural properties of Ex, Re and lignite were studied in detail with ^{13}C NMR and FT-IR to reveal the different reactivities and yields.

The project is financially supported by the National Natural Science Foundation of China (No. 21676019 and 21076138).

Keywords: Alkali-oxygen oxidation; Carboxylic acids; Lignite; Two stage

---

* Corresponding author; E-mail: houyc@tynu.edu.cn; Tel./Fax: +86 136191281359.

501230 CFD Modeling of Immiscible Liquid-Liquid Flow in a Large Scale Crude Oil Storage Tank Equipped with Side-Entering Mixers

Abstract:

The periodic internal cleaning of crude oil storage tanks at terminals and refineries, needed when the rate of Basic Sediment and Water (BSW) accumulation is at a concentration above 0.5 wt%, consists in a demand of great impact for the maintenance team. This results in low overall storage efficiency due to the equipment unavailability and high environmental impact, such as the discharge of the sludge inside the tank and replacement of the bottom plates due to corrosion as a result of the presence of brine. Mineral solid present in petroleum; high molecular weight organic sediments, aqueous brine drops and corrosion residues participate in an undesirable process of formation of complex solid compounds with high hardness and density. Compacted sediment, associated with the degradation of organic compounds inside storage tanks leads to the formation of a thick sludge layer, which by definition is a multi-phase water-oil-solid system of great stability, difficult to remove and / or process and with low chemical treatability. The mixing process is an important physical operation to avoid sludge formation, achieving better homogenization of the fluid (blending). Settling is common when dispersed and continuous phases are of different density and when agitation provides only minimal circulation throughout the vessel. It is known that the mechanical mixer is the more efficient mixing device in terms of the average tank speed per power unit. Furthermore, the side-entry impellers are considered to be a low capital investment and of simple installation. This computational study has been carried out using the Computational Fluid Dynamics package OpenFOAM, which gives an estimate of the liquid-liquid flow pattern, considering the characterization of the mixture using side-entering impellers in large scale circular tanks. The power number and simulated flow of a single impeller in a cylindrical storage tank were validated using empirical correlations. Sensitivity analysis of the mixture by multiple impellers in cylindrical storage tank has been also performed to study the effect of the spacing between the adjacent impellers and the angle of inclination for each of the impellers in the performance of the mixture. It also important to determine the minimum speed for drop suspension in other to keep drops suspended, avoiding this way sludge formation. This of course depends on phase density differences, agitation speed, impeller size, and its location within the vessel.

Keywords: Mixing; crude oil; side-entering mixers; Computational fluid dynamics; immiscible liquid-liquid; suspension of drop

484123 Reaction Kinetics of 1-Methylnapthalene Hydrocracking over Metal/Beta Catalyst to BTX

Abstract:

Light cycle oil (LCO, in diesel boiling point range), mainly composed of two-aromatic ring compounds, is produced in large amounts through FCC process in oil refineries. Since the demand for heavy fuel oil and diesel is in steady reduction recently, it is highly desirable to convert this fraction into high-value products, i.e., light alkyl aromatic hydrocarbons such as BTX (benzene, toluene and xylene), for which demands are steadily increasing.

To produce BTX, one aromatic ring of LCO must be selectively preserved to form one phenyl ring with alkyl chains. Selective hydrocracking (HC) is common ring opening reactions used for effective conversion of multiple-aromatic ring components to one-aromatic ring components, and bifunctional catalysts such as metals supported on acid zeolites have mostly been used for this purpose. However, there is hardly report on the inherent correlation between catalyst characteristics and reaction selectivity because of the high complexity of the LCO hydrocracking. The metallic catalytic centers and acid catalytic sites have hydrogenation function and cracking function respectively, and they cooperate during the selective hydrocracking of LCO. A reaction kinetic model is useful for the quantitative evaluation of the function of the metallic catalytic sites and the acid catalytic sites in the selective hydrocracking process, and therefore can be used to guide the preparation of good selective LCO hydrocracking catalysts.

In this work, we studied the hydrocracking kinetics of 1-methylnaphthalene as a model molecule of the LCO. The hydrocracking experiments were conducted using metals/Beta molecular sieve as the catalyst (Metals = NiW, NiMo, CoMo, CoW, W, Mo) and a six-lump reaction kinetic model was proposed. A mathematic kinetic model was established, and all kinetic constants and the activation energy of every step reaction in the reaction model were obtained. There is a good consistency between the kinetic model and experimental data, indicating that the kinetic model are reliable.

492249 CO_2 Micro Foam and Its Application in Enhanced Oil Recovery

Abstract:

The diameter of foam is the key to determine foam`s physical properties. Based on that, I tried to enhance the EOR (enhanced oil recovery) property of $CO_2$ foam by decreasing it`s diameter. Several kinds of foaming agents (C-14 BE, CTMAB, SDBS and Polyoxyethylene alkyl amine), foam stabilizers (Polyacrylamide, Xanthan Gum and Carbopol Aqua SF-1) and preparation methods (airflow method and waring-blender method) were used to filtrate the best preparation conditions of CO_2 micro foam. Besides, temperature, concentration and the water`s degree of mineralization are also considered. After that, I examined the basic physical properties, structure and the diameter of CO_2 micro foam by Ross-Miles method and AFM (Atomic Force Microscope). With these consequences, I tried to explain the causes of these special characteristics of CO_2 micro foam in theory.

On the other hand, I used CO_2 micro foam to do low-permeability core flooding experiments to get displacement efficiency and the situation of residual oil (microscopic imaging of core slice) which are the evaluation indicators of EOR effect. The two evaluation indicators clearly expressed the exact EOR effect of CO_2 micro foam.

I also prepared normal CO_2 foam and air foam as a comparison. The comparative items, including structure, stability, foaming capacity and EOR effect, characterized the differences between CO_2 micro foam, normal CO_2 foam and air foam. At last, the CO_2 micro foam prepared by C-14 BE and polyacrylamide compound has the best physical property and EOR effect.

484979 Kinetic Modelling of the Combustion of Aliphatic Hydrocarbons

Abstract:

Kinetic Modelling of the Combustion of Aliphatic Hydrocarbon

1KAMALU C.I.O., 2ANARAH A.C., 3OKOH E., 4OKERE P.C.,*5OBIJIAKU J.C., 6UZONDU F.N.

***3Phone: +234(0)8036602533 *3Email: okohelechi1968@yahoo.com**

1,2,5,6 Department of Chemical Engineering, Federal University of Technology, Owerri, Nigeria.

3 Department of Oil and Gas Technology, Federal Polytechnic, Ekowe, Bayelsa State, Nigeria.

4 NNPC/PPMC, Port Harcourt Depot, AlesaEleme, Rivers State, Nigeria.

ABSTRACT

A general kinetic model which describes the combustion characteristics of aliphatic hydrocarbons at any given temperature, involving the variations of conversion, carbon number and combustion time was developed for the combustion of aliphatic hydrocarbons. A combustion quality tester (CQT) was designed, fabricated, and then used to carry out combustion experiments on CNG. Data was sourced from coherent works from the internet, for certain experiments that could not be carried out due to required equipment sophistication. Experimental data on CNG combustion were curve-fitted using the derived kinetic models to ascertain their goodness of fit. It was found that the curves fitted well at 97.5%, 97.9%, 97.6% and 98.7% goodness of fit for combustion temperatures of 385K, 390K, 395K and 400K respectively. The models agree vehemently with experimental data, and show that reaction rate of the fuel increases with temperature, and the overall reaction rate with temperature is an inverse function of combustion time. Next, ignition quality tester (IQT) was used to investigate pressure and temperature effects on ignition delay, effect of fuel additive (2-EHN) on cetane number and ignition delay, and also measurement of cetane number according to ASTM D6890 standards. Test was conducted under steady state conditions at constant pressure and temperature of 145psi and 828K respectively for cetane number experiments. Data were curve-fitted at 98.18%, 99.64%, 99.43% 99.82% and 99.19% goodness of fit for cetane number-%additive vol., cetane number-ignition delay, ignition delay-%additive vol., ignition delay-pressure and ignition delay-temperature experiments. Data agreed well with derived models. Ignition delay time of the fuel gives an inverse function with temperature, pressure, cetane number and percentage vol. additive concentration. Fuel additive (2-EHN) has a profound impact on cetane number and increases it up to about 8 points beyond which there is no appreciable increase no matter the volume of additive added. The present research finding can find use in combustion laboratories and automotive industries, to correlate certain variable parameters instead of habitually resorting to experiments

Recent Publications

1. Alatorre G.G., B¨ohm H., Atakan B., Kohse-H¨oinghaus K. (2001). Experimental and Modelling study of 1-Pentene Combustion at Fuel-Rich conditions.Journal of Physical Chemistry. 215:981-995.
2. Burcat A., Olchanski E., Sokolinski C. (1999).2Methyl-Pentane Ignition Kinetics in a Shock-Tube.Journal of Combustion Science and Technology.147: 1-37.
3. Carriere T., Westmoreland P.R., Kazakov A., Stein Y.S., Dryer F.L. (2002). Modelling Ethylene Combustion from Low to High Pressure.Proceedings of the Combustion Institute.
4. Castaldi M.J., Marinov N.M., Melius C.F., Huang J., Senkan S.M., Pitz W.J., Westbrook C.K. (1996). Experimental and Modelling Investigation of Aromatic and Polycyclic Aromatic Hydrocarbon Formation in a Premixed Ethylene Flame.Proceedings of the Combustion Institute.
5. Chen C.J., Bozzelli J.W. (2000). Thermochemical Property, Pathway and Kinetic Analysis on the Reactions of AllylicIsobutenyl Radical with O_2: An elementary reaction mechanism for isobutene oxidation. Journal of Physical Chemistry. 104: 9715-9732.

Biography **Okoh Elechi.** is a lecturer in the department of oil/gas processing technology, Federal Polytechnic Ekowe, Nigeria he holds a Bachelor degree in Chemical Engineering from the University of Port Harcourt, Rivers State, Nigeria and he is presently pursuing a Masters degree in Chemical Engineering from the Niger Delta University, Amassoma Bayelsa State Nigeria.

Email: okohelechi1968@yahoo. com

Notes/Comments:

Predicting the Properties of Petroleum Blends

Tareq A. Albahri
Chemical Engineering, Kuwait University, PO Box 5969, Safat 13060, Kuwait,
dr.albahri@ku.edu.kw

Hessa A. AlMulla
Manufacturing Optimization Group, Kuwait National Petroleum Company (KNPC), Kuwait.

ABSTRACT

- Developed very accurate correlations to predict the properties of petroleum fractions blends useful for implementation in optimization models and advanced control strategy systems.
- Correlations are for Viscosity, Smoke point, Freezing point, Reid vapor pressure, Aniline point and Cloud point.
- The correlations gave an absolute average deviation of 2.6 mm^2/s for viscosity, 0.015 mm for smoke point, 0.05 K for freezing point, 6.47 kPa for RVP, 0.32 K for aniline point and 0.9 K for cloud point, respectively.

Blending for Viscosity

In cSt.

$$(VBI)_i = \frac{0.183 + 0.458\log(v)_i}{1.105 + 0.305\,log\,(v)_i} \quad (1)$$

$$(VBI)_m = \sum X_{vi}\,(VBI)_i \quad (2)$$

$$\log(v)_m = \frac{-0.315 + 1.796\,(VBI)_m}{1 - 1.264\,(VBI)_m + 0.45\,(VBI)_m^{\;2}} \quad (3)$$

where m designates the mixture or blend, i designates the blending component, X_{vi} is the volume fraction of the blending component i, VBI is the viscosity blending index, and $(VBI)_m$ blending index of the blend, and $(v)_m$ is the viscosity of the blend. 235 blends with viscosity values between $0.25 < v < 564.96$ mm^2/s measured at temperatures from 20 to 150°C. The average absolute deviation (AAD) was 2.6 mm^2/s.

In SUS either of the following equally accurate eq.s (4) followed by one of eq.s (5)

$$VBI_i = 0.6068 + 0.04797(v)_i - \frac{0.80755}{(v)_i^2} \quad (4\text{-a})$$

$$VBI_i = \frac{-97993438 + 103852082\,(v)_i}{1 + 118310078\,(v)_i - 3052999(v)_i^2} \quad (4\text{-b})$$

$$VBI_i = 0.5909\,((v)_i\; - 1.4414)^{0.25372} \quad (4\text{-c})$$

$$(VBI)_m = \sum X_{vi}\,(VBI)_i \quad (5)$$

$$\log(v)_m = \frac{5.2602 + 34.83\,(VBI)_m^{\;3.9746}}{3.7246 + (VBI)_m^{\;3.9746}} \quad (6\text{-a})$$

$$\log(v)_m = 21.0553\; - 19.6471\exp[-0.446654\,(VBI)_m^{\;3.9315}] \quad (6\text{-b})$$

$$\log(v)_m = 1.4123 + \frac{33.4177(VBI)_m^{\;3.9746}}{3.7246 + (VBI)_m^{\;3.9746}} \quad (6\text{-c})$$

In SFS

$$VBI_i = 0.5008\,(VBI)_i^{\;0.345} \quad (7)$$

$$(VBI)_m = \sum X_{vi}\,(VBI)_i \quad (8)$$

$$\log(v)_m = 13.07836 - 12.6188\,exp\,[-0.7749(VBI)_m^{\;3.9578}] \quad (9)$$

Blending for Smoke Point

$$(SP)_m = \sum X_{vi}\,(SP)_i \quad (10)$$

$(SP)_m$ is the smoke point of the blend, and $(SP)_i$ is the smoke point of component i. The experimental values using 33 different crude blends in the range between $7.63 < SP < 22.87$ mm. The AAD is 0.015 mm which is highly accurate.

Blending for Freezing point

$$(FPBI)_i = 3.23 * 10^{-6} x\; 1.067^{FPK_i} \quad (11)$$

$$(FPBI)_m = \sum X_{vi}\,(FPBI)_i \quad (12)$$

$$(FPK)_m = 193.798 + 15.379\; x\; Ln((FPBI)_m) \quad (13)$$

FPK_i is the freezing point of component i in K, and $(FPBI)_i$ is the freezing point blending index of component i, and FPK_m is the freezing point of the blend in K,

33 different blends in the range between 207.86 and 327.93 K gave AAD 0.05 K.

Blending for Reid Vapor Pressure (RVP)

The RVP index for each component $(VPBI)_i$ is given by equation (14). The RVP index of the blend is calculated through equation (14). The blend's RVP is then calculated using equation (16)

$$(VPBI)_i = RVP_i^{1.35} \quad (14)$$

$$(VPBI)_m = \sum X_{vi}\,(VPBI)_i \quad (15)$$

$$RVP_m = (VPBI)_m^{0.7} \quad (16)$$

33 sets of petroleum systems in the range between $16.89 < RVP < 71.87$ kPa gave an AAD of 6.48 kPa.

Blending for Aniline Point

$$APBI_i = -4852.435 + 14.765APK_i + \frac{73{,}778{,}706}{APK_i^2} \quad (17)$$

$$APBI_m = \sum X_{vi}\,APBI_i \quad (18)$$

$$APK_m = \frac{(4{,}093{,}333 + 9{,}025\,(APBI_m)^{0.77911})}{(16{,}731 + (APBI_m)^{0.77911})} \quad (19)$$

where APK_i is aniline point of blending component i in K, $APBI_i$ is the aniline point blending index of component i, and APK_m is the blend's aniline point in K.

31 crude blends in the range between $320 < APF < 366.5$ K gave an AAD of 0.32 K.

Blending for Cloud Point

$$CPBI_i = 1.501x10^{-4}\left(1 + \frac{CPK_i}{602.08}\right)^{34.4947} \quad (20)$$

$$(CPI)_m = \sum X_{vi}\,(CPBI)_i \quad (21)$$

$$CPK_m = \frac{-854.7 + 2626.94\,x\,(CPI)_m^{0.0709}}{8.236 + (CPI)_m^{0.0709}} \quad (22)$$

where $(CPBI)_i$ is the cloud point blending index of each component i, CPK_i as the cloud point of component i in K, $(CPI)_m$ is the cloud point blending index of the blend, and CPK_m is the blend's cloud point in K.

33 blends in the range between $201.4 < CPF < 323.56$ K gave an AAD of 0.9 K.

492988 On-Sun Demonstration of Hydrogen Production Via Solar Thermochemical Water Splitting

Abstract:

A sustainable hydrogen economy based on the low-cost production of H_2 from renewable resources has the potential to drastically transform the energy sector. While solar energy is the most abundant renewable energy resource, the capture, storage, and distribution of it remains a challenge. Solar thermochemical water splitting (STWS) provides a promising route for efficient utilization of this dispersed resource because it allows for use of the entire solar spectrum to convert water to a renewable energy carrier and fuel, H_2. Simple two-step reduction/oxidation (redox) cycles are preferred because the production of oxygen and hydrogen can be spatially or temporally separated, eliminating the need for high-temperature gas separation, while also avoiding the design complications and inefficiencies of multistep processes. In this process, oxygen is generated by the reduction of a metal oxide during a high temperature step in an inert atmosphere. In the second step, the reduced metal oxide is exposed to high temperature steam to reoxidize the material and produce hydrogen fuel. Although the STWS reaction schemes appear straightforward, their implementation in an economically viable process is extremely demanding. In order to commercially realize this technology, several challenges must be overcome including an energy efficient reactor design, stable reactor and heat exchanger materials, and robust redox active particles. In this work we present results showing near-isothermal on-sun hydrogen production utilizing dual lab-scale fluidized beds. The use of fluidized beds that are cycled between oxidation and reduction negates the need for energy intensive moving solids. Active hercynite particles were redox cycled at the High Flux Solar Furnace at the National Renewable Energy Laboratory and production rates of H_2 exceeding 500 umol/g were measured. The effect of environmental variability on the incident solar radiation and hydrogen production is evaluated during on-sun testing. This work demonstrates the viability of solar thermal water splitting leveraging continuous operation of both reduction and oxidation at near-isothermal conditions.

502795 Optimal Aspect Ratio and Recirculation Rate for LNG Storage Tanks in a Regasification Terminal

Abstract:

In the present work, a rigorous process model of an LNG regasification terminal is developed on Aspen HYSYS. The model is then used to study the design and operation of regasification terminals. It is observed that a near-unity H/D ratio results in a minimum average boil-off gas generation for a wide range of tank volumes and operating parameters. The typical H/D ratio of 0.5 used in the industry results in 1 to 5% increased average boil-off gas generation. It is also observed that the average boil-off gas generation decreases monotonically with recirculation rate. This is understood to be the composite effect of decreasing heat ingress from the recirculation line and a decrease in in-tank pump duty. However, excessively low recirculation rate results in two-phase flow in the recirculation line. Based on these observations, an equation for selecting an optimal recirculation rate is developed. The equation is general, and only requires as input the heat transfer characteristics of the recirculation line. A separate figure predicts average compressor power as a function of tank volume, heat transfer characteristics of the recirculation line and send-out rate. The present study further concludes that the average boil-off gas generation arises from five effects (1) heat ingress into the tank, (2) decreasing liquid level, (3) vapor expansion, (4) heat ingress into the recirculation line and (5) pump duty. The concepts and results presented in this study should prove useful in the design and operation of regasification terminals.

Gas Production from the SCOOP (or How I formed an Oil Company)

Richard L. Long, Jr. PhD, P.E.
Professor Emeritus, NMSU
AIChE Annual National Meeting
San Francisco, Ca.
November 7, 2013
AIChE Annual NationalMeeting
Update: November 1, 2017
Minneapolis, Minn.

Public Lands Available in the United States in 2013

- BLM currently oversees about 264 million acres of public lands in the U.S.(about 1/8th of all land in the United States).
- BLM does occasionally sell parcels of land under certain conditions(see Government Sales of Public Lands)at fair market values (including mineral rights – but it is not always guaranteed that the mineral rights are accurate. When new technology becomes available, an opportunity in mineral rights may arise, which was not there before.).
- In 1976 the federal government abolished all remaining traces of the Homestead Act of 1862. All future sales must be at the fair market value.
- New technology developments in the oil industry: horizontal drilling (since the 1980s) and hydraulic fracturing(since the 1950s).
- In the 1980s most of the oil companies closed up shop and moved out of Oklahoma to Houston and went international(except for Continental Resources , Inc.(remained in Enid, Oklahoma)) and a few other companies. These latter companies, partly due to the long time support of George Mitchell of Mitchell Energy, developed technology that allowed horizontal drilling and made possible the exploitation of the source rocks, especially the shale deposits of the U.S.
- The global oil price went from $3/barrel in the 1960s to $100/barrel today.
- Today, till recently, George Mitchell was the largest shareowner in Devon Energy, a second generation oil company in Oklahoma City.
- In 2005, Continental Resources figured out how to recover oil (at a price) from the Bakken shale in North Dakota and now in the SCOOP (Woodford Shale) in Oklahoma.

Company History

- In 2008 a lease acquisition company approached a sibling requesting a lease. Unbeknownst to me, I happened to inherit a small mineral right. Originally, it was not worth anything. The land and the mineral rights attached were sold to a family member to use as a farm. The insurance company from whom my relative purchased it concluded that the mineral rights were worth just a token or nothing at all and decided to include the mineral rights to give my relatives a token incentive to acquire the land and farm it for the war effort (WWII). They did so and my understanding from the family was that other family members had actually inherited it. So I concluded that I did not have an inheritance right. I did spend a small amount of money about 10 years prior to this to try to find out the details of the documentation to this in the county courthouse, but to no avail. After the lease acquisition company approached us, I began to get more serious about this and decided to be sure that I did own the rights described. I retained some mineral rights' lawyers in the area. After some time I acquired some of the details in the courthouse records. After further time I acquired all known records of the mineral rights owners. And I obtained 3 independent letter opinions from attorneys defining the mineral rights that I own.
- With that, I retained one of the attorneys, formed an LLC, and filed appropriate paperwork with the county courthouse and filed with the state of Oklahoma.
- Ultimately, the lease acquisition company disappeared and I was exposed to the energy company who wanted to lease the mineral rights.
- I then negotiated with the energy company a joint operating agreement (JOA). I then elected to participate with a portion of the mineral rights and lease the remainder(this is in accord with Oklahoma law). A well was drilled (2008). It did produce and still produces gas to this day. It is a small, but steady income. The decline curve initially was of the "hyperbolic type", though it has since lined out.

Geology

- Reservoir: oil and gas storage in rock.
- Key parameters: (1)porosity: void fraction

  (2)permeability: ease of flow

Ideal reservoir: high porosity

high permeability

Reservoir Considerations

- Usually, reservoir is above the source rock
- **Source rock**: the source rock is where the oil and gas are formed: usually, a clay-like material – often times a shale rock; expression: oil and gas are formed in the "kitchen".
- **Characteristics of shale**: low porosity and low permeability (a "tight formation").
- **New Technology**: horizontal drilling and hydraulic fracturing: in effect, this can change the "effective" porosity and permeability.

Horizontal Drilling and Hydraulic Fracturing

- Private development of these two technologies and their effective combination was developed over time partly by George Mitchell of Mitchell Energy, now deceased. He was convinced that he was on the right track and it proved so.
- The major oil companies (Exxon, etc.) tried to recover oil from shale type source rocks, but were largely unsuccessful at the then current price of oil(example: TOSCO and the joint ventures of the majors in the 1960s, paying particular attention to the Green River Basin, and also, a small effort by the Sandia National Labs up until the 1980s. I personally did do research for a company looking at the kerogen in the Green River shale play and determined the kinetics and modeling production in the play.
- Horizontal drilling: Key development was the down hole mud motor development in which the drill bit turned without turning the drill pipe, somewhat like a Lazy Susan joined with a drill bit, with mud flow producing the torque.
- Hydraulic fracturing: fracturing of porous rock by water with "proppants" – sand particles and added chemicals etc. This is an apparent improvement in "effective" permeability.

Shale Plays

- The organic material in shale plays is known as "kerogen". If heat and pressure is applied, kerogen decomposes to crude oil and gas. Shale is usually low permeability, or a "tight reservoir."
- The major oil companies (e. g. Exxon, etc.) in the 1960s tried to extract oil and gas from shale oil, especially in the **Green River Basin**, at a crude oil price(recall TOSCO and other related joint ventures).They could do it, but at a much higher price than the market price. In the Green River Basin the kerogen is low sulfur("sweet") and higher nitrogen- a good quality oil.
- In 1981, the **Barnett shale** (Ft.Worth) was discovered by Mitchell Energy. In 1997,Mitchell switched from "gel frac" to "slickwater frac", lowered the cost and increased the recovery. In 1998, Mitchell drilled two horizontal wells that were technical successes but economic failures.
- Mitchell sold Mitchell Energy to Devon Energy in 2002, becoming the largest share owner in Devon. He is now deceased.
- The **Bakken shale** play has been very successful in producing oil. Continental Resources (**not** Conoco) has been successful in the Bakken and is now moving to the Woodford shale(SCOOP).They are now drilling horizontal and also doing hydraulic fracturing. Harold Hamm is the founder and principal owner.

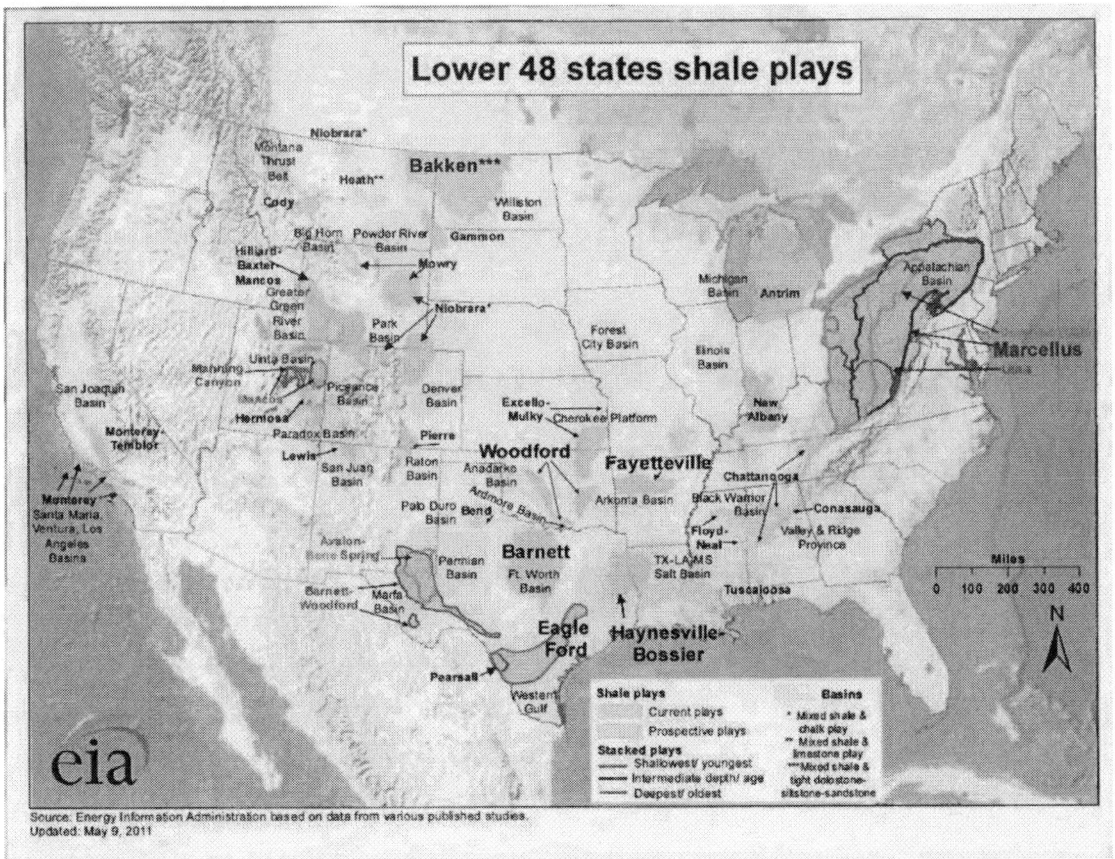

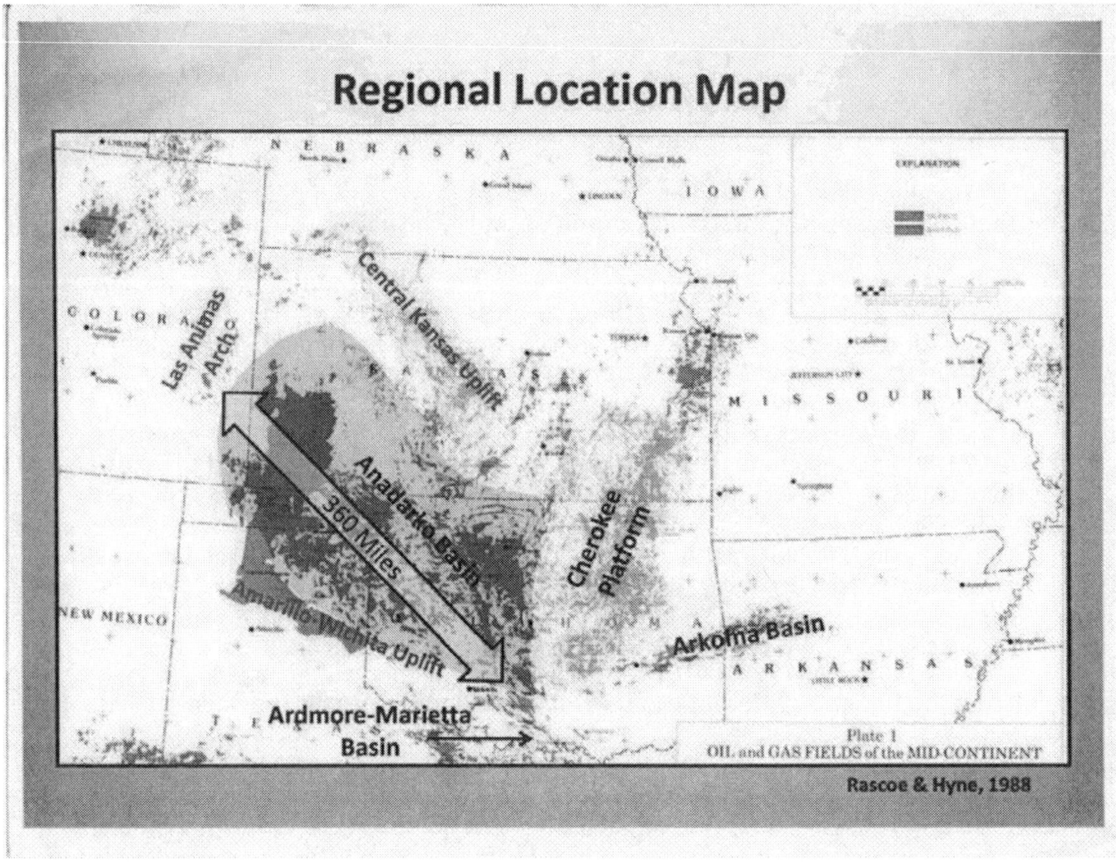

Company Mineral Rights Purchase Offers

- A total of 5 offers has been made to purchase mineral rights.
- An estimate of 40%/ yr compounded growth after taxes has been offered for the mineral rights for the last 5 years.
- Capital gains looks very encouraging.
- Goal: invest for cash flow and capital gains.

Perspective

- At least one very preliminary offer for company financing has been made.
- I have taught engineering economics several times and have a basic understanding of micro-economic decision making.
- Basically, if financing is to be considered, you have to have the truth. No tricks, no treachery.
- Location of mineral rights: in the SCOOP.
- Look at investor presentations for 2013 and guidance for 2014 at CLR website.

Lessons

- When a herd of elephants comes into field mouse territory, your task, if you are a field mouse, is to avoid getting stepped on.
- "It's not personal, it's just business" – The Godfather.
- Plan for the worst. Hope for the best.

Shale Oil and Gas
("life reflects art")

"Happy Days"

- "The Fonz" –aka-
  Harold Hamm(founder, CLR)

Hamm in 1964

"Happy Days"

- "The Nerd" –aka-
  Richard Long (1965)

Questions?

Update

2017

Lease/Participation Agreements

- In 2008, a lease was signed with Unit Petroleum, Inc.
- In 2008, a Joint Operating Agreement (JOA) was signed with Unit Petroleum, Inc.
- In 2014, a lease was signed with Jackfork, LLC.
- Other participation agreements pending, 2017.

References

- Money in the Ground, 4rth edition, 2006, by John Orban, III, Meridian Press, Oklahoma City, Ok.
- Oil and Gas Law in a Nutshell, 5th edition, 2009, by John S. Lowe, Thomson Reuters, West Nutshell Series.
- "Tomorrowland", Walt Disney Films, 2015.
- "The Godfather", Paramount Films, 1972.
- "Gasland", HBO Documentary Films, 2010.
- "Fracknation", Hard Boiled Films, 2012.

References (Continue)

- Guide to Safety in the Laboratory for Chemical Engineers- copyright©1987, Dept of ChE, NMSU, by George Whitmyre, Dept of ChE, University of Delaware, and Richard L. Long, Dept of ChE, NMSU.
- Guide to Writing and Problem Solving for Chemical Engineers – Kinko's Publishing Group, ISBN 1- 55577-032,©Copyright 1985, by Long, R., Barna, B., Bridges, C.W., Rakow, A.L., and Wilson, D. B., used at 8 universities.

Questions?

504591 The Relationship between Olefin Hydrogenation and Octane Number Loss in FCC Gasoline Hydrodesulfurization

Abstract:

Whatever how the developing of clean-gasoline production technology, nowadays hydrodesulfurization (HDS) still possesses the main position as the most effective approach for large scale production of clean gasoline in industry. HDS technology has a lot of advantages in efficient desulfurization and convenience operation, however, it still has a bottleneck problem[1]. During desulfurization, some of olefin are inevitable hydrogenation, which leading to huge loss of octane number[2]. Does each kind of olefin saturated has the same effect on octane number of gasoline? What is the relationship between type, carbon number, and structure of olefin and octane number of gasoline? For these questions, people are still not well understand. Furthermore, content of olefins in gasoline will be further cut down in the very near future. Which kinds of olefin should be remove from gasoline? More deeply studies need to be done before handling these olefins in gasoline.

In this study, based on the detailed olefin compositions and distributions in three different kinds of FCC gasoline feedstock obtained by gas chromatography[3], we studied the relationship between carbon numbers, structures, and content of olefins and loss of octane number of caused by olefins hydrogenation by combined calculation and experimental methods.

During the study, the olefins studied were quantitatively hydrogenated by artificial calculation method to obtain the effect of different carbon numbers olefins on the octane number of gasoline. As the results shown, the saturation of C5-C8 olefins has the greatest impact on the octane number of gasoline. In addition, olefins content was also one of the major factors affecting octane number of gasoline. Therefore, C5-C7 olefins were one of the major contribution components in gasoline, which considers the influence of olefins content in gasoline. It means that the protection of C5-C7 olefins was key to reduce the octane number loss in HDS. Beyond that, the relationship between olefins structures and octane number of gasoline was also studied. And it shown that the RON loss caused by the n-olefins hydrogenation was larger than i-olefins hydrogenation. Thus, in order to keep the octane number of the gasoline, the protection of C5-C7 olefins was more critical.

Acknowledgement

This research is supported by National Natural Science Foundation of China (Grants 21336011, 21476260, 21236009, and U1162204). And also supported by the Science Foundation of China University of Petroleum-Beijing (Grants 2462015YQ0316).

References

[1] Brunet, S., Mey, D., Perot, G., et al. On the hydrodesulfurization of FCC gasoline: a review[J]. *Applied Catalysis A: General*, 2005. 278(2): 143-172.

[2]Ishutenko D. I., Nikulâ€™shin P. A., Konovalow V. V., et al. Selective hydrotreating of cat cracked gasoline over a Kâ€"CoMoS/Al_2O_3 catalyst[J]. *Kinetics and Catalysis,* 2015. 56(6): 747â€"757.

496180 Inherent Safety Assessment in Natural Gas Liquefaction Process

Abstract:

The natural gas liquefaction process is typically divided as cascade processes and mixed refrigerant (MR) process according the cooling methodology. Both processes uses the refrigerant to cool down the natural gas, which traditionally consists of mixture of hydrocarbon. However, the hydrocarbon has a potential risk of a flammability. Thus there has been some approach to change the refrigerant from hydrocarbon to nonflammable refrigerant. N_2O is a nonflammable and nontoxic material which can be used as a refrigerant to overcome the safety problem of the liquefaction process which is particularly important in offshore platform. The aim of this study is to compare the inherent safeness of the liquefaction process for the conventional hydrocarbons and N_2O by using inherent safety index. Using the inherent safety index enables to quantitatively assess the safeness of the both processes, and further can be extensively applied to the other chemical processes.

501943 A Comparative Assessment of Sol-Gel and Solid State Syntheses Derived Redox Materials for H_2 Production *Via* Thermochemical Water-Splitting Process

Abstract:

Redox materials of the type $M_{1x}M_{2y}Fe_2O_4$ (M_1 or M_2 = Zn, Mn, Co, Ni) were prepared by the solid state and sol-gel syntheses routes. In the solid state synthesis route, metal oxide precursors were mixed in the required amounts in water and ball milled for 36 hours. Resultant powered material was calcined in air at higher temperatures to achieve the desired phase formation. In the sol-gel synthesis approach, metal salt precursors were sonicated in ethanol and gel formation was achieved using propylene oxide. The gel was aged, dried and finally calcined in air. The solid state diffusion properties of these redox ferrites were studied using the variable temperature impedance spectroscopy whereas the morphological properties were investigated using scanning/transmission electron microscopy (SEM/TEM), powdered X-ray diffraction (XRD), and Brunner-Emmett-Teller (BET) specific surface area analyzer. A solid state or sol-gel derived redox material was loaded inside a continuous Inconel tubular reactor and thermochemical water-splitting reaction was performed at isothermal conditions. Materials syntheses, characterization and H_2 volume generation over multiple thermochemical cycles will be presented.

501733 Overview of Impact of Electrode Catalyst Loadings on Thin Flexible Fuel Cell (TFFC) Performance

Abstract:

Thin Flexible Fuel Cell (TFFC) is a flat and flexible device without any rigid support capable of being integrated into portable electronics. TFFC uses hydrogen and oxygen as fuel to provide high volumetric and gravimetric energy densities capable of competing with battery technology as a power source. The group is currently investigating a TFFC with an air-breathing design that maximizes the cathode interface with ambient air. The current challenges of making TFFC commercialized are the cost and performance. Enhanced cell performance is critical for a competitive cost of the device. The group investigated the impact of catalyst loading on the cell performance to increase the overall performance of the device. The catalyst loading of the cathode electrode was varied while keeping the anode catalyst loading constant and vice versa. The effects of the various catalyst loadings were characterized using various analytical techniques. Polarization curves and Electrochemical Impedance Spectroscopy (EIS) were used to separate and quantify the sources of losses for the TFFC. Power curves provided the maximum power density of the device. Cyclic voltammetry (CV) was implemented to characterize the electrochemically active surface area (ECSA) and utilization of the various catalyst layers. Rotating Disc Electrodes (RDE) tests were performed for a direct assessment of the catalyst. We will report the results of these assessments and summarize the resulting guidance for optimizing catalyst loading on similar fuel cell structures.

497849 NiMo-Ceria-Zirconia Catalytic Reforming Layer for Solid Oxide Fuel Cells Running on Isooctane

Abstract:

Solid oxide fuel cells (SOFCs) are a promising energy conversion device because they can efficiently and directly convert the chemical energy of fuels into electrical energy. Due to the high operating temperatures (typically 600-1000 °C), SOFCs hold particular promise because of their ability to use a variety of complex liquid fuels, such as conventional liquid transportation fuels (e.g., gasoline, diesel-like fuels and jet fuels) and next generation liquid bio-fuels (e.g., biodiesel), either by using external reforming systems or directly feeding these fuels into their inexpensive transition metal-based anodes

Nickel-based anodes are commonly used in SOFCs owing to their low cost, good chemical stability, and excellent catalytic activity toward hydrogen oxidation and reforming of small hydrocarbon molecules. However, the Ni-based anodes are well known for promoting severe surface carbon deposition. Excessive formation of carbon species on the anode leads to a rapid deterioration of the cell performance by physically blocking access of the reactants to the active catalyst sites. To operate the Ni-based SOFCs by feeding liquid isooctane fuel, a catalytic layer can be applied over the Ni-based anode to act as an internal micro-reformer. The micro-reforming layer is designed to convert the mixture of isooctane and air to synthesis gas (hydrogen and carbon monoxide), which is believe to suppress the carbon deposition over the Ni-based anode.

In this work, a NiMo-Ceria-Zirconia (NiMo-CZ) catalyst was used as a micro-reforming layer for SOFCs running on isooctane. The catalyst layer was applied on top of the conventional anode supported single cell with a configuration of Ni-yttira-stabilized zirconia (YSZ) anode, YSZ/$Ce_{0.8}Sm_{0.2}O_{1.9}$ bi-layer electrolyte and $La_{0.6}Sr_{0.4}Co_{0.2}Fe_{0.8}O_{3-Î´}$ cathode. Our results showed that application of the novel catalyst layer was an effective way to enhance the electrochemical oxidation of complex hydrocarbons at the anode. At 750 Â°C, the single cell exhibited a low polarization resistance of 1.36 Î© cm^2 and a good maximum power density of 405 mW cm^{-2} in isooctane/air. At the current density of 500 mA cm^{-2}, the cell voltage presented a fairly low degradation rate of 3.0 mV h^{-1} during over 12 h stability test. The excellent electrochemical performance suggests the high catalytic activity of the NiMo-CZ catalyst for reforming isooctane and suppressing performance degradation of the single cell in the isooctane fuel.

500402 Homogeneous Reaction Kinetics of Carbohydrates with Viologen Catalysts in Biofuel Cell Applications

Abstract:

The production of renewable energy from biomass has been demonstrated using processes such as Fischer-Tropsch synthesis, direct fermentation of biomass sugars, biomass conversion to syngas followed by fermentation, and utilization of biomass carbohydrates in biofuel cells.Â Processes have been used to produce either electrical power, heat, liquid biofuels, or hydrogen. However, limitations and challenges for commercial viability still exist with all of these processes. Challenges inhibiting biofuel cell processes include low power output, low biological catalyst durability, and low carbohydrate conversion. Recently, a potentially transformative breakthrough has occurred with the discovery of a homogeneous non-biological viologen catalyst that facilitates carbohydrate conversion to electricity.

The viologen catalyst serves as a mediator or â€œelectron shuttleâ€ to extract electrons from the carbohydrate to form a reduced viologen. The reduced viologen is then oxidized at the anode of a biofuel cell to generate power. The combination of homogeneous viologen reduction by the fuel coupled with oxidation of the viologen at the electrode provides a continuous electron shuttle, enabling energy to be extracted from carbohydrates. For a biofuel cell process, the viologen catalyst provides greater durability relative to that achievable with biological catalysts.

Our initial studies have shown that viologens are effective in shuttling electrons from a wide variety of carbohydrates (C1 to C6 compounds). The homogeneous reaction rates and the efficiency of electron extraction were found to vary with reaction conditions such as pH, temperature, and reactant concentrations. Optimization of the reaction rates and efficiency are critical towards developing a viable biofuel cell using biomass carbohydrates as the feedstock.

Studies were conducted to determine the homogeneous rate law that describes the reaction between the carbohydrate and viologen. Specifically, studies were performed using NMR and spectroscopic techniques to develop a mechanistic description of the reaction rate applicable for a wide range of conditions. Studies of viologen stability were also performed to quantify potential viologen decomposition conditions that could compromise carbohydrate conversion efficiency. This work will present the mechanistic model of the carbohydrate-viologen reaction and provide insights towards optimizing reactor conditions that enhance reaction rates and carbohydrate conversion efficiency.

500124 Hydrogen Fueled Micro-Fuel Cell with Microfluidic Channels on a PFSA Plane

Abstract:

Microfluidic devices have shown considerable promise in a wide range of applications from medical screening to portable energy sources. Micro fuel cells (Î¼FCs) have been attracting much attention as a leading candidate for prospective portable power sources and battery replacements [1-4]. Their ability to create an efficient and clean source of energy combined with the ease of portability makesÂ Î¼FCs available to meet the needs of various portable electronic applications in the future [4, 5]. To this purpose, making more efficient devices with cost effective processes and materials is crucial.

Many efforts on fuel cell miniaturization are focused on silicon-based techniques as silicon is the most common substrate in MEMS technology [6, 7]. However, combining silicon devices with polymeric fuel cells at mm or sub-mm scale presents many challenges, none of which have been solved in a completely satisfactory manner [8]. Furthermore, in recent years due to material property issues associated with PDMS such as bulk absorption of small molecules and evaporation through the device, there has been a tendency towards the employment of thermoplastics for microfluidic systems [9].

Distinct from previous attempts on fabricating in-planeÂ Î¼FCs in other studies, in this work we present a novel On-Membrane Micro Fuel Cell design where the micro-flow channels for fuel and oxidant input are fabricated by using a combination of laser micromachining and hot-bonding process. In preparation of the device, a laser patterned acrylic sheet was pressed onto the Nafion substrate by a gas-cushion hot-bonding and the inlet tubes were mounted on the acrylic sheet. The micro device performance was characterized by I-V polarization curves and impedance spectroscopy under dry and humidified conditions. The combined in-plane and through-plane flux of protons in the membrane can be compared to literature data.

With this fabrication technology high aspect ratio structures can be fabricated over large surface areas, which prompts a commercially successful manufacturing of polymer-based micro-components.

Refrences:

1. Sundarrajan, S., Allakhverdiev, S. I., & Ramakrishna, S. (2012). Progress and perspectives in micro direct methanol fuel cell. *International Journal of Hydrogen Energy*, *37*(10), 8765-8786.

2. Taylor, A. D., Lucas, B. D., Guo, L. J., & Thompson, L. T. (2007). Nanoimprinted electrodes for micro-fuel cell applications. *Journal of Power Sources*, *171*(1), 218-223.

3. Z. Yuan, J. Yang, Y. Zhang and X. Zhang, "The optimization of air-breathing micro direct methanol fuel cell using response surface method," *Energy*, vol. 80, pp. 340-349, 2015.

4. Kamarudin, S. K., Daud, W. R. W., Ho, S. L., & Hasran, U. A. (2007). Overview on the challenges and developments of micro-direct methanol fuel cells (DMFC).*Journal of Power Sources*, *163*(2), 743-754.

5. Cao, J., Xu, J., Chen, Z., Wang, W., Huang, Q., & Zou, Z. (2013). A Silicon-based micro direct methanol fuel cell stack with a serial flow path design.*International Journal of Energy Research*, *37*(4), 370-376.

6. Shah, K., Shin, W. C., & Besser, R. S. (2004). A PDMS micro proton exchange membrane fuel cell by conventional and non-conventional microfabrication techniques. *Sensors and Actuators B: Chemical*, *97*(2), 157-167.

7. Lu, G. Q., Wang, C. Y., Yen, T. J., & Zhang, X. (2004). Development and characterization of a silicon-based micro direct methanol fuel cell.*Electrochimica Acta*, *49*(5), 821-828.

8. Omosebi, A., & Besser, R. S. (2013). Fabrication and performance evaluation of an in membrane micro-fuel cell. *Journal of Power Sources*, *242*, 672-676.

9. Sackmann, Eric K., Anna L. Fulton, and David J. Beebe. "The present and future role of microfluidics in biomedical research." *Nature* 507.7491 (2014): 181-189.

496142 Design and Analysis of Tri-Generation System Powered By Seaweed Biogas

Abstract:

Over the last two decades, fuel cells have been intensely researched. Biogas-powered solid oxide fuel cells (SOFCs) have become prospective devices for stationary power generation. In this work, simulation and economic analysis were performed for a 3 MW biogas-fueled SOFC system. The composition of the anode outlet of the proposed SOFC system contains CO, H_2, H_2O, and CO_2 which is comparable to typical high temperature water-gas-shift (WGS) feed to co-produce hydrogen. The conversion of CO in water-gas shift reaction was 88% with the hydrogen concentration of 35 mol %. Pressure swing adsorption is used to purify the hydrogen. In order to utilize the heat from the WGS reaction, organic rankine cycle (ORC) was used. To our knowledge, this has been the first work on design and analysis of an SOFC integrated into a combined heat, hydrogen and power (CHHP) system.

501490 Multi-Functional Mixed Oxide Catalysis in Cascade Chemistry to Convert Ethanol to High Value Oxygenates

Abstract:

The development of alternative cleaner, safer, and environmentally friendly technologies is one of the priorities in chemistry for sustainable future. Cascade catalysis, which combines various chemical steps without separating the intermediates, can play a vital role in improving the sustainability of chemical processes via process intensification with high economic benefit. At PNNL we focus on developing the low cost multi-functional mixed oxide catalyst for the conversion of the bioethanol to high value compounds via cascade catalysis. Particularly, we are interested in synthesizing high molecular weight oxygen containing compounds that can be used as fuel additives or intermediate for infrastructure compatible fuel as well as chemical commodity. Recent work conducted at PNNL has demonstrated a novel pathway to selectively convert ethanol to higher alcohols at ~50% yields and >80% selectivity with stable catalyst life times above 600 hours, using a copper based hydrotalcite catalyst in a continuous flow reactor. In this presentation we will discuss the detailed reaction mechanism and the role of mixed metal oxide in the ethanol conversion to high value oxygenates.

494643 Acetaldehyde Condensation-Cyclodehydration to Aromatics over Mg-Al Oxides

Abstract:

Acetaldehyde is a constituent of many biomass-derived feedstocks, but its high reactivity and tendency to polymerize make it challenging to upgrade into compounds for incorporation into chemical and fuel infrastructures. Mg-Al oxide catalysts derived from calcined layered double hydroxides are used in a multiple condensation-cyclodehydration reaction of acetaldehyde to produce benzene and methylbenzaldehydes, the latter of which are precursors to terephthalic acid. By varying the Mg/Al ratio in the catalysts, conversion and selectivity profiles are impacted as a function of time on stream (TOS). Carbon selectivity towards methylbenzaldehydes reaches 33% over the catalyst with Mg/Al=3. Despite deactivation due to the formation of non-volatile carbonaceous deposits, selectivity towards methylbenzaldehydes stabilizes at 20% after 18 hours TOS. Changes in selectivity with deactivation, as well as the difference in selectivity over the varying Mg/Al catalysts, indicate different mechanisms are activated over different site types.

Calibration-Free Methods for GC Quantification of Bio-Derived Chemicals

Charlie Spanjers, Ph.D.
Product Innovation Engineer
Activated Research Company

AIChE, 11/2/2017

How would YOU determine the composition of a truly unknown sample with GC?

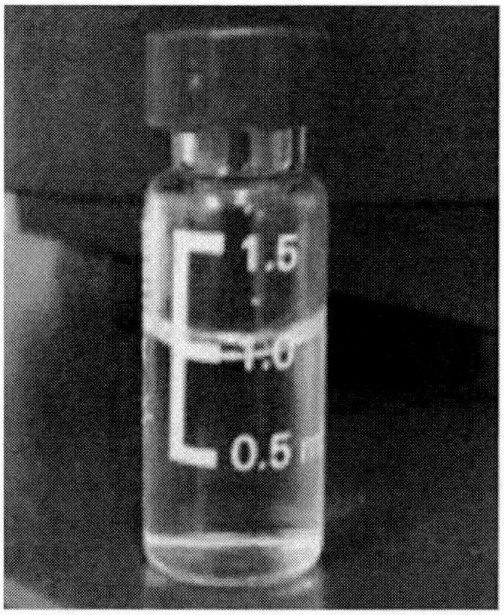

?

How would YOU determine the composition of a truly unknown sample with GC?

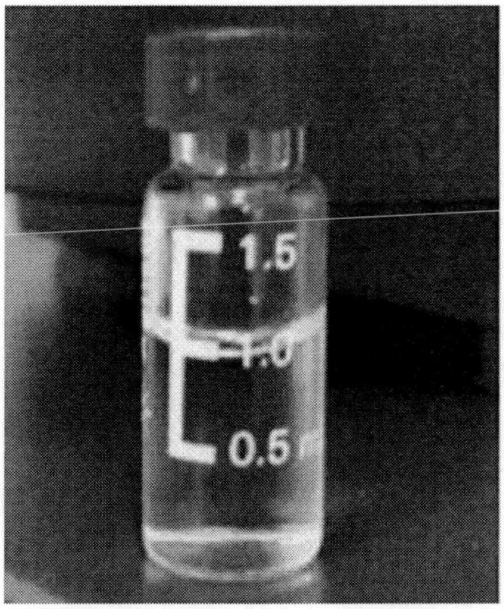

1. Identify the components with a MS
2. Purchase standards
3. Prepare calibration samples
4. Run the calibration standards
5. Quantify using calibration results

How would YOU determine the composition of a truly unknown sample with GC?

1. Identify the components with a MS
2. Purchase standards — Costly
3. Prepare calibration samples — Error-prone
4. Run the calibration standards — Time-consuming
5. Quantify using calibration results

How would YOU determine the composition of a truly unknown sample with GC?

Is there a system that can eliminate these steps?

YES!

1. Identify the components with a MS
2. ~~Purchase standards~~
3. ~~Prepare calibration samples~~
4. ~~Run the calibration standards~~
5. Quantify using calibration results

The Solution: Simultaneous Compound Identification and Quantification

Compound Identification

Quantification

Solution: Post-Column Oxidation/Methanation Gives Uniform Molar Response

Catalytic microreactor converts organic compounds to methane after separation and prior to detection in the FID

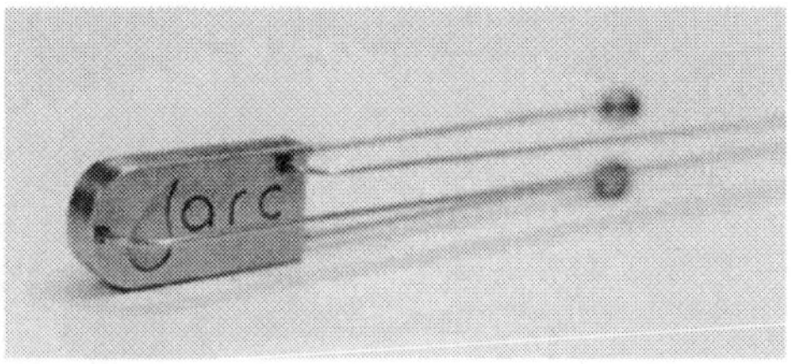

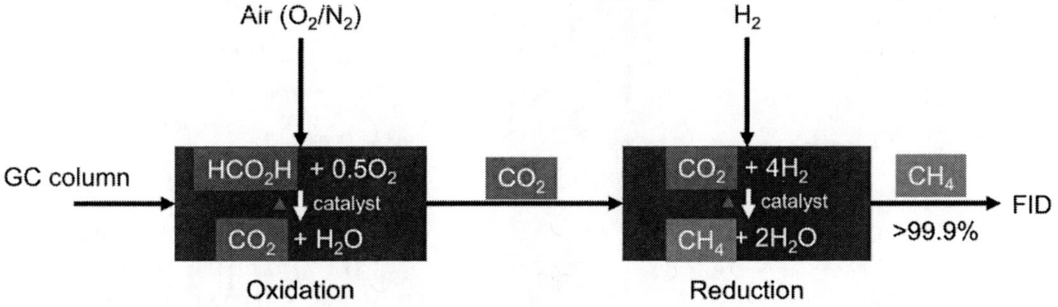

The Polyarc/FID is a carbon counting detector

Polyarc System: How it Works

Post-column and pre-detector conversion to methane.

GC with flame ionization detection (GC/FID)

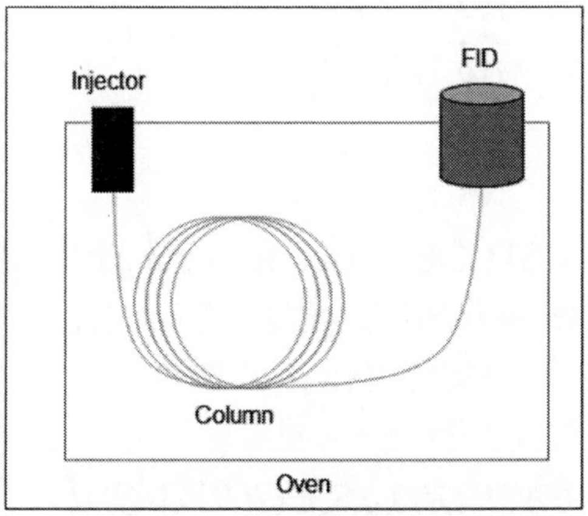

GC with Polyarc® and FID

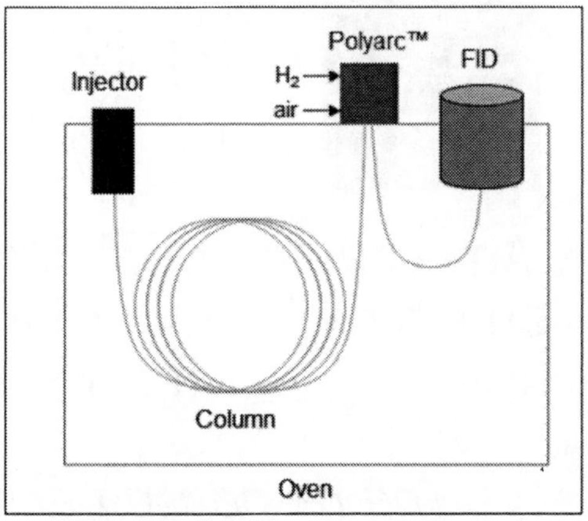

https://www.activatedresearch.com/technology/

Accurate Single Injection Identification and Quantification

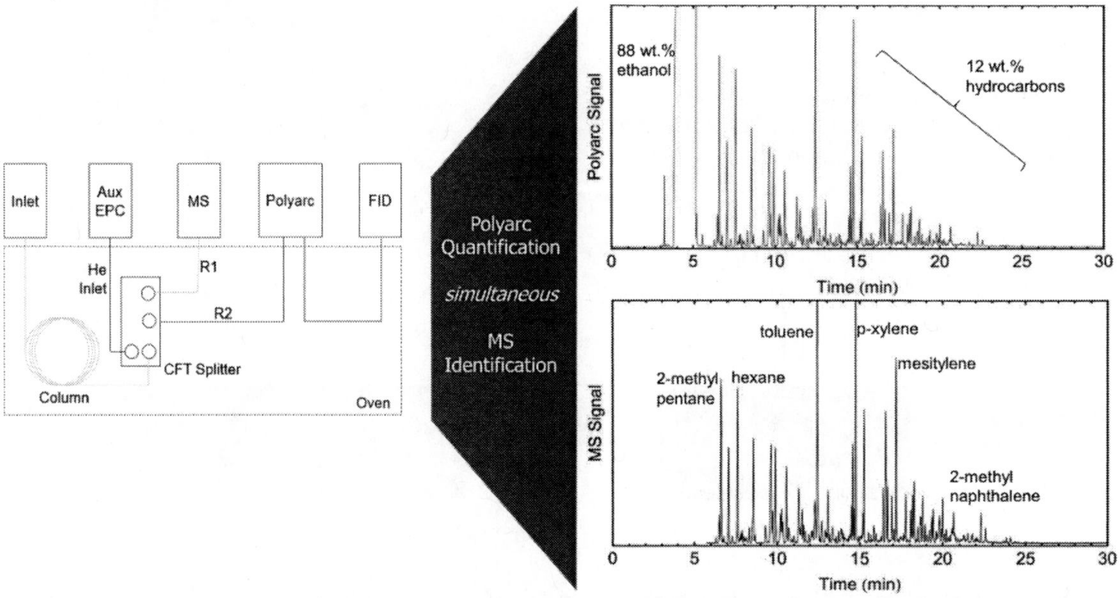

E85 Sample
*Authored by the National Renewable Energy Laboratory and ARC.

Accurate Single Injection Identification and Quantification

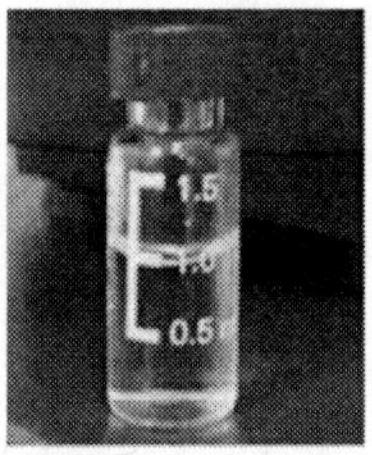

$$C_A = C_S \left(\frac{Area_A}{Area_S}\right)\left(\frac{Mw_A}{Mw_S}\right)\left(\frac{\#C_S}{\#C_A}\right)$$

Any compound of interest can be quantified using an internal standard and the integrated peak areas from a single injection.

The response per carbon atom is equivalent for all organic molecules.

Accurate Single Injection Identification and Quantification of Pyrolysis Products

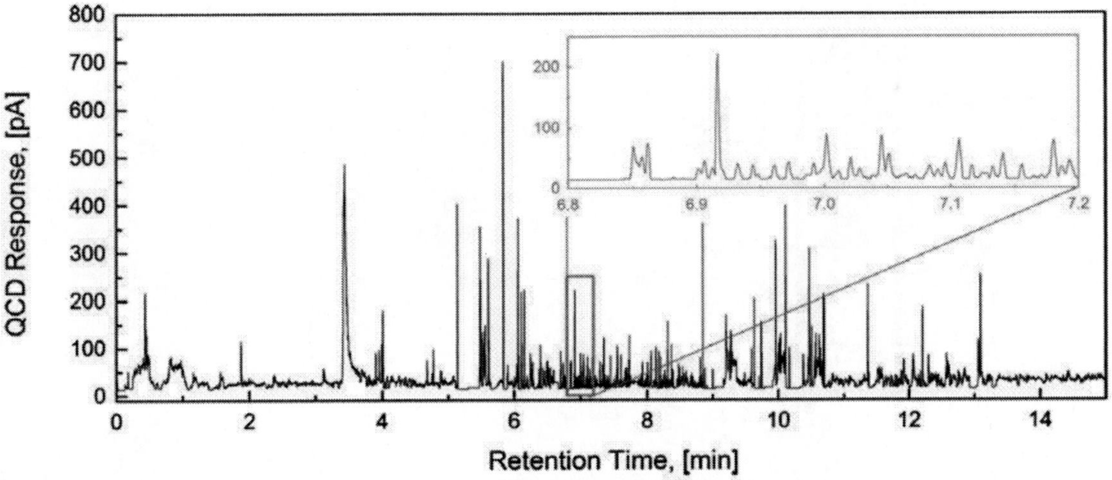

Lab Chip, 2015, **15**, 440-447

Accurate Single Injection Identification and Quantification of Biomass-Derived Isoprene

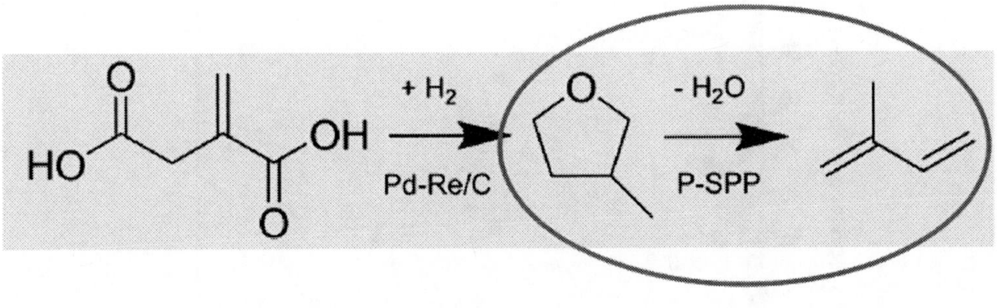

ACS Catal. 2017, *7*, 1428−1431

Accurate Single Injection Identification and Quantification of Biomass-Derived Isoprene

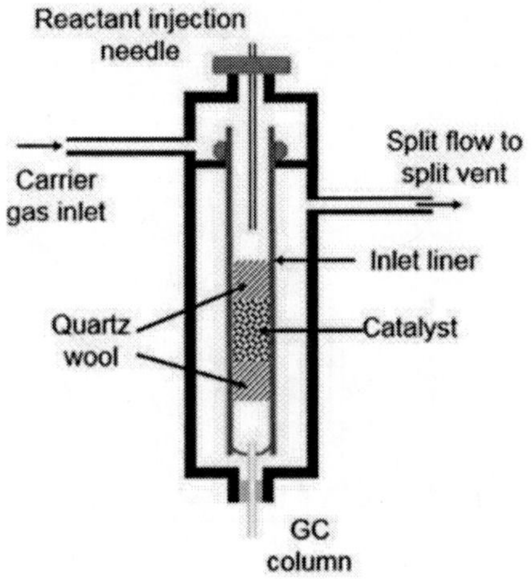

+

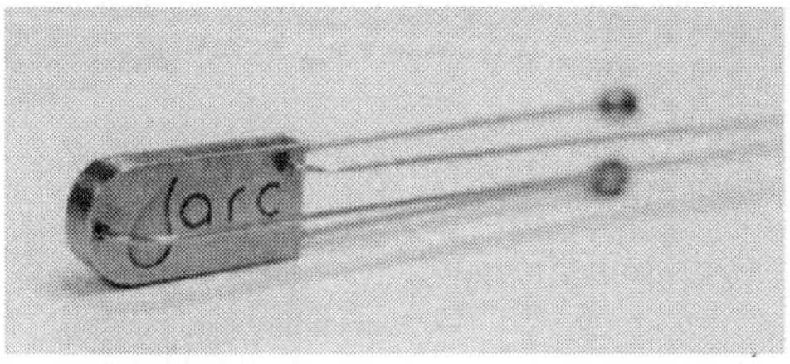

ACS Catal. 2017, 7, 1428–1431

Accurate Single Injection Identification and Quantification of Biomass-Derived Isoprene

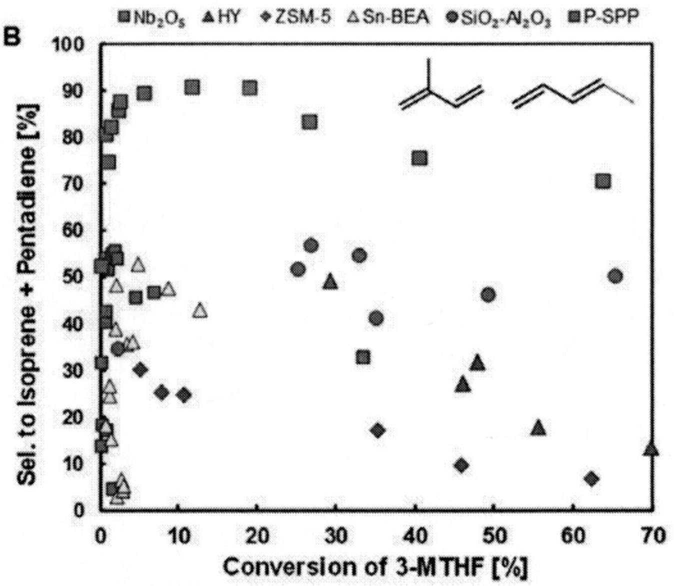

ACS Catal. 2017, 7, 1428–1431

Accurate Single Injection Identification and Quantification of Biomass-Derived Isoprene

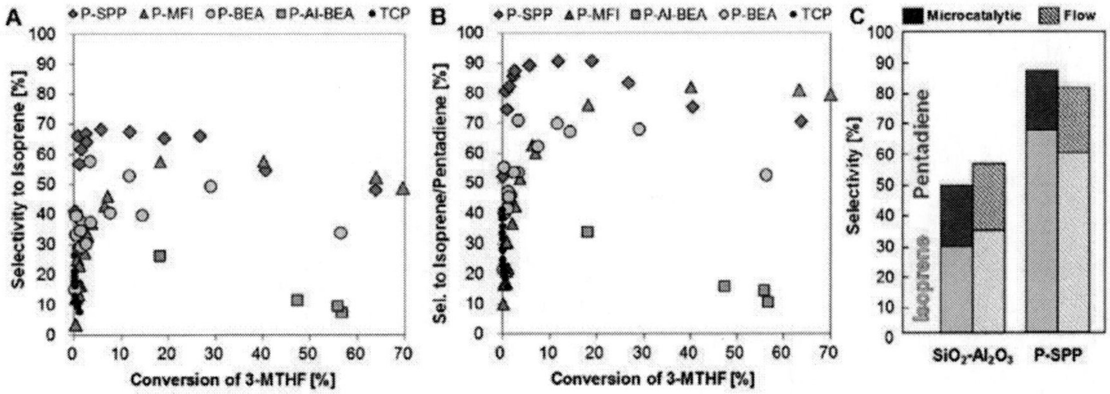

ACS Catal. 2017, 7, 1428–1431

Accurate Single Injection Identification and Quantification of Biomass-Derived Butadiene

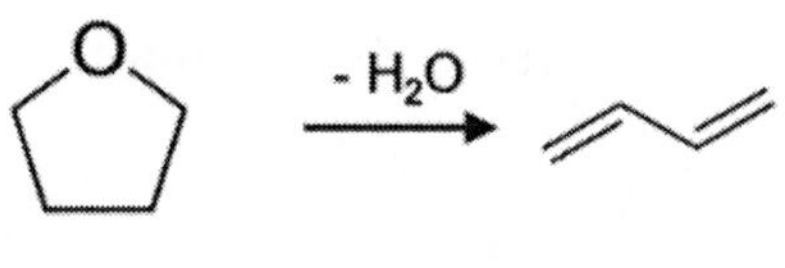

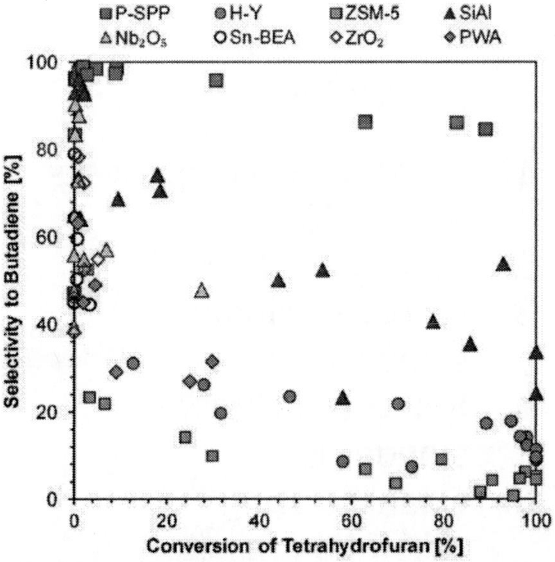

ACS Sustainable Chem. Eng., **2017**, 5 (5), pp 3732–3736

Accurate Single Injection Identification and Quantification of FAMES

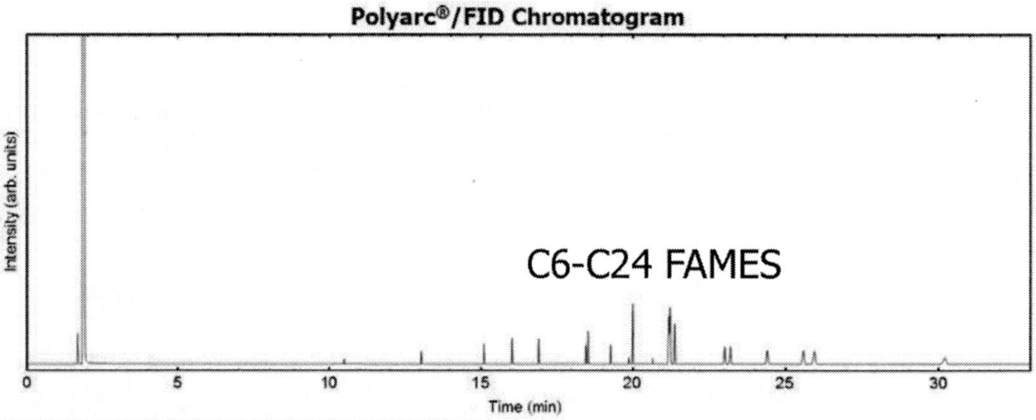

Analyte	Mw (g/mol)	#C	Ret time (min)	Area	Conc. (wt. %)
Pentane	72	5	1.862	3.21·10[illegible]	**95***
C6:0	130	7	10.506	1.08·10[illegible]	**0.041**
C8:0	158	9	13.022	2.58·10[illegible]	**0.093**
C10:0	186	11	15.096	4.07·10[illegible]	**0.142**
C11:0	200	12	16.029	5.55·10[illegible]	**0.190**
C12:0	214	13	16.909	5.61·10[illegible]	**0.190**
C14:1	240	15	18.454	4.26·10[illegible]	**0.140**
C14:0	242	15	18.529	7.16·10[illegible]	**0.237**
C15:0	256	16	19.276	4.29·10[illegible]	**0.141**
C16:1	268	17	19.867	1.48·10[illegible]	**0.048**
C16:0	270	17	19.995	1.43·10[illegible]	**0.467**
C17:0	284	18	20.667	1.46·10[illegible]	**0.047**
C18:1/2/3	~294	19	21.235	3.36·10[illegible]	**1.069**
C18:0	298	19	21.384	1.19·10[illegible]	**0.385**
C20:1	324	21	23.018	7.28·10[illegible]	**0.231**
C20:0	326	21	23.200	7.39·10[illegible]	**0.238**

<2% Average Error for 20 FAMES Using Single Injection

Comparison of Old Method and New Method for Quantification of Unknowns

Old Method

1. Identify the components with a MS
2. Purchase standards
3. Prepare calibration samples
4. Run the calibration standards
5. Quantify using calibration results

→

New Method

1. Single Injection for Identification and Quantification

Will Simultaneous Identification/Quantification Methods Become the New Standard?

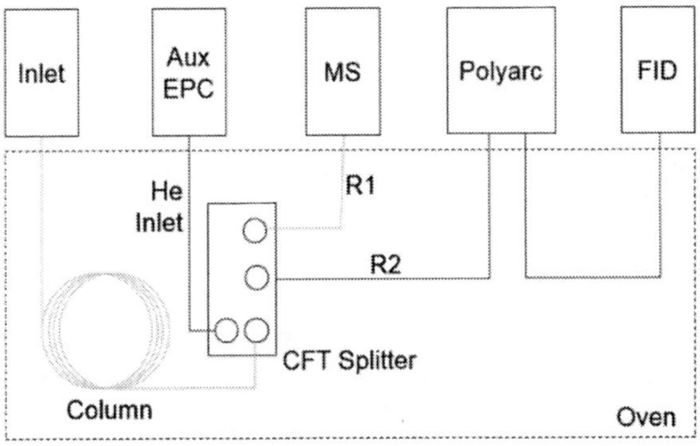

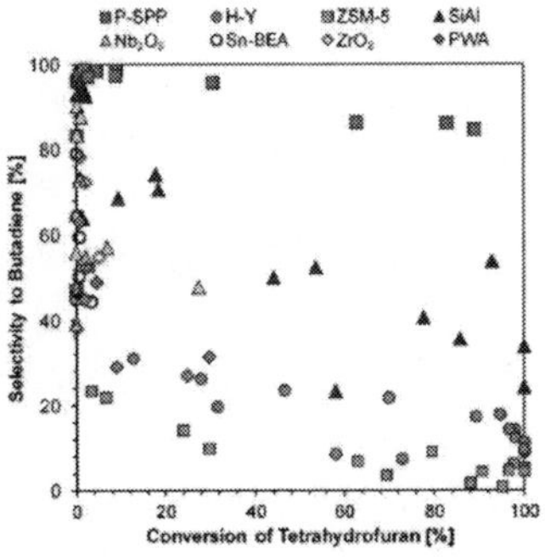

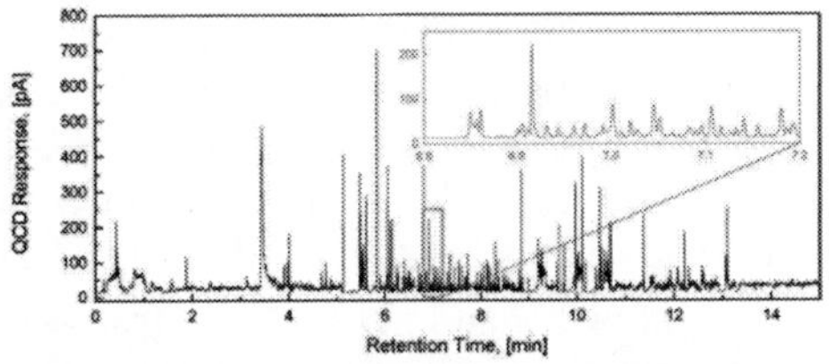

Questions?

500525 A Novel Method of Producing Levulinic Acid at High Concentrations and Yields from Corn Stover Hydrolysate

Abstract:

A novel method of producing levulinic acid at high concentrations and yields from Corn Stover hydrolyzate

Ravikumar Gogar,[1] Sridhar Viamajala,[1] Patricia Relue,[2] and Sasidhar Varanasi[1]

Department of Chemical and Environmental Engineering, [1] Department of Bioengineering, [2] University of Toledo, Toledo, OH 43606

Levulinic acid is a promising platform molecule that can be used to produce transportation fuels, resins, polymers, pharmaceuticals, and solvent precursors from biomass. In this work, dilute acid pre-treated corn stover was used as feed to obtain saccharified biomass hydrolysate. We developed a novel hybrid enzyme- and chemo-catalytic technology that isomerizes, isolates, and dehydrates the sugar to levulinic acid in high yields and concentrations. First, a *simultaneous-isomerization-reactive-extraction* (SIRE) step was used to convert glucose from biomass hydrolysate, using GXI enzyme at 60^{0} C and pH = 8.5, to fructose at higher yield (88%). This approach, which extracts the fructose into an immiscible octanol phase, as it is formed, not only overcomes the unfavorable equilibrium of the isomerization reaction, but also separates the sugar from the pretreatment-derived impurities in the hydrolyzate. Aryl Boronic Acids (ABA) and Aliquat® 336 are added to the octanol phase to facilitate the extraction of fructose selectively via preferential binding to ABA. Fructose was then transferred from the octanol into an immiscible phase of an acidic ionic liquid and aqueous mixture (at pH = 6) via back-extraction. Fructose was then converted in the IL-water phase, at facile reaction conditions (T = 95 C, P =1 atm), to Levulinic acid at high yield (63 mol%); levulinic acid was obtained at significant loading of 6.4%, thereby, reducing energy cost for its separation from the reaction mixture. The proposed process offers advantage of recyclable streams and catalyst, near complete utilization of feed by overcoming glucose to fructose equilibrium, and lower energy cost by providing levulinic acid at higher concentration, thereby making the process cost competitive.

492940 Hydrodeoxygenation of Acetic Acid As a Model Compound for the Aqueous Phase Catalytic Pyrolysis Oils

Abstract:

Catalytic pyrolysis of biomass generates organic, aqueous, gaseous and solid fractions. The organic fraction can be easily hydrotreated to produce hydrocarbons, but the aqueous phase that contains between 10 to 25% soluble organics can pose challenges in wastewater treatment. The aqueous fraction from the catalytic pyrolysis of Pinyon Juniper wood was characterized for its organic content. The fraction contained about 15 wt% organic compounds determined from Karl Fischer analysis. The organic fractions were further characterized using gas chromatography and mass selective detection (GC/MS). The analysis showed that the dissolved organics were composed of acetic acid, ketones, aldehydes, and phenolic compounds. In this study we investigated the hydrodeoxygenation (HDO) of 15 wt.% acetic acid solution to represent aqueous phase Pinyon Juniper catalytic pyrolysis oil (APPJCPO). HDO experiments were carried out at different temperatures (150, 250, 350, and 450 °C) using Ni/SiO_2-Al_2O_3 catalyst in a high pressure Parr reactor. HDO of acetic acid produced acetaldehyde, ethanol, ethyl acetate, carbon dioxide, methane, ethane, and coke at different concentrations depending on the reaction temperature. Reaction pathways of acetic acid HDO were proposed based on analysis of the products. The final products of acetic acid HDO were methane, water at reaction temperature of 450 °C. HDO of APPJCPO was carried out at 450 °C. During HDO the pH of APPJCPO increased from 2.97 to 6.93. The final products of HDO of APPJCPO were water and methane. After HDO experiments, the catalyst was partially deactivated due to coke formation. This study provided insight in the reaction network of acetic acid HDO and suggests that HDO is a promising technique to overcome toxicity and corrosion of aqueous phase pyrolysis oil.

493018 C_5+ Ketones Synthesis from Ethanol over Mixed Metal Oxides

Abstract:

Currently, bioethanol accounts for almost 90% of the global biofuel production. Driven by the latest innovations in cellulosic biomass conversions and syngas fermentations, world ethanol production is expected to increase. The energy independence and security act requirement to produce 36 billion gallons/year biofuel by 2022 could drive the conversion technologies further to generate more ethanol well beyond the existing E15 blending wall requirement. Therefore, it is anticipated that the excess ethanol produced will become available as a platform molecule for the production of value-added chemicals and infrastructure compatible fuels in the near future and beyond. Recent work conducted at PNNL has demonstrated a novel pathway for selectively converting ethanol to C_5+ ketones with minimal loss of carbon using a low cost multi-functional mixed oxide catalyst with >80% yield to C_5+ ketones. The uniqueness of this chemistry is the condensation of acetone with acetaldehyde to form pentanone rather than the self-condensation between acetone(s). This helps to improve the final product yield and reduces the overall carbon loss. This presentation will discuss the uniqueness of the reaction mechanism and the catalyst development for the ethanol conversion to C_5+ ketones.

AUTHOR INDEX

Abdelrahman, Omar A. 28
Agblevor, Foster 283
Akhfash, Masoumeh 235
Al Ghafri, Saif 222, 235
Albahri, Tareq 249
Almulla, Hessa 249
Al-Shankiti, Ibraheam 250
Amar, Vinod S. 264
Amghizar, I. 137
Arami-Niya, Arash 235
Ausserleitner, Julia 171
Bahari, Meisam 267
Banerjee, Dwijen 136
Bastos, Jaci C. S. C. 125
Basu, Somnath 113
Beims, Ramon F. 179
Besser, Ronald S. 265, 268
Bhattacharyya, Debangsu 175, 234
Bingham, Hilary 267
Bkour, Qusay 266
Botton, Vanderleia 179
Caballero, Jonatan D. A. 199, 201
Cai, Tianxing 176
Cao, Ying-Qian 246
Carter, Scott 267
Cavataio, Giovanni 185
Chaisoontornyotin, Wattana 135
Chauvel, Paul 111
Chen, Sheng-Li 246
Chen, Xiaoyin 202
Churchill, Stuart W. 30
Cross, Collin 87
Czernik, Caitlin 250
Dauenhauer, Paul J. 28
David, Aditi 243
De Lasa, Hugo 140
Delgass, W. N. 199, 201
Dhiman, Saurabh 243
Ding, Yanfen 161
Djokic, Marko R. 137
Dong, Lei 172
Dong, Lixia 262
Effendi, Ivannie 269
Effendy, Surya 251
Elechi, Okoh 248
Ender, L. 179
Epling, William 200
Erickson, Kayla 67
Farooq, S. 251
Fasahati, Peyman 269
Favero, Claudio V. B. 33
Fonseca, Reynaldo 245
Fontoura, Diener V. R. 245
Foust, Thomas D. 182
Gangar, Bijal 204
Gao, Jinsen 158, 262
Glotzer, Sharon C. 29
Go, Kang S. 178
Gogar, Ravikumar 282
Gong, Jing 161
Gonzalez, Juan M. 201
Gounder, Rajamani 199, 201
Gray, Michel 270, 284
Guillén-Gosálbez, Gonzalo 34
Guo, Mond 270, 284
Gye, Hye-Ri 263
Ha, Su 266
Hammerschlag, Berndt 171
Harb, John 267
Harold, Michael 183
Harper, Michael R. 2, 25
Hayashi, Yojiro 240
Hayashida, Kei 240
Herceg, Eldad 111
Hill, James C. 30
Hoepfner, Michael P. 135
Hoskins, Amanda 250
Hou, Xiaoxue 266
Hou, Yucui 238, 241, 242, 244
Huang, Huirong 161
Huang, Ling-Xiang 246
Huber, Anna 171
Hughes, Thomas 235
Ibrahim, Dauda 34
Im, Soo I. 178
Isogami, Hisashi 240
Jahromi, Hossein 283
Jensen, Klavs F. 271
Jiang, Yuan 175, 234
Jiao, Fuyu 235
Jobson, Megan 34, 35, 98
Johns, Michael L. 235
Johnson, Glenn 243
Karimi, Iftekhar A. 251
Karinshak, Kyle 183
Khan, Mohd S. 251
Khurana, Ishant 199, 201
Kim, Changhwan 203
Kim, Dae-Hyun 263
Kim, Do H. 203
Klein, Michael T. 26
Kumar, Ashok 201
Kurokawa, Hideaki 240
Kushnerick, Doug 2, 25
Lail, Marty 186
Larson, Ronald G. 33
Ledezma-Martinez, Minerva 35
Lee, Chul-Jin 263
Lee, Hyokyoung 203
Lee, J. 203

AUTHOR INDEX

Lee, Ki B. 178
Lewis, Randy S. 267
Li, Hui 201
Li, Jia 174
Li, Jie 34
Li, Ping 202
Li, Wenbin 244
Li, Yongdan 202
Lin, Wensheng 236, 237
Liu, Bin 180
Liu, J. J. 269
Liu, Qing 238
Liu, Shuangxing 247
Liu, Zewei 172
Livingston, Dana A. 111
Long, Richard L. 252
Lopez-Zamora, Sandra M. 140
Loyola-Fuentes, J. 98
Luo, Kaiwei 262
Lusardi, Marcella 271
Mahmoodi, Seyed R. 265, 268
Marin, Guy B. 137
Martignoni, Waldir P. 139
Martinez, Tania S. 139
May, Eric 222, 235
Mayer, Matthew 265
Meier, Henry F. 125, 139, 179
Mevawala, Chirag 234
Miao, Guang 173
Miller, Jeffrey T. 199, 201
Millican, Samantha L. 250
Millington, Dinara 126
Moir, Michael 3
Molina, Alejandro 140
Musgrave, Charles B. 250
Mylona, Sofia 235
Nagl, Roland 171
Nava, Glixon M. 136
Nduagu, Experience I. 126
Netter, Judy 250
Neurock, Matthew 27
Ng, Samson 135
Nguyen, Quoc T. 31
Nho, Nam S. 178
Niu, Muge 241
Noriler, Dirceu 125
Norton, M. G. 266
Nunhez, J. R. 245
Oakley, Jordan 235
Olsen, Tim 112
Paolucci, Christopher 199, 201
Papavassiliou, Dimitrios V. 31
Parekh, Atish A. 199, 201
Park, Junwoo 178
Peles, Yoav 32
Peng, Zeheng 161
Pfromm, Peter H. 180
Pilehvari, Ali A. 204
Pławsky, Joel L. 32
Premchand, Kiran 183
Pucher, Peter 171
Puszynski, Jan A. 264
Qiu, Wei 237
Ramasamy, Karthikeyan K. 270, 284
Reardon, Kenneth F. 182, 239
Relue, Patricia 282
Ren, Shuhang 241, 244
Ren, Xiaoling 172
Riazi, M. R. 138
Ribeiro, Fabio H. 199, 201
Ristic, Nenad D. 137
Rosa, Leonardo M. 125, 139
Ruan, Binhui 236
Ryou, Youngseok 203
Sani, Rajesh 243
Sano, Tadashi 240
Schneider, William F. 199, 201
Schuler, Bruno 1, 24
Schwaiger, Nikolaus 171
Schwank, Johannes W. 202
Seneviratne, Kumarini 235
Seo, Chang Y. 202
Shan, Nannan 180
Sharf, Dilamara R. 179
Shende, Rajesh V. 264
Shih, Arthur J. 199, 201
Shirazi, Saeid A. 182, 239
Siahvashi, Arman 222
Siebenhofer, Matthaeus 171
Silva, Marcela K. 125
Simionatto, E. L. 179
Sloley, Andrew W. 50, 68
Smith, Robin 35, 98
Sow, Alpha 126
Spanjers, Charles S. 272
Spogis, Nicolas 245
Srinivasan, Chiranth 31
Stahnke, Carina 125
Stefanov, Zdravko 111
Subramaniam, Senthil 284
Treusch, Klara 171
Umeozor, Evar 126
Utzig, Jonathan 139
Van Den Berg, Frans 14
Van Geem, Kevin M. 137
Varanasi, Sasidhar 282
Viamajala, Sridhar 282
Villa, A. L. 201
Volk, Michael 136
Wallace, Mark 250

AUTHOR INDEX

Wang, Qian....238
Wang, Wei....161
Watanabe, Masatomo....240
Watt, Gerald D....267
Waturuocha, Amaka....136
Weimer, Alan W.....250
Wiggers, Vinicyus R....179
Wilburn, Monique S.....200
Wilson, Shawn....181
Wisecarver, Keith....136
Woodcock, Cory....32
Wu, Jason....185
Wu, Liqiong....173
Wu, Tao....246
Wu, Weize....238, 241, 244
Wu, Ying....173
Xiao, Jing....172, 173
Xu, Chunming....158, 262
Xu, Jie....246
Xu, Qiang....176
Xu, Yiling....176
Yang, Fan....238, 241, 244
Yang, Yuan....135
Ye, Feiyan....173
Yeh, Bryan....177
Yezerets, Aleksey....199, 201
Yu, Xiangfei....32
Yuan, Guimei....246
Zhang, Jeffrey....212
Zhang, Jingzhou....135
Zhang, Yuhao....158, 262
Zhang, Yunlong....2, 25
Zhao, Kai....266
Zhao, Liang....158, 262
Zhao, Siyue....237
Zheng, Qinghe....186
Zhenlei, Wang....176

AIChE
120 Wall Street, FL 23
New York, NY 10005-4020

ISBN 978-1-5108-5803-9

CURRAN ASSOCIATES INC.
proceedings.com